Advances in
INORGANIC CHEMISTRY
AND
RADIOCHEMISTRY

Volume 19

CONTRIBUTORS TO THIS VOLUME

Ingrid Fischler

Oskar Glemser

Jochen Jander

Ernst A. Koerner von Gustorf

Luc H. G. Leenders

R. Mews

Robin N. Perutz

Karl-Heinz Tytko

Advances in
INORGANIC CHEMISTRY
AND
RADIOCHEMISTRY

EDITORS

H. J. EMELÉUS

A. G. SHARPE

University Chemical Laboratory
Cambridge, England

VOLUME 19

1976

ACADEMIC PRESS   New York San Francisco London

A Subsidiary of Harcourt Brace Jovanovich, Publishers

ACADEMIC PRESS, INC.
111 Fifth Avenue, New York, New York 10003

United Kingdom Edition published by
ACADEMIC PRESS, INC. (LONDON) LTD.
24/28 Oval Road, London NW1

LIBRARY OF CONGRESS CATALOG CARD NUMBER: 59-7692

ISBN 0-12-023619-2

PRINTED IN THE UNITED STATES OF AMERICA

CONTENTS

Recent Chemistry and Structure Investigation of Nitrogen Triiodide, Tribromide, Trichloride, and Related Compounds

JOCHEN JANDER

Aspects of Organo-Transition-Metal Photochemistry and Their Biological Implications

ERNST A. KOERNER VON GUSTORF, LUC H. G. LEENDERS, INGRID FISCHLER, AND ROBIN N. PERUTZ

Nitrogen–Sulfur–Fluorine Ions

R. MEWS

Isopolymolybdates and Isopolytungstates

Karl-Heinz Tytko and Oskar Glemser

LIST OF CONTRIBUTORS

Numbers in parentheses indicate the pages on which the authors' contributions begin.

INGRID FISCHLER (65), *Institut für Strahlenchemie im Max-Planck-Institut für Kohlenforschung, Mülheim/Ruhr, West Germany*

OSKAR GLEMSER (239), *Anorganisch-Chemisches Institut der Universität Göttingen, Göttingen, West Germany*

JOCHEN JANDER (1), *University of Heidelberg, Heidelberg, West Germany*

ERNST A. KOERNER VON GUSTORF* (65), *Institut für Strahlenchemie im Max-Planck-Institut für Kohlenforschung, Mülheim/Ruhr, West Germany*

LUC H. G. LEENDERS† (65), *Institut für Strahlenchemie im Max-Planck-Institut für Kohlenforschung, Mülheim/Ruhr, West Germany*

R. MEWS (185), *Institute of Inorganic Chemistry, University of Göttingen, Göttingen, West Germany*

ROBIN N. PERUTZ‡ (65), *Institut für Strahlenchemie im Max-Planck-Institut für Kohlenforschung, Mülheim/Ruhr, West Germany*

KARL-HEINZ TYTKO (239), *Anorganisch-Chemisches Institut der Universität Göttingen, Göttingen, West Germany*

* Deceased.
† Present address: Bell Laboratories, Murray Hill, New Jersey.
‡ Present address: Department of Chemistry, University of Edinburgh, Edinburgh, Scotland.

RECENT CHEMISTRY AND STRUCTURE INVESTIGATION OF NITROGEN TRIIODIDE, TRIBROMIDE, TRICHLORIDE, AND RELATED COMPOUNDS

JOCHEN JANDER

University of Heidelberg, Heidelberg, West Germany

I. Introduction

Although new investigations on inorganic nitrogen compounds of chlorine, bromine, and iodine have been reviewed comparatively recently (*28*, *78*), it seems appropriate to discuss the chemistry and structures of nitrogen triiodide, -chloride, and -bromide once more since highly interesting and, in part, surprising results have emerged in the course of the past 2 years.

II. Nitrogen Triiodide

A. Polymeric Structure and IR Spectrum of Nitrogen Triiodide-1-ammonia $(NI_3 \cdot NH_3)_n$

With the elucidation of the structure of nitrogen triiodide-1-ammonia (*58*, *77*), a structural principle, which is widely applicable in both inorganic and organic nitrogen–iodine chemistry, was unexpectedly discovered.

Before this key concept is considered more closely (Sections II, B–F), the structure itself will be described (Fig. 1). It contains a polymeric N—I framework rather than isolated NI_3 units. This is made up of NI_4 tetrahedra joined by common iodine atoms (I1, I1′) into chains of tetrahedra with infinite —N—I—N—I— chains, which are built into sheets of tetrahedra by I—I contacts [I3—I2 neighboring chain (3.36 Å)] between chains lying above one another along the *c*-axis. In many respects the structure of the chain of NI_4 tetrahedra resembles that of SiO_4 tetrahedra in metasilicates. The NI_4 tetrahedra are slightly distorted and have two pairs of equal N—I distances [N1—I1, I1′ (2.30 Å) and N1—I2, I3 (~2.15 Å)]. One iodine atom of each tetrahedron is loosely attached to an ammonia molecule [I2—N2 (2.53 Å)] that projects into the corrugated space between the sheets of tetrahedra (see Fig. 1).

A special characteristic of this structure, in addition to the NI_4 tetrahedra, is the linear or almost linear groups N1—I1′—N1′, N1—I2—N2, and N1—I3—I2 (neighboring chain). This illustrates a

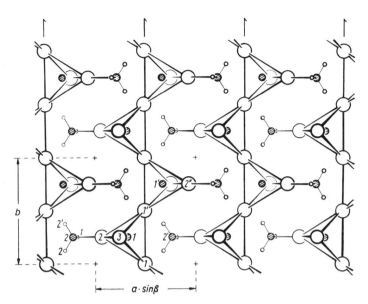

FIG. 1. Projection of the monoclinic crystal structure of nitrogen triiodide-1-ammonia on the (001) plane. Large circles, I atoms; shaded circles, N atoms; small circles, H atoms (58).

tendency, which has been observed many times (100, 151, 63) for covalently bonded iodine atoms, to form a second bond. In fact, it is responsible for the polymeric character of nitrogen triiodide-1-ammonia. The bond in these linear groups of 3 atoms is best described in terms of molecular orbital theory as a 3 center-4 electron bond, using a p orbital of the central iodine and an sp^3 hybrid orbital of each nitrogen, or a p orbital of the second iodine atom (20). It is notable that the N1—I1′—N1′ group is completely symmetrical, that is, there is no longer a difference between the intramolecular and the intermolecular bond.

Table I shows the IR spectra for the isotopic species $^{14}NI_3 \cdot ^{14}NH_3$, $^{14}NI_3 \cdot ^{14}ND_3$, and $^{15}NI_3 \cdot ^{15}NH_3$. Raman spectra cannot be recorded, even with laser radiation, since the brown–black ammoniate is too rapidly decomposed. Assignment of the bands is difficult because of the polymeric structure of the compound. It was done, after eliminating bands associated with the ammonia in the adduct, by making the simplifying assumption that the polymeric N—I structure is made up of slightly distorted tetrahedra with C_{2v} symmetry, the vibrations of which are not coupled with one another through the relatively heavy

TABLE I

INFRARED SPECTRUM OF NITROGEN TRIIODIDE-1-AMMONIA (98)[a]

$NI_3 \cdot NH_3$ (cm^{-1})	$NI_3 \cdot ND_3$ (cm^{-1})	$^{15}NI_3 \cdot {}^{15}NH_3$ (cm^{-1})	Assignment on basis of the tetrahedral model (see text)
3346 w		[b]	ν H—N
3243 w		[b]	ν H—N
1600 w (broad)		[b]	δ H—N—H
1095 m		[b]	δ H—N—H
558 m	556 ss	540 m	ν_{as} N—I2,3
514 m (shoulder)	410 ss	516 m	NH_3 rocking
488 s	491 ss	470 s	ν_s N—I2,3
382 ss	375	371 ss	ν_{as} N—I1
257 ss	237 ss	250 ss	ν $H_3N\cdots$I2
175 w	173 w	174 w	ν_s N—I1
148 s	146 s	148 ss	δ I1—N—I2,3
142 ss	83 ss	148 ?	δ $H_3N\cdots$I2—N
133 ss	131 ss	136 ss	δ I1—N—I1
113 w	113 w	114 w	δ I1—N—I2,3
75–78 ss	68 m	77 ss	δ I2,3—N—I2,3

[a] Intensities: ss, very strong; s, strong; m, medium; w, weak.

[b] The NH bands of $^{15}NI_3 \cdot {}^{15}NH_3$ are the same, to within the accuracy of the measurements, as those of $^{14}NI_3 \cdot {}^{14}NH_3$.

iodine atoms (I1, I1'). Nine normal vibrations would then be expected, eight being IR-active; all these vibrations were found (98).

The structural and spectroscopic data make it possible to calculate approximate force constants for the various N—I bonds in nitrogen triiodide-1-ammonia (98). In the N1—I1'—N1' chains, the value for f_{N-I} is 0.81 and that for the N1—I2 or N1—I3 bonds is 1.01 mdyn/Å; for the I2—N2 bond, f_{N-I} is 0.58 mdyn/Å. Table II shows the force constants with the corresponding bond lengths, together with force constants and bond lengths for other nitrogen–iodine compounds. Although force constants were calculated on the basis of different approximations, the expected inverse relationship between the force constant and the bond length is in general fulfilled. This lends support to the values calculated for the force constants. The force constant of the $[I(uro)_2]^+$ cation shows a greater deviation, which may be explained by the fact that the simplifying assumption of a rigid sphere for the ligands on iodine made for the ions $[I(py)_2]^+$, $[I(3\text{-pic})_2]^+$, and $[I(uro)_2]^+$, is no longer correct in the case of urotropin. According to Siebert (140), a value of 1.6 mdyn/Å, taking account of the Goubeau

TABLE II

DISTANCES, STRETCHING FORCE CONSTANTS, AND BOND ORDERS OF
THE N—I BOND

Compound	r_{N-I} (Å)	f_{N-I} (mdyn/Å)	N_{N-I} (140)	References
cis-OSNI	2.03[a]	1.93	1.21	31
$NI_3 \cdot NH_3$: N1—I2,3	2.15	1.01	0.63	58, 98
$[I(py)_2]^+$	2.16 ± 0.1	1.08	0.68	64, 56
$[I(3\text{-pic})_2]^+$	2.24	1.02	0.64	118, 56
$(CH_3)_3N \cdot I_2$	2.27	0.87	0.54	142, 38
$[I(uro)_2]^+$	2.30	1.13	0.71	125, 88
$NI_3 \cdot NH_3$: N1—I1	2.30	0.81	0.51	58, 98
$NI_3 \cdot NH_3$: N2—I2	2.53	0.58	0.36	58, 98

[a] Sum of covalent radii according to Pauling (121).

factor (51), is put forward as a theoretical standard for the valence force constant of the N—I bond in nitrogen triiodide (with a free electron pair on nitrogen). From this the bond orders shown in Table II may be calculated (140).

In Table III, valence force constants and bond orders for iodides of elements in Group V of the periodic system are compared. Whereas bond orders from phosphorus to bismuth are about unity (monomeric iodides with a predominantly σ-bond character), the values for nitrogen show a significant deviation: Nitrogen triiodide is, indeed, a polymer and has complex bonding relationships (see the foregoing).

The reaction enthalpy for the decomposition reaction of nitrogen triiodide-1-ammonia,

$$2NI_3 \cdot NH_3 \longrightarrow 2NH_3 + N_2 + 3I_2 \tag{1}$$

TABLE III

STRETCHING FORCE CONSTANTS AND BOND ORDERS OF THE
IODIDES OF THE FIFTH MAIN GROUP ELEMENTS

Compound	f_{N-I} (mdyn/Å)	f_1 (mdyn/Å)	N	Ref.
$NI_3 \cdot NH_3$	1.01	1.6	0.63	98
	0.81		0.51	
	0.58		0.36	
PI_3	1.21	1.03	1.18	141
AsI_3	1.05	0.95	1.11	141
SbI_3	0.81	0.75	1.08	107
BiI_3	0.65	0.71	0.92	107

has been determined with a bomb calorimeter as $\Delta H = -42.1 \pm 2$ kcal/mole (1). This value has been used to calculate the N—I bond energy (1), although apparently without taking into account the lattice energy and the polymeric character of nitrogen triiodide. (The hypothetical monomeric triiodide has three N—I bonds per unit formula NI_3, whereas the polymer shows four such bonds, of two different lengths, per unit formula.)

B. Exchange Studies on Nitrogen Triiodide-1-ammonia with ND_3-Ammonia and Ammonia-^{15}N

From the structure of the ammoniate, a relatively easy exchange of the ammonia with other bases (see Section II, E) or isotopically labeled ammonia is to be expected. In experiments involving the condensation of ammonia-^{15}N onto nitrogen triiodide-1-pyridine in order to prepare labeled $NI_3 \cdot {}^{15}NH_3$, the expected product was, surprisingly, not obtained:

$$NI_3 \cdot Py + {}^{15}NH_{3(liq)} \quad \rightleftharpoons\!\!\!/ \quad NI_3 \cdot {}^{15}NH_3 + Py \qquad (2)$$

Instead, $^{15}NI_3 \cdot {}^{15}NH_3$ was produced, with exchange also occurring in the NI_4 tetrahedra (N1, Fig. 1). The reverse reaction also fails to give the desired product: Rather, it gives $NI_3 \cdot NH_3$ (97):

$$^{15}NI_3 \cdot Py + NH_{3(liq)} \quad \rightleftharpoons\!\!\!/ \quad {}^{15}NI_3 \cdot NH_3 + Py \qquad (3)$$

Since nitrogen triiodide-1-ammonia is only very slightly soluble in liquid ammonia (ca. 1 mg/ml at $-75°C$) and no dissolution was observed in the experiments, an exchange mechanism other than that involving solution seems likely. To clarify this point exchange experiments between solid nitrogen triiodide-1-ammonia and gaseous ammonia were made and followed by IR spectroscopy (29, 30). They are summarized in Table IV.

The exchange experiments show in the first place that, in absence of a solvent, not only the adduct ammonia (N2, Fig. 1) but also nitrogen in the NI_4 tetrahedra (N1, Fig. 1) can be exchanged. Furthermore, the completeness of the exchange is seen to depend on three factors.

1. The quantity of ammonia available. Experiments 1–3 show that in a static ammonia atmosphere corresponding to a molar ratio of about $NI_3 \cdot NH_{3(solid)} : 2NH_{3(gas)}$, no exchange of the adduct ammonia occurs, whereas, if for the same time a greater excess of ammonia is available in a flow system, exchange of the adduct ammonia is complete. Lack of exchange in a stationary ammonia atmosphere cannot be attributed to too low a diffusion rate of the gaseous ammonia since it has a low molecular weight and a high diffusion coefficient.

TABLE IV

EXCHANGE EXPERIMENTS BETWEEN SOLID NITROGEN TRIIODIDE-1-AMMONIA AND GASEOUS AMMONIA AT $-5°C$ AND ATMOSPHERIC PRESSURE

No.	Starting products[a]		Reaction conditions (hr)	Final product[a]: Nitrogen triiodide-1-ammonia
	Nitrogen triiodide-1-ammonia	Ammonia		
1	$NI_3 \cdot ND_3$, 2	NH_3	Flow, 24	$NI_3 \cdot NH_3$, 4
2	$NI_3 \cdot NH_3$, 2	ND_3	Static, 20	$NI_3 \cdot NH_3$, 4
3	$NI_3 \cdot NH_3$, 2	ND_3	Flow, 24	$NI_3 \cdot NH_3/ND_3$, 4; only a little NH_3
4	$^{15}NI_3 \cdot ^{15}NH_3$, 2	ND_3	Flow, 5	$^{14/15}NI_3 \cdot ^{15}NH_3/ND_3$, 4; $^{14}NI_3{:}^{15}NI_3 = 1{:}2$; $^{15}NH_3{:}ND_3 = 1{:}1$
5	$^{15}NI_3 \cdot ^{15}NH_3$, 2	NH_3	Flow, 0.5	$^{14/15}NI_3 \cdot ^{14/15}NH_3$; $^{14}NI_3{:}^{15}NI_3 = 1{:}2\text{-}3$
6	$^{15}NI_3 \cdot ^{15}NH_3$, 2	NH_3	Flow, 2	$^{14/15}NI_3 \cdot ^{14/15}NH_3$; $^{14}NI_3{:}^{15}NI_3 = 1{:}1$
7	$^{15}NI_3 \cdot ^{15}NH_3$, 2	NH_3	Flow, 2.25	$^{14/15}NI_3 \cdot ^{14/15}NH_3$; $^{14}NI_3{:}^{15}NI_3 = 1{:}1$
8	$^{15}NI_3 \cdot ^{15}NH_3$, 2	NH_3	Flow, 18	$^{14/15}NI_3 \cdot ^{14/15}NH_3$; $^{14}NI_3{:}^{15}NI_3 = 1{:}1$
9	$^{15}NI_3 \cdot ^{15}NH_3$, 2	NH_3	Flow, 67	$^{14/15}NI_3 \cdot ^{14/15}NH_3$; $^{14}NI_3{:}^{15}NI_3 = 1{:}1$; $^{14}NH_3{:}^{15}NH_3{:}\text{mainly}^{14}NH_3$
10	$NI_3 \cdot NH_3$, 1	NH_3	Flow, 3.5	$NI_3 \cdot NH_3$, 2
11	$^{15}NI_3 \cdot ^{15}NH_3$, 1	NH_3	Flow, 2	$^{14/15}NI_3 \cdot ^{14/15}NH_3$, 2; $^{14}NI_3{:}^{15}NI_3 = 1.5{:}1$
12	$^{15}NI_3 \cdot ^{15}NH_3$, 3	NH_3	Flow, 3	$^{15}NI_3 \cdot ^{14/15}NH_3$; $^{14}NH_3{:}^{15}NH_3 = 1{:}1$

[a] (1) Poorly crystallized $NI_3 \cdot NH_3$ from dilute solutions of ammonium chloride and iodine chloride by rapid mixing at room temperature; (2) moderately crystallized $NI_3 \cdot NH_3$ from dilute solutions of ammonium chloride and iodine chloride by slow dropwise addition of the latter at 0°C; (3) well-crystallized $NI_3 \cdot NH_3$ from 2 by standing 20 hr in an atmosphere of gaseous ammonia at $-5°C$; (4) only NH_3/ND_3 exchange detected, no H/D exchange.

2. The duration of the exchange experiment. Experiments 5 and 6 show that nitrogen in the NI_4 tetrahedra is more completely exchanged in an experiment lasting 2 hr than after half an hour. Similarly, exchange of the adduct ammonia has proceeded further after 24 hr than after 5 hr (Experiments 3, 4, and 9).

3. The state of the nitrogen triiodide-1-ammonia crystals. If the experiment is prolonged by several hours, there is no further increase in nitrogen exchange in the NI_4 tetrahedra, whereas exchange of adduct ammonia goes to completion in this time period (Experiments 6–9). Experiment 10 provides the key to this observation, showing that the ammoniate sinters or recrystallizes when exposed to gaseous ammonia: Poorly crystallized material with a high surface area is transformed into material that is visibly more crystalline and of lower surface area. Sintering processes occurring by a gas–solid interaction are already known and have been closely studied for various examples (67). Experiments 6, 11, and 12 show that the state of the crystals and their surface area play an important part in the exchange process: Poorly crystalline nitrogen triiodide-1-ammonia of high surface area is able to exchange three-fifths of the compound; moderately crystalline material, a half; and highly crystalline material of low surface area, none of the nitrogen in the NI_4 tetrahedra in comparable times. Exchange of the adduct ammonia, on the other hand, is little influenced by the state of the crystals.

Consequently, we have developed the following picture of the exchange process on a poorly crystalline sample of high surface area.

1. Gaseous ammonia first comes into equilibrium with the surface by an adsorption–desorption process.

2. The adsorbed ammonia (a) by diffusion comes into equilibrium with adduct ammonia between the puckered sheets of the structure and (b) by surface iodine–hydrogen exchange comes into equilibrium with imperfectly formed, energy-rich chains or planes of NI_4 tetrahedra.

3. In parallel with processes 1 and 2, sintering of the ammoniate takes place with a reduction in surface. As a result, process 2b is gradually stopped whereas 2a is at most retarded.

Complete exchange of nitrogen in the NI_4 tetrahedra (N1, Fig. 1) in liquid ammonia (97), which has been mentioned previously, can for the time being not be better explained than by postulating a solution mechanism. Complete solution, to be sure, does not seem to be necessary. It is sufficient to assume a superficial solution equilibrium that retards recrystallization and surface area reduction of the nitrogen triiodide-1-ammonia.

C. NITROGEN TRIIODIDE-3-AMMONIA $(NI_3 \cdot 3NH_3)_n$ AND ITS IR SPECTRUM

Nitrogen triiodide-3-ammonia has an IR spectrum at $-60°C$ (9) which is very similar to that of the monoammoniate (Table V) (89). The similar position of the intense NI skeletal vibrations is especially striking. The spectrum, therefore, leads to the view that NI_4 tetrahedra are again present in this compound and that they resemble those in the monoammoniate in structure and dimensions. In all probability, the NI_3 molecules form infinite chains of linked NI_4 tetrahedra as before. Whether the chains are linked through I—I contacts as for nitrogen triiodide-1-ammonia (see Section II, A) or whether there are isolated chains with the I2 and I3 atoms (Fig. 1) fully coordinated by ammonia cannot be determined from the IR spectrum. The multiple absorptions in the I—NH_3 vibrational region (boxed in Table V) leave the possibility open that I—I contacts in the triammoniate are suppressed in favor of I—NH_3 contacts. On the other hand, the observation that the crystal form does not change in going from the 3-ammoniate to the 1-ammoniate (91, 57) indicates that there is no change in the NI skeleton and that the additional ammonia molecules of the 3-ammoniate are held in lattice cavities, possibly by hydrogen bonds.

D. NITROGEN TRIIODIDE AMMONIATES FORMED BELOW $-75°C$ (MONOIODAMINE-1-AMMONIA AND MONOIODAMINE FORMED BELOW $-85°C$)

Red–brown (well-crystallized) or black (finely divided) nitrogen triiodide-1-ammonia may be prepared not only directly from aqueous solution (41) but also via green iridescent nitrogen triiodide-3-ammonia as an intermediate in, for example, the reaction of iodine with liquid ammonia (78, 28). In this case one is restricted by the melting and boiling points of ammonia to the temperature range of $-75°C$ to $-35°C$. It has been shown earlier by the most diverse experimental methods that only the two nitrogen triiodide ammoniates referred to exist above $-75°C$ (28, 91).

The existence of further ammonia adducts stable at lower temperatures was established by the reaction of iodine bromide at $-75°$ to $-85°C$ with a mixed solvent consisting of two parts by volume of chloroform with one of ammonia (30, 110, 55). This mixture remains liquid to $-95°C$. In addition to soluble ammonium bromide, the reaction yields a red precipitate with a nitrogen-to-iodine ratio of 1.5–2.5:1, i.e., a mean of about 2:1. Its low-temperature IR spectrum

TABLE V

Infrared Spectrum of Nitrogen Triiodide-3-ammonia Compared with That of Nitrogen Triiodide-1-ammonia[a]

NI$_3$·NH$_3$ at −60°C (cm⁻¹)	NI$_3$·3NH$_3$ at −60°C (cm⁻¹)	NI$_3$·3ND$_3$ at −60°C (cm⁻¹)	NI$_3$·ND$_3$ (cm⁻¹)	Assignment on basis of the tetrahedral model
3346 w				ν H–N
3243 w				ν H–N
1600 w (broad)				δ H–N–H
1095 m				δ H–N–H
558 m	579 s	574 ss	556 ss	ν_{as} N–I2,3
514 m (shoulder)	530 ? w 512 m (shoulder)	433 ss	410 ss	NH$_3$ rocking
488 s	506 s	505 s	491 ss	ν_s N–I2,3
382 ss	385 ss	369 w	375	ν_{as} N–I1
257 ss	203–268 ss	200–244 ss	237 ss	ν H$_3$N···I2
175 w	172 w	164 w	173 w	ν_s N–I1
148 w	141 w	142 w	146 s	δ I1–N–I2,3
142 ss	128 ss 113 w (shoulder) 101	125 m 110 m 90 ? m 87 ? s	83 ss	δ H$_3$N···I2–N
133 ss	133 m (shoulder)	136 s[b] 131 s[b]	131 ss	δ I1–N–I1
113 w	107 m	104 m	113 w	δ I1–N–I2,3
75–78 ss	81–83 ss	77–81 ? ss	68 m	δ I2,3–N–I2,3

[a] Intensities: ss, very strong; s, strong; m, medium; w, weak.
[b] Crystal field splitting?

TABLE VI

INFRARED SPECTRUM OF NITROGEN TRIIODIDE-CA. 5-AMMONIA AND NITROGEN TRIIODIDE-CA. 2-AMMONIA COMPARED WITH NITROGEN TRIIODIDE-3-AMMONIA AND NITROGEN TRIIODIDE-1-AMMONIA [a]

$NI_3 \cdot NH_3$ (cm^{-1})	$NI_3 \cdot \sim 2NH_3$ ca. $-100°C$ (cm^{-1})	$NI_3 \cdot \sim 5NH_3$ ca. $-100°C$ (cm^{-1})	$NI_3 \cdot 3NH_3$ $-60°C$ (cm^{-1})	Assignment on basis of the tetrahedral model
3346 w				ν H—N
3243 w				ν H—N
1600 w (broad)				δ H—N—H
1095 m				δ H—N—H
558 m	574 m	579 s	579 s	ν_{as} N—I2,3
	543 s			
514 m (shoulder)		515 m (shoulder)	530? w	NH$_3$ rocking
			512 m (shoulder)	
488 s	502 m	506 ss	506 s	ν_s N—I2,3
	483 m			
382 ss	380 ss	385 ss	385 ss	ν_{as} N—I1
257 ss	248 s	200–270 ss	203–268 ss	ν H$_3$N$\cdots$I2
175 w	174 w	168 w	172 w	ν_s N—I1
148 w	148 w	141 w	141 w	δ I1—N—I2,3
	141 w			
142 ss	124 s	128 s	128 s	δ H$_3$N$\cdots$I2—N
		113 w	113 w (shoulder)	
		101 w	101 w	
133 ss	132 w	132 m (shoulder)	133 m (shoulder)	δ I1—N—I1
113 w	112 w	107 m	107 m	δ I1—N—I2,3
	105, 103 w			
75–78 ss	89, 81 s	81–83 ss	81–83 ss	δ I2,3—N—I2,3
	79			

[a] Intensities: ss, very strong; s, strong; m, medium; w, weak.

(Table VI) corresponds almost exactly with that of nitrogen triiodide-3-ammonia, the almost identical position of the intense NI skeletal vibration being especially noteworthy. This leads us to suppose that the red compound is a pentammine, $NI_3 \cdot \sim 5NH_3$ with the same polymeric $(NI_3)_n$ framework.

The relationship between the triammine and the pentammine is also brought out by the observation that the red pentammine is converted into the green triammine above $-75°C$ either in liquid ammonia or in the mixed solvent referred to above. The green compound is also transformed into the red one in the mixed solvent at $-75°$ to $-85°C$.

Below $-85°$ to $-90°C$, iodine bromide and the mixed solvent do not give the red pentammine but a second red compound that likewise has a nitrogen-to-iodine ratio of 1.5–2.5:1, although the low-temperature IR spectrum is different. Above $-85°C$ it is transformed into the first red compound, nitrogen triiodide-5-ammonia (30, 55). We believe the second red compound to be monoiodamine-1-ammonia, which is discussed in Section II, L. The existence of two red compounds was not known in earlier publications (78, 28, 91, 81, 27, 79, 80), but one can deduce from the temperature during the preparation which compound was obtained.

If a sample of the red pentammine is pumped for several days at $-90°C$ at 10^{-5} Torr or covered with ether and held for 4 hr at $-75°C$, a black substance is produced with a nitrogen iodine ratio 1.1–1.4:1, that is, approximately 1:1. Its low-temperature IR spectrum (Table VI) shows, in addition to bands of the polymeric 5- or 3-ammoniate, those of the polymeric 1-ammoniate (with some small deviations). We consider the black compound to be a nitrogen triiodide with the mean composition $NI_3 \cdot \sim 2NH_3$ in which $NI_3 \cdot 3NH_3$ and $NI_3 \cdot NH_3$ structural units are present; the NI_3 skeleton should be polymeric and made up of NI_4 tetrahedra as in the 1-, 3-, and 5-ammoniates (110).

Removal of ammonia from the red monoiodamine-1-ammonia in the same way does not result in a black compound with the composition $NI_3 \cdot \sim 2NH_3$; instead we obtain a second black compound with a reproducible nitrogen-to-iodine ratio of 1:1 and a different low-temperature IR spectrum (30). We believe this to be monoiodamine. Authors of earlier publications did not realize that two black compounds existed, and they were, therefore, not differentiated. It is, however, possible to decide which of the two was handled from the temperature during the preparation of the red compound from which ammonia was removed (78, 28, 81, 27, 80).

It is impossible to decide from the IR spectrum how the additional ammonia molecules in the pentammine are bonded to the $(NI_3)_n$ framework (cf. what was said in Section II, C about additional ammonia molecules in the triammine). The ease and speed with which red pentammine is re-formed from the black diammine and ammonia, which is in contrast to the difficult and slow formation of green triammine from the monoammine, leads us to suppose that both of the new polymeric nitrogen triiodide ammines may have a disordered lattice structure with a large surface area and possibly only short $(NI_3)_n$ chains of varying length. The poor reproducibility of the analytical results points to the same conclusions.

E. Adducts of Nitrogen Triiodide with Aprotic N, O, S, and P Bases

As was shown in Section II, B ammonia in nitrogen triiodide-1-ammonia is relatively easily exchanged with isotopically labeled ammonia. A corresponding exchange without change in the polymeric N—I framework is also possible with aprotic N, O, S, and P bases. So far the reaction with tertiary N bases has been most fully investigated.

1. Adducts with Tertiary N Bases

These adducts are usually formed by direct reaction of preformed nitrogen triiodide-1-ammonia with excess of base, with or without additional solvent (usually water), and ammonia is set free. It is almost always possible to work at room temperature or 0°C. It seldom happens that the desired adduct is obtained by adding the base (pyridine, quinuclidine, urotropine) to the reaction mixture in which nitrogen triiodide is being made from ammonia and iodine monochloride or potassium triiodide. The compounds obtained and their properties are shown in Table VII.

The formulas of these adducts are, as a rule, established by the analytically determined ratio of total nitrogen to iodine, which in most cases is 2:3 within experimental error. In the case of nitrogen triiodide-1-pyridine, 1-picoline-4, and 1-quinoline, it is possible to differentiate between inorganic and organic nitrogen: the ratio $N_{inorg}:I:N_{org}$ is 1:3:1, as would be expected (77, 147). In reactions with urotropine and quinuclidine, different ratios were found [urotropine:$N_{inorg}:I:N_{org}$ = 1:5:4 (82); total analysis of quinuclidine gives C 13.2, H 2.2, N 3.7, and I 80.2%; $N_{inorg}:I:N_{org}$ = 1:4:1 (146, 90)]. In fact, urotropine forms an adduct with the abnormal composition $NI_3 \cdot I_2 \cdot (CH_2)_6N_4$. Since this is based on a changed polymeric structure, it is discussed

TABLE VII

Adducts of Nitrogen Triiodide with Tertiary N Bases

Base	Formula	Color	Stability	References
Pyridine	$NI_3 \cdot C_5H_5N$ $NI_3 \cdot 3C_5H_5N$?	Dark brown Green	25°C decomposition −30°C decomposition	77, 62, 147, 7, 61 77, 7, 146
Picoline-4	$NI_3 \cdot CH_3C_5H_4N$		25°C decomposition	62, 147
Picoline-3	$NI_3 \cdot CH_3C_5H_4N$	Brown-black	25°C decomposition	146, 90
Picoline-2	$NI_3 \cdot CH_3C_5H_4N$	Brown-black		146, 90
Lutidine	$NI_3 \cdot (CH_3)_2C_5H_3N$	Brown-black	25°C decomposition	146, 90

	Formula	Color	Stability	Ref.
Collidine H_3C (structure with CH_3, CH_3)	$NI_3 \cdot (CH_3)_3C_5H_2N$	Brown-black	0°C decomposition	146, 90
Quinoline (structure)	$NI_3 \cdot C_9H_7N$ $NI_3 \cdot 3C_9H_7N$?	Green-black	25°C decomposition Light-sensitive 25°C decomposition	147 72
Pyrazine (structure)	$NI_3 \cdot C_4H_4N_2$	Brown-black	25°C decomposition	145
Trimethylamine	$NI_3 \cdot (CH_3)_3N$	Black	0°C rapid decomposition	146, 90
Triethylamine	$NI_3 \cdot (C_2H_5)_3N$?		0°C inst. decomposition	146, 90
Quinuclidine (structure)	$N:I:C_7H_{13}N$ $= 1:4:1$	Black		146, 90
Urotropine (structure)	$NI_3 \cdot I_2 \cdot (CH_2)_6N_4$	Red	25°C stable	82, 124

later in Section II, I. The nature of the compound formed by quinucli-dine has not yet been clarified.

X-Ray powder photographs of nitrogen triiodide-1-pyridine and -1-picoline show both compounds to form monoclinic crystals like those of the 1-ammonia, and to have almost the same angle β as well as b- and c-axes of the unit cells ($\beta = 93°-98°$; $b = 7.5-7.6$, and $c = 6.3-6.4$ Å) (62). This indicates that the sheets of NI_4 tetrahedra that extend in the bc plane of the $(NI_3)_n$ framework (cf. Section II, A and Fig. 1) remain practically unchanged in going from the 1-ammoniate to nitrogen triiodide-1-pyridine or 1-picoline-4. On the other hand, the progressive lengthening of the a-axis when ammonia is replaced by pyridine or pico-line-4 ($a = 7.1, 10.6, 11.3$ Å) shows the steadily increasing space needed by the base bound to I2 between the puckered sheets of tetrahedra.

Finally, it is shown by a full X-ray structural investigation of the pyridine adduct (Fig. 2) (61) that the adducts of nitrogen triiodide with tertiary N bases are structurally very similar to nitrogen triiodide-1-ammonia (Section II, A). Chains of tetrahedra with infinite —N—I—N—I— linkages showing a characteristic translational period of 7.5 Å parallel to the b-axis again occur. The adduct pyridine mole-cules are bonded through nitrogen (N2) to I2. Table VIII summarizes the corresponding distances and angles. They emphasize the structural

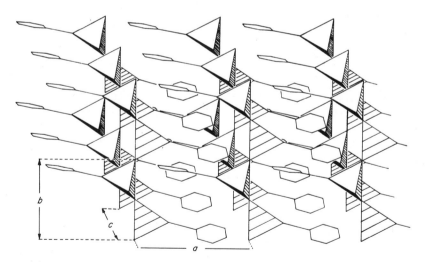

FIG. 2. Spatial representation of the monoclinic structure of nitrogen triiodide-1-pyridine; projection approximately on the (001) plane. In spite of a somewhat different presentation, the analogy of the chain structure of NI_4 tetrahedra with the structure shown in Fig. 1 is apparent. The notation of the N and I atoms of the polymeric N—I framework in Fig. 1 is, therefore, applicable to Fig. 2 (61).

TABLE VIII

DISTANCES AND ANGLES IN NITROGEN TRIIODIDE-1-AMMONIA
AND -1-PYRIDINE

Distance or angle measured	$NI_3 \cdot C_5H_5N$	$NI_3 \cdot NH_3$
I3—I2 (neighboring chain)	3.93 Å	3.36 Å
N1—I1,I1′	2.36 Å	2.30 Å
N1—I2,I3	∼2.10 Å	∼2.15 Å
I2—N2	2.59 Å	2.53 Å
N1—I1′—N1′	180°	180°
N1—I2—N2	174°	176°
N1—I3—I2 (neighboring chain)	153°	172°

analogy but also show that in the pyridine adduct the cohesion of the separate chains of tetrahedra to the sheets of tetrahedra in the bc plane is not as strong as in the ammonia adduct. The angle N1—I3—I2 (neighboring chain) is distorted to 153° by the bulky pyridine ring and formation of the normally linear 3 center–4 electron bond between the 3 atoms is made more difficult. The iodine–iodine distance (I3—I2, neighboring chain) between the chains of tetrahedra is, as a result, increased from 3.36 to 3.93 Å and can be looked on as only a very weak I—I contact (the sum of the van der Waals radii of 2 iodine atoms is 4.30 Å).

The nitrogen triiodide–tertiary N-base adducts, for which no full X-ray structural analysis is available, may be identified by their IR spectra, besides their dark color, their tendency to explode on shock or friction, and their analyses. The upper part of Table IX shows the spectra of these adducts to exhibit the three strong NI-stretching vibrations at ca. 380, ca. 490, and ca. 560 cm^{-1}, which also occur in the spectrum of nitrogen triiodide-1-ammonia and are characteristic of the NI_4 tetrahedra that make up its polymeric N—I framework. The fourth, less intense, NI-stretching vibration at ca. 170 cm^{-1} and the INI deformation vibrations are, on the other hand, less often firmly assigned, as they partly overlap frequencies of the N base and may also be difficult to observe because of the poor quality of the spectra. Further proof for the formation of a nitrogen triiodide adduct with ammonia replaced by tertiary N base is provided by the absence in the IR spectrum of all the vibrational frequencies of ammonia, especially the very characteristic rocking frequency at 514 cm^{-1}. In their place intense bands associated with the N base occur. These may be displaced, as in their iodine adducts, or may overlap with the bands of the N—I framework.

TABLE IX

INFRARED SPECTRA OF THE N—I FRAMEWORK FOR ADDUCTS OF NITROGEN TRIIODIDE WITH N, O, S, AND P BASES COMPARED WITH NITROGEN TRIIODIDE-1-AMMONIA[a]

Adduct	ν_{as} N—I2,3	ν_s N—I2,3	ν_{as} N—I1	ν_s N—I1	δ I1—N— I2,3	δ I1—N—I1	δ I1—N— I2,3	δ I2,3—N— I2,3	Refs.
NI3·NH3	558	488	382	175	148	133	113	75–78	98
NI3·pyridine	567	486	372	164	(135)	127	82	70 sh	98
NI3·3pyridine?	575	500	384	ca. 175		127	72 sh	70	7, 77
NI3·picoline-3	560	504	374						146
NI3·picoline-2	556	495	369						146
NI3·lutidine	562	(481)	388						146
NI3·collidine, −40°C	564	500	394						146
NI3·3quinoline?	566	510	380			137			77, 72
NI3·pyrazine	564	503	384						145
NI3·trimethylamine	(562)	(486)	(386)						146
x-NI3·y-dinitrophenolate?	553	482	375						53
NI3·5.5pyridine-N-oxide, −40°C	(551)	489	368						53
NI3·2.5tetrahydrofuran, −40°C	563	~500	388	162	142	130			53
NI3·2.5dioxane-1.4	573	509	377	(163)	145	134	103	82	53
NI3·2.5thiophane, −40°C	563	506	375						53
NI3·0.5dithiane, −40°C	558	480	369	169 sh					53

[a] Values expressed per [cm^{-1}].

The IR spectrum of the compound with quinuclidine (see above and Table VII) differs considerably from the spectra shown in Table IX; consequently, the structure of this compound cannot yet be predicted.

Decomposition of the nitrogen triiodide–tertiary N-base adducts generally leads to the more stable adducts of iodine itself with the tertiary base. Thus, nitrogen triiodide-1-picoline-3 or 1-trimethylamine in aqueous suspension at 0°C is transformed after a time into red-brown diiodine-1-picoline-3, $I_2 \cdot pic$-3, or yellow diiodine-1-trimethylamine, $I_2 \cdot N(CH_3)_3$. The color and IR spectrum of the former lead us to suppose that it is bispicoline-3-iodonium triiodide, $[I(pic$-$3)_2]^+I_3^-$. Attempts to obtain crystals of the compound with quinuclidine, the nature of which has not been elucidated (see above and Table VII) resulted, apparently because of the duration of the experiments, in yellow diiodine-1-quinuclidine, $I_2 \cdot quin$, or blue-black needle-shaped crystals of quinuclidinium pentaiodide, $[quin \cdot H]^+I_5^-$. The latter has an interesting structure with infinite iodine chains made up of I_3^- ions and I_2 molecules (90) and is transformed at 25°C within weeks into a red-brown substance, presumably quinuclidinium triiodide, $(quin \cdot H)^+I_3^-$. Clearly, formation of the I_2–base adduct from the NI_3–base adduct is more favored as the base becomes stronger. In the case of adduct $NI_3 \cdot N(C_2H_5)_3$, decomposition in aqueous suspension at 0°C yields iodoform in a few minutes—possibly via adduct $I_2 \cdot N(C_2H_5)_3$.

2. Adducts with Aprotic O, S, and P Bases

A dinitrophenol or dioxane adduct may be obtained from aqueous solution by the general synthetic methods given in Section II, E, 1. For the latter, it is necessary to add hydrochloric acid in order to bring about replacement of the strongly basic ammonia by weakly basic dioxane. Experiments designed to replace ammonia in nitrogen triiodide-1-ammonia by excess of the basic oxalate or acetate anion were unsuccessful. The other adducts shown in Table X are either markedly more temperature-sensitive or are derived from bases that are insoluble in water, so that their synthesis from aqueous solution at $\geqslant 0°C$ is impossible. In these cases they are obtained by addition of the base during the synthesis of nitrogen triiodide in methylene chloride at $-40°C$. It is necessary to make sure that there is always a deficiency of ammonia, so that it does not compete as an adduct molecule with the more weakly basic O, S, or P base. Even under these favorable conditions, however, thiophene is not incorporated into the nitrogen triiodide framework.

TABLE X

Adducts of Nitrogen Triiodide with Aprotic O, S, and P Bases

Base	Formula	Color	Stability	References
2,4-Dinitrophenol	$x\text{-}NI_3 \cdot y\text{-}(O_2N)_2C_6H_3ONH_4$?	Black	25°C decomposes	53
Pyridine-N-oxide	$NI_3 \cdot \sim 5.5C_5H_5NO$	Black	$> -40°C$ decomposes	53
Tetrahydrofuran	$NI_3 \cdot \sim 2.5C_4H_8O$	Black	$> -40°C$ decomposes	53
1,4-Dioxane	$NI_3 \cdot \sim 2.5C_4H_8O_2$		25°C decomposes	53
Thiophane	$NI_3 \cdot \sim 2.5C_4H_8S$	Black	$> -40°C$ decomposes rapidly	53
1,4-Dithiane	$NI_3 \cdot \sim 0.5C_4H_8S_2$	Black	$> -40°C$ decomposes	53
Triphenylphosphine	$NI_3 \cdot \sim 0.5(C_6H_5)_3P$	Black	$> -40°C$ very unstable	53

In almost all instances it may be shown by an analytical determination of the N:I ratio that true nitrogen triiodide adducts are formed; a ratio of 1N:3I is found, as expected, although products obtained from methylene chloride are, from the synthesis, still contaminated with ammonium chloride. The proportion of O, S, or P base found by analysis (relative to iodine) or by gas chromatography has so far been determined for only one sample and is, therefore, not certain; variations in the value obtained from base to base could be an indication of the large surface area and the much distorted lattice structure of the nitrogen triiodide sample (cf. Section II, D).

Infrared spectra in the lower part of Table IX are also indicative of the formation of adducts with O, S, and P bases, as are analytical data, black color and shock or friction sensitivity of the compounds. Three strong N—I stretching vibrations at ca. 380, ca. 490, and ca. 560 cm^{-1} again occur for these compounds, although vibrations associated with ammonia are missing and in their place we found vibrations associated with the adduct base, which are in part displaced.

The compounds studied decompose much more readily to iodine or iodinated compounds than the adducts with tertiary N bases described above. Thus, there is an accelerated decomposition of an aqueous suspension of nitrogen triiodide-1-ammonia at room temperature in presence of tetrahydrofuran, thiophane, or triphenylphosphine. The reaction product from the last of these is triphenylphosphine diiodide, Ph_3PI_2; it is not possible first to detect or isolate nitrogen triiodide adducts with the three bases. Only by lowering the temperature to $-40°C$ can one isolate the corresponding adducts by adding the corresponding base to the reaction mixture for the synthesis of nitrogen triiodide in nonaqueous solution. On warming, there is gas evolution from the tetrahydrofuran adduct, and iodine separates, whereas, on warming the solution used for synthesizing the thiophane adduct, a red-violet crystalline substance is obtained (C 16.9, H 2.9, S 11.1, I 67.7, and N 2.2%). Its structure is not known at present. Under the most favorable conditions at low temperature, triphenylphosphine gives a very unstable adduct in low yield, the greater part of the triphenylphosphine being converted to Ph_3PICl.

It cannot be excluded from these observations that the rapid decomposition with nitrogen evolution of an aqueous suspension of nitrogen triiodide-1-ammonia at room temperature in the presence of p-nitrophenol (producing yellow 2,6-diiodo-4-nitro- phenol), cyanide, or thiocyanate (producing iodine with iodine cyanide or thiocyanate as intermediates) stems from the formation of very unstable adducts of these bases with nitrogen triiodide, which cannot be isolated. In the case of *p*-nitrophenol, this assumption is supported by the isolation of an adduct with the weaker "base" 2,4-dinitrophenol under the same conditions (see above and Table X).

F. Polymeric Nitrogen Triiodide Structural Analogs from Diiodomethylamine, CH_3NI_2, and N,N'-Tetraiodoethylenediamine, $I_2NCH_2CH_2NI_2$

These compounds show the same structural analogy to nitrogen triiodide as all of the compounds so far described, although at first sight this would not be expected from their formulas (*81, 84, 148, 85*). Red-brown microcrystalline diiodomethylamine may be obtained from methylamine and iodine or iodine chloride in aqueous solution at 0°C or from the anhydrous amine and iodine or *N*-iodosuccinimide at lower temperatures. Its adducts with pyridine (brick red), trimethylamine (reddish yellow), and methylamine [reddish yellow; this is not CH_3NHI, as was previously supposed (*78, 133*) but $CH_3NI_2 \cdot CH_3NH_2$ (*28, 81*)] are prepared by pouring or condensing the base onto diiodomethylamine or nitrogen triiodide-1-ammonia at $-15°$ to $-75°C$. The red-violet N,N'-tetraiodoethylenediamine is made by iodination of the diamine with nitrogen triiodide-1-ammonia or iodine chloride at 0°C (cf. Section II, L, 2). The formulas are established by N:I ratio analysis.

The most probable structure for monomeric diiodomethylamine (without adduct molecules) would have a trigonal pyramid with C_s symmetry for the C—N—I framework; six infrared active normal vibrations would then be expected. In a polymeric structure it is likely that the C—N—I framework would, like that of nitrogen triiodide-1-ammonia, be made up of I2CNI1I1' tetrahedra with C_s symmetry. These would be lined up through common I1,I1' atoms into infinite chains (Fig. 3). In this case, nine infrared active normal vibrations would

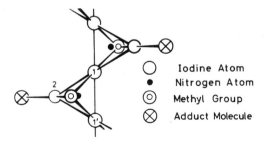

◯	Iodine Atom
●	Nitrogen Atom
◎	Methyl Group
⊗	Adduct Molecule

Fig. 3. Section of the chain of tetrahedra for the polymeric diiodomethylamine adduct (*84*).

be expected if we neglect vibrational coupling between the separate tetrahedra through the relatively heavy iodine atoms (I1,I1′) as we did for nitrogen triiodide-1-ammonia.

The compounds and the IR spectra associated with their C—N—I frameworks are shown in Table XI. In so far as they have been measured, nine absorptions are found for the diiodomethylamines, with little difference between the parent compound and its adducts. We must, therefore, conclude that (a) the diiodomethylamines have a polymeric structure and (b), as for nitrogen triiodide, there is no essential structural difference between the amine and its adducts.

Table XI also shows clearly the close relationship between the spectra of the diiodomethylamines and nitrogen triiodide-1-ammonia. Replacement of a heavy iodine atom by a light methyl group results in a frequency increase for the four stretching vibrations. For the bending vibrations, an increase occurs only where I—N—C angles are involved (instead of I—N—I). In addition, the diiodomethylamines exhibit an infrared-active torsional vibration that is missing for nitrogen triiodide-1-ammonia. The relationship between the spectra shows the principle underlying the polymeric structure to be the same in both substances.

X-Ray structural analysis of diiodomethylamine-0.5-pyridine, which is still incomplete, shows there to be a monoclinic unit cell with a translational period of 7.50 Å parallel to the b-axis characteristic of the infinite —N—I—N—I—N— chains. It establishes the identity of the polymer structural type in this compound and nitrogen triiodide-1-ammonia. Figure 4 shows a view of the structure in the direction of these infinite —N—I—N—I—N— chains. One can recognize two tetrahedra of an infinite chain of tetrahedra; in fact, they lie obliquely under one another.

It is evident from Table XI that, for the C—N—I framework of $N,N′$-tetraiodoethylenediamine, there are nine absorptions if C—C vibrations are omitted and the CN-stretching vibration, which appears as a doublet, is counted as one. Only unimportant displacements in relation to diiodomethylamine and its adducts with N bases are observed. As for them, the infrared spectrum may be explained in terms of I2CNI1I1 tetrahedra with C_s symmetry that are arranged in one direction in infinite chains with common I1 atoms and are bound together in the other direction through methylene groups (Fig. 5). Vibrational coupling between separate tetrahedra through the relatively heavy iodine atoms I1 may again be neglected. Thus for $N,N′$-tetraiodoethylenediamine the same polymer structural principle prevails as in nitrogen triiodide-1-ammonia and in the other compounds that have been discussed.

TABLE XI

INFRARED SPECTRUM OF THE C—N—I (OR N—I) FRAMEWORK OF DIIODOMETHYLAMINE, ITS ADDUCTS WITH N BASES, AND OF N,N'-TETRAIODOETHYLENEDIAMINE COMPARED WITH NITROGEN TRIIODIDE-1-AMMONIA

CH_3NI_2 (cm^{-1})	$CH_3NI_2 \cdot Py$ (cm^{-1})	$CH_3NI_2 \cdot \frac{1}{2}Py$ (cm^{-1})	$CH_3NI_2 \cdot (CH_3)_3N$ (cm^{-1})	$CH_3NI_2 \cdot CH_3NH_2$ (cm^{-1})	I_2NCH_2—CH_2NI_2 (cm^{-1})	Proposed assignment	$NI_3 \cdot NH_3$ (cm^{-1})	Assignment on basis of the tetrahedral model
970	980	(960)	975	990	1045 990	ν N—C	558	ν_{as} N—I2,3
545	544	{540[a] 525[a]}	518	535	520	ν N—I2	488	ν_s N—I2,3
415	404	385[b]	382	414	405	ν_{as} N—I1	382	ν_{as} N—I1
346	(349)	(350)	320	345	303[c]	ν_s N—I1	175	ν_s N—I1
193[d]	e	e	e	194[d]	187	δ I1—N—C	148	δ I1—N—I2,3
162[d]	e	e	e	169[d]	173	τ Torsion		
133[d]	e	e	e	{135[a,d] 131[a,d]}	141	δ I1—N—I1	133	δ I1—N—I1
113[d]	e	e	e	125[d]	131	δ I1—N—I2	113	δ I1—N—I2,3
234	230	231	225	226	232	δ I2—N—C	75	δ I2,3—N—I2,3

[a] Crystal field splitting possible.
[b] A further weak band occurs at 470 cm^{-1}.
[c] H_2O band?
[d] Measured at −60°C.
[e] Not measured.

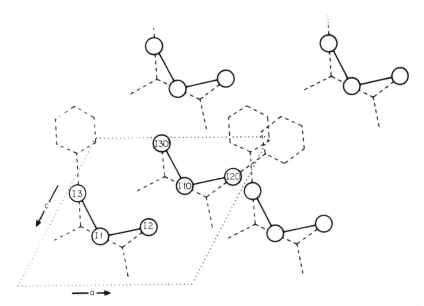

FIG. 4. Projection of the proposed structure for monoclinic diiodomethyl-amine-0.5-pyridine on the (010) plane. The iodine positions, shown by full lines, are definite; final predictions for the nitrogen and carbon positions cannot yet be made (dashed lines) (*148*).

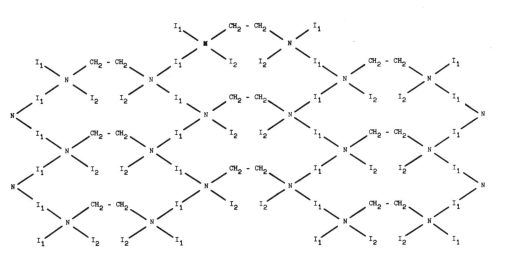

FIG. 5. Proposed structure for N,N'-tetraiodoethylenediamine (*85*).

The properties of diiodomethylamine and of N,N'-tetraiodo-ethylenediamine so far known are in accord with the polymeric structure. Both are difficultly soluble in all available solvents or dissolve with decomposition. The deep color of these compounds is indicative of iodine–iodine contacts between the chains of tetrahedra as proved for nitrogen triiodide-1-ammonia (see Section II, A). If these contacts are disturbed by adduct formation with N bases, as in nitrogen triiodide-3-ammonia (Section II, C) or in the adducts of diiodomethylamine with pyridine, trimethylamine, or methylamine (see above), the color becomes lighter.

G. Modification of the Polymeric Structural Principle of Nitrogen Triiodide by Partial Replacement of Iodine: Iododimethylamine, $(CH_3)_2NI$

When three iodine atoms are bonded to nitrogen, as in the hypothetical monomer of nitrogen triiodide, the compound stabilizes itself by polymerization as we have seen in Section II, A: The N atom in the monomer with its free pair of electrons acts as an electron donor, with one of the iodine atoms as acceptor, in an acid–base reaction. The same tendency to stabilize by polymerization also arises if only 2 iodine atoms are bonded to nitrogen, the third being replaced by a methyl group (Section II, F illustrates this with the example of diiodomethylamine, its N-base adducts, and N,N'-tetraiodoethylenediamine). When 1 atom of iodine and two methyl groups are bonded to a nitrogen atom, polymerization according to the same principle is still conceivable, but obviously does not take place (84, 146, 81).

Yellow microcrystalline iodomethylamine is prepared from the amine and iodine or iodine chloride in aqueous solution at 0°C or, better, because of the temperature sensitivity, in aqueous methanol at $-35°C$. In a very elegant preparative method the anhydrous amine is iodinated with nitrogen triiodide-1-ammonia, which may be replaced by N-iodosuccinimide or iodine, at $-55°$ to $-75°C$ (cf. Section II, L, 2). Crystalline iododimethylamine may also be made in chloroform at $-60°C$ by the following reaction:

$$(CH_3)_3SiN(CH_3)_2 + ICl \longrightarrow (CH_3)_2NI + (CH_3)_3SiCl \qquad (4)$$

Its formula may be established by N:I ratio analysis. It is soluble in dimethylamine, methylene chloride, and chloroform and does not exist in a brown form (59) as was previously assumed (120).

The solubility and light color show that stable polymerization and iodine–iodine contacts are unlikely, and the IR spectrum confirms this fact. One would expect six, infrared, active normal vibrations for the monomer (trigonal pyramid with C_s symmetry), whereas for the polymer there would be eight associated with the C—N—I framework, neglecting vibrational coupling between the separate tetrahedra (CCNII tetrahedra with C_{2v} symmetry). The spectrum (Table XII) does

TABLE XII

INFRARED SPECTRUM OF THE C—N—I
FRAMEWORK OF IODODIMETHYLAMINE
(SOLID)

Band position (cm^{-1})	Proposed assignment
1030	ν N—C
884	ν N—C
467	ν N—I
266	δ C—N—C
166	δ I—N—C
149	δ I—N—C
92⎱	Lattice vibrations?
67⎰	ν N$\cdots$I— ?

in fact show eight bands, but only one N—I stretching vibration at 467 cm^{-1}. In addition, in view of the replacement of 2 iodine atoms of the NI$_4$ tetrahedral unit by lighter methyl groups, the two bands below 100 cm^{-1} are at too low values to be deformation frequencies and must be attributed to lattice vibrations or, at the most, very weak intermolecular N$\cdots$I contacts. Thus the IR spectrum is more in keeping with a monomolecular structure. Proposed assignments for the remaining six bands correspond with this model. A structure is suggested in Fig. 6. Only by X-ray studies will it be possible to settle finally whether or not there are weak intermolecular N$\cdots$I contacts.

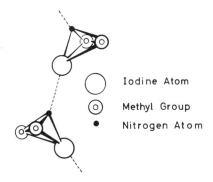

Iodine Atom

Methyl Group

Nitrogen Atom

FIG. 6. Proposed structure for iododimethylamine (84).

H. Iodine Adducts of Diiodomethylamine and Iododimethylamine

Stabilization of nitrogen triiodide and diiodomethylamine by polymerization to —N—I—N—I—N— chains raises the question of whether the electron-donor nitrogen atom in the iodamines is also able to add on molecular iodine as an electron acceptor and thus to stabilize the iodamines. Are there compounds such as I_3N—I_2—NI_3, I_3N—I_2, CH_3I_2N—I_2—NI_2CH_3, CH_3I_2N—I_2, $(CH_3)_2IN$—I_2—$NI(CH_3)_2$, or $(CH_3)_2IN$—I_2?

A urotropine-stabilized compound of formula I_3N—I_2 is reported in Section II, I. It may be mentioned at this point that it is possible to prepare iodine complexes of diiodomethylamine and iododimethylamine (99). Bright red 2-diiodomethylamine-1-diiodine, $(CH_3NI_2)_2 \cdot I_2$, and ochre-colored 2-iododimethylamine-1-diiodine, $[(CH_3)_2NI]_2 \cdot I_2$, are obtained from the free amine and excess iodine in aqueous solution at 0°C. Bright red-brown iododimethylamine-1-diiodine, $(CH_3)_2NI \cdot I_2$, may be prepared from iododimethylamine and an excess of ethereal iodine solution at -25°C, although it is not possible to obtain diiodomethylamine-1-diiodine, $CH_3NI_2 \cdot I_2$, in this way.

Analysis of these three compounds gives N:I values of 1:3 (2-diiodomethylamine-1-diiodine), 1:2 (2-iododimethylamine-1-diiodine), and 1:3 (iododimethylamine-1-diiodine). Both of the first two complexes are appreciably more stable thermally than the corresponding simple nitrogen–iodine compounds and may be stored for weeks at -30°C without decomposition.

According to detailed investigations that have already been mentioned (63), complexes of this type always have a linear structure, which is in full accord with results referred to above. Either a 2:1 complex, corresponding with D···X—X···D, or a 1:1 complex corresponding with D···X—X is formed (D = donor atom; X—X = elemental halogen). Which complex is formed depends on the donor or acceptor strength of the two partners forming the complex. For stronger donor–acceptor interaction, the 1:1 type appears to be preferred, whereas weaker interaction leads to the 2:1 type. A closely similar relationship is found for the complexes considered here: The strong donor trimethylamine forms only the 1:1 complex, whereas for diiodomethylamine, which is a weaker donor, only the 2:1 complex results. With iododimethylamine, which is of intermediate strength, both the 1:1 and the (less stable) 2:1 complexes are able to form. Figure 7 shows the structures proposed for the three iodo complexes (according to Ref. 63 and the results referred to above).

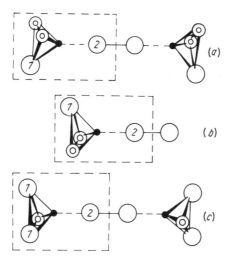

FIG. 7. Proposed structures for 2-iododimethylamine-1-diiodine (a), iododi-methylamine-1-diiodine (b), and 2-diiodomethylamine-1-diiodine (c) (*99*).

Infrared spectra may be discussed on the basis of the structures proposed in Fig. 7, assuming CCNI1I2 or I1I1NCI2 tetrahedra with C_s symmetry and neglecting vibrational coupling through the heavy iodine atoms (*99*). This leads to assignments for the vibrational frequencies of 2-diiodomethylamine-1-diiodine with greater ease than in the case of the two iodine adducts of iododimethylamine. Final verification of these proposed structures will depend on X-ray structural analysis.

I. MODIFICATION OF THE POLYMERIC STRUCTURAL PRINCIPLE OF NITROGEN TRIIODIDE BY BULKY ADDUCT BASES: NITROGEN TRIIODIDE-1-DIIODINE-1-UROTROPINE, $NI_3 \cdot I_2 \cdot C_6H_{12}N_4$

It is impossible to insert urotropine into the puckered cavities of polymeric nitrogen triiodide, using the methods mentioned in Section II, E, because the molecule is too bulky for this to occur. An adduct compound, which analysis shows to be $NI_3 \cdot I_2 \cdot$ urotropine is first formed in addition to diiodine urotropine, $I_2 \cdot$ urotropine, when part of the nitrogen triiodide-1-ammonia has decomposed, setting free iodine. The principle underlying its structure is, however, different from that in the nitrogen triiodide adducts discussed so far (*82, 124*).

Brick red nitrogen triiodide-1-diiodine-1-urotropine is the most stable nitrogen triiodide adduct so far described; it explodes only on

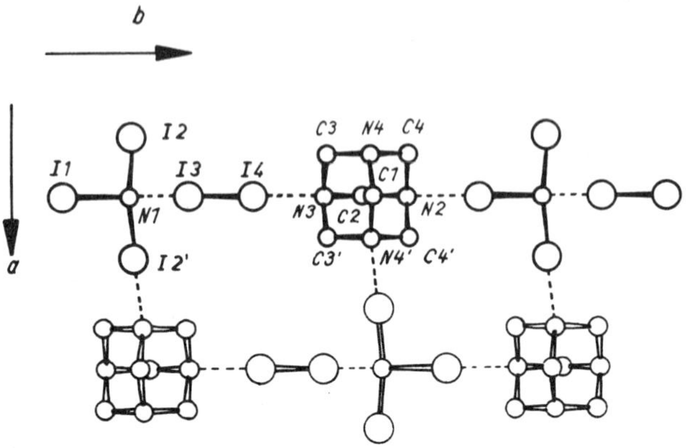

FIG. 8. Projection of the rhombic structure of nitrogen triiodide-1-diiodine-1-urotropine on the (001) plane. Intermolecular bonds are shown by dashed lines (*124*).

impact, under pressure, or when heated. It is also insoluble in all the usual solvents, and, on treatment with an alkaline aqueous ammonia-(pyridine) solution, nitrogen triiodide-1-ammonia (-1-pyridine) is formed instantly.

The crystal structure (Fig. 8) shows the nitrogen triiodide and urotropine molecules to be approximately tetrahedrally surrounded by 4 molecules of another sort. Iodine molecules make up one of the four links between nitrogen triiodide and urotropine. The geometrical arrangement and distances (Table XIII) indicate that there is marked intermolecular bonding between the iodine atoms of nitrogen triiodide and nitrogen atoms in urotropine (I1—N2, I2—N4', I2'—N4). In

TABLE XIII

DISTANCES AND BOND ANGLES IN NITROGEN TRIIODIDE-1-DIIODINE-1-UROTROPINE

N1—I1	2.14 Å	N1—I3	2.47 Å
N1—I2	2.14 Å	N3—I4	3.23 Å
		N2—I1	2.58 Å
I3—I4	2.81 Å	N4'—I2	2.57 Å
Sum of the covalent radii according		N + I	2.03 Å
to Pauling (*121*)		I + I	2.66 Å
Sum of the van der Waals radii		N + I	3.65 Å
according to Pauling (*121*)		I + I	4.30 Å
N1—I1—N2	175°	N1—I3—I4	175°
N1—I2—N4'	179°	I3—I4—N3	178°

addition, the nitrogen atom of the triiodide shows intermolecular bonding to the iodine molecule (N1—I3), which, in turn, is similarly bonded to the urotropine molecule (I4—N3). The resemblance to the structure of iodoform-1-urotropine is apparent (21).

The intermolecular contacts shown result in tetrahedral coordination of 4 iodine atoms (sp^3 hybridization) round the nitrogen of the triiodide (N1). Although the principle underlying the structure of the new compound differs from that of the compounds discussed in Sections II, A–F and H, the tetrahedral structural element [NI_4, $NI_3(CH_3)$, $NI_2(CH_3)_2$] occurs once more. There are, of course, two pairs of almost equal N—I distances in the NI_4 tetrahedra of nitrogen triiodide-1-ammonia and -1-pyridine, whereas NI_4 tetrahedra in nitrogen triiodide-1-diiodine-1-urotropine have three such distances that are almost equal and one that is longer.

In spite of the changed principle underlying the structure, twofold coordination of the iodine atom is maintained as a structural element: All four iodine atoms I1, I2, I3, and I4 have twofold coordination and their bond angles are approximately 180° (3 center–4 electron bond; cf. Section II, K).

J. MODIFICATION OF THE POLYMERIC STRUCTURAL PRINCIPLE OF NITROGEN TRIIODIDE BY CHANGE IN THE sp^3 HYBRIDIZATION AT NITROGEN: N-DIIODOFORMAMIDE, $HCONI_2$

In diiodoformamide, a red, needle-shaped crystalline substance slowly decomposing at room temperature (49), there are two fundamental changes in comparison with the compounds discussed so far: Nitrogen shows sp^2 instead of sp^3 hybridization and oxygen is an additional donor. It is thus no longer possible to have tetrahedral coordination round nitrogen, and the acceptor, iodine, can attain its coordination number of 2 only through donation from oxygen.

The crystal structure (123) (Fig. 9) shows in fact that, of the structural elements so far encountered, only the almost linear twofold coordination of iodine is operative. Each diiodoformamide molecule contains two acceptor (the iodine atoms) and two donor functions (both free electron pairs on oxygen), which is not unlike the situation in formamide itself (103). The molecules combine to form staggered chains along the a-axis through I2 and O, the N—I2—O group being almost linear (Table XIV). In addition, each molecule possesses a free donor and acceptor site, which lie on different sides of the chain and through which the chains are joined up into sheets lying at right angles to the b-axis. Complete linearity of the N—I1—O group is, however,

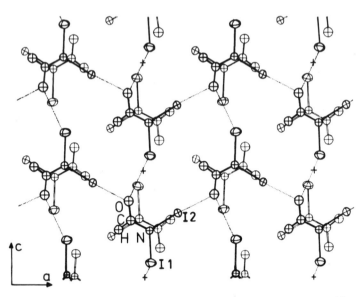

FIG. 9. Projection of the rhombic structure of N-diiodoformamide on the (010) plane. Atoms shown in heavy print lie at $y = 0.25$; the other atoms shown are at $y = -0.25$. Intermolecular bonds for $y = 0.25$ are shown by dotted lines (123).

impossible on geometrical grounds. The structure observed represents the best possible arrangement of all O—I1 bonds between neighboring chains. The unfavorable form of the grouping results in a weakening of the intermolecular O—I1 bond compared with the O—I2 bond (Table XIV). Distances between atoms of different layers are almost always greater than the sum of the van der Waals radii. The position of the layers in relation to one another is determined by optimum packing of the large iodine atoms.

TABLE XIV

DISTANCES AND BOND ANGLES IN N-DIIODOFORMAMIDE

N—I1	2.04 Å	I1—N—I2	119°
N—I2	2.10 Å	N—I2—O	171°
O—I2	2.57 Å	N—I1—O	151°
O—I1	3.13 Å		
I1—I2	3.56 Å		

Sum of the covalent radii according to Pauling (121)	O + I	1.99 Å
Sum of the van der Waals radii according to Pauling (121)	O + I	3.55 Å

K. Linear Twofold Coordination of Iodine as an Essential Structural Element in Solid Nitrogen–Iodine Compounds

That linear twofold coordination of iodine is the essential structural feature for stabilizing nitrogen–iodine compounds is again illustrated in Table XV (124). It occurs in all structures that have been investigated so far. Distances $d1$ and $d2$ never exceed the sum of the van der Waals radii, and the bond angle at the central iodine atom always lies in the region of 180°.

Bond lengths $d1$ and $d2$, which change inversely with bond strength, depend on the donor strengths of $X1$ and $X2$ and on the geometry of the group $X1$—I—$X2$. For example, the stronger and shorter $d1$ becomes because of the high donor strength of $X1$ and a favorable geometry, the weaker and longer $d2$ will be. It is, however, not possible at present to make a quantitative comparison of $d1$ with $d2$ because of the paucity of structural data and the variable nature of both the donor $X[N, O, I]$ and the geometry. In the case of interhalogen compounds and adducts of iodine halides, an attempt has been made to introduce reduced radii $R1$ and $R2$ for the iodine atom in the two bonds of the $X1$—I—$X2$ group and then to establish experimentally a relationship between the two radii (65). Only some of the nitrogen–iodine compounds satisfy this relationship, however, and, because of the paucity of data, it is not possible to determine what is responsible for the deviations.

In describing the bond, a simplified MO model is used, namely, the 3 center–4 electron bond, which has led to good results for the interhalogen compounds. Qualitative predictions based on this model are as well fulfilled for nitrogen–iodine compounds as for iodine adducts.

A bond angle of ~180° at iodine is also encountered together with angles of ~90° in structural studies of polyvalent organic iodine compounds such as phenyl iodine dichloride, $C_6H_5ICl_2$, and diphenyl-iodonium chloride, $(C_6H_5)_2ICl$ (4). These angles suggest that 3 center–4 electron bonds are again involved, with the participation of only p orbitals of the iodine atom. It may, therefore, be expected that in recently discovered polyvalent iodine–nitrogen compounds, such as the benziodazoles (I) (153, 5) and the iododichloridamines (II) (23), there

(I)   (II)   (III)

TABLE XV

TWOFOLD COORDINATION OF IODINE IN THE CRYSTAL STRUCTURE OF NITROGEN–IODINE COMPOUNDS STUDIED[a]

Compound	Grouping $X1 \xrightarrow{d1} I \xrightarrow{d2} X2$	$d1$ (Å)	$d2$ (Å)	Angle (°)	Section	Figure
$(NI_3 \cdot py)_n$	N1—I1'—N1'	2.36	2.36	180	II, E	2
	N1—I2—N2	~2.10	2.59	174		
	N1—I3—I2 (neighboring chain)	~2.10	3.93	153		
$(NI_3 \cdot NH_3)_n$	N1—I1'—N1'	2.30	2.30	180	II, A	1
	N1—I2—N2	~2.15	2.53	176		
	N1—I3—I2 (neighboring chain)	~2.15	3.36	172		
$(I \cdot uro_2)I_3$	Uro—I—uro	2.30	2.30	177	II, A	
$(I(3\text{-pic})_2)BF_4$	3-pic—I—pic-3	2.24	2.24	180	II, A	
$(I \cdot py_2)I_3 \cdot 2I_2$	Py—I—py	2.16 ± 0.1	2.16 ± 0.1	180	II, A	
$(NI_3 \cdot I_2 \cdot uro)_n$	N1—I2—N4'	2.14	2.57	179	II, I	8
	N1—I1—N2	2.14	2.58	175		
	I4—I3—N1	2.81	2.47	175		
	I3—I4—N3	2.81	3.23	178		
$(HCONI_2)_n$	N—I2—O	2.10	2.56	171	II, J	9
	N—I1—O	2.04	3.12	151		

[a] Sum of van der Waals radii according to Pauling (121): N + I, 3.65; O + I, 3.55; I + I, 4.30 Å.

will also be bond angles of ca. 180° and ca. 90° with bonds of the 3 center–4 electron type. [This expectation has recently been fulfilled by N-chloro-3-aza-3H,2,1-benzoxiodol-1-yl chloride (III) (*112*).]

L. Behavior of Nitrogen Triiodide-3-ammonia in Solution, Especially in Liquid Ammonia

1. Nature of Species in Solution: Experimental Evidence for Occurrence of Monoiodamine-1-ammonia in Liquid Ammonia

Nitrogen triiodide-1-ammonia has a very low solubility in all solvents that have so far been studied. It is to be assumed or has been proven that this solubility depends on a reaction between iodine, acting as an acceptor, and the solvent, acting as a competing base:

$$NI_3 \cdot NH_3 + 3OH^- \underset{}{\overset{H_2O}{\rightleftharpoons}} 2NH_3 + 3OI^- \quad (28) \tag{5}$$

$$NI_3 \cdot NH_3 + 3ROH \underset{}{\overset{ROH}{\rightleftharpoons}} 2NH_3 + 3ROI \quad (78) \tag{6}$$

$$NI_3 \cdot NH_3 + 3R_2NH \underset{}{\overset{R_2NH}{\rightleftharpoons}} 2NH_3 + 3R_2NI \quad (\text{Section II, G}) \tag{7}$$

In liquid ammonia, nitrogen triiodide-1-ammonia is at once converted into nitrogen triiodide-3-ammonia, which is likewise poorly soluble in this solvent (ca. 1 mg/ml at $-75°C$). What species are present in the liquid ammonia solution?

Our working hypothesis is that, by analogy with Eqs. (5)–(7), a small amount of monoiodamine-1-ammonia, $H_2NI \cdot NH_3$, is formed in the equilibrium and that this has twofold coordinated iodine with a bond angle of $\sim 180°$ [Eq. (8), vertical], as do the solid nitrogen–iodine compounds considered in Sections II, A–K.

So far this hypothesis lacks proof, but the following experimental findings are not inconsistent with it.

1. Solid nitrogen triiodide-3-ammonia is completely stable in an atmosphere of ammonia at $-25°C$, whereas it decomposes slowly under liquid ammonia even at $-75°C$. There must, therefore, be an unstable nitrogen–iodine compound in solution that is in equilibrium with nitrogen triiodide-3-ammonia (*91*).

2. The greater ease of decomposition of the dissolved compound may be explained by the free pair of electrons on the nitrogen atom of

$$[H_3N-I-NH_3]^+ + NH_3 + NH_2^- \rightleftharpoons [H_3N-I-\overline{NH_2}] + 2NH_3 \rightleftharpoons [H_2\overline{N}-I-\overline{NH_2}]^- + NH_4^+ + NH_3 \qquad (8)$$

colorless   yellow in solution   red
red in solid state

$(\div 3) \Vert (\times 3)$

$$[H_3N-I-\overline{NIH}] + [H_3N-I-\overline{NH_2}] + 8NH_3$$

$(\div 3n) \quad (\times 3n)$

$(\div n) \Vert (\times n)$

$$[H_3N-I-\overline{NI_2}] + 10NH_3$$
$$[NI_3\cdot NH_3]_n + 10n\,NH_3$$
brown

$$[H_2\overline{N}-I-\overline{NH_2}]_n^{n-} + nNH_4^+ + 9n\,NH_3$$
black

$$[NI_3\cdot 3NH_3]_n + 8n\,NH_3$$
green

$> -75°C \Vert < -75°C$

$$[NI_3\cdot \sim 5NH_3]_n + 6n\,NH_3$$
red

acidic   neutral   basic

monoiodamine-1-ammonia, which makes it easier for iodine to be split off as iodide ion with simultaneous nitrogen evolution:

$$[H_3N-I-\overline{N}H_2] \underset{(\div 3)}{\overset{(\times 3)}{\rightleftharpoons}} 3I^- + 3NH_4^+ + NH_3 + N_2 \tag{9}$$

3. Suspensions of nitrogen triiodide-3-ammonia in liquid ammonia are able to add I and NH_2 groups to $C=C$ double bonds as in halogen-hydrin reactions (*80*) (Section II, L, 2).

4. By chilling saturated solutions of nitrogen triiodide-3-ammonia with liquid nitrogen, a red solid compound with an N:I ratio of 2:1, which is not identical with nitrogen triiodide-~5-ammonia, discussed in Section II, D, may be obtained. It is much more readily decomposed and has an intense IR band at 480 cm^{-1} in the N—I stretching region (*80, 55*).

Our working hypothesis further predicts that the dissolved mono-iodamine-1-ammonia will give up a proton in a basic solution containing amide and be transformed into the monoiodamine ion $[H_2NINH_2]^-$ [Eq. (8), horizontal].

This hypothesis is also not inconsistent with the following experimental observations:

1. Addition of potassium or sodium amide to a suspension of nitrogen triiodide-3-ammonia in a molar ratio of $3K(Na)NH_2 : 1NI_3 \cdot 3NH_3$ yields a red-colored solution from which nitrogen triiodide-3-ammonia separates again on addition of acid (ammonium iodide). Addition of less amide leads to black salts of low solubility, which have been more fully investigated in the case of the silver compound (obtained in an analogous reaction). With excess amide, they go over to the red solutions (*92*).

2. The extremely ready decomposition of the dissolved red substance to iodide and nitrogen, even at $-75°C$, may be explained by the presence of two pairs of free electrons on the nitrogen atom of the monoiodamine ion:

$$[H_2\overline{N}-I-\overline{N}H_2]^- \underset{(\div 3)}{\overset{(\times 3)}{\rightleftharpoons}} 3I^- + 4NH_3 + N_2 \tag{10}$$

3. A polarographic reduction stage may be identified in basic solutions of the red compound, which does not appear in acid or neutral solutions [Eq. (8), horizontal] and disappears when the red compound decomposes (*109*).

According to our hypothesis, the small amount of monoiodamine-1-ammonia is transformed almost completely in acid solutions containing ammonium salts into the diammineiodonium ion, $[H_3N-I-NH_3]^+$

[Eq. (8), horizontal], the iodide of which is in equilibrium with the iodine ammines:

$$I_2 + x\text{-}NH_3 \; \rightleftharpoons \; I_2 \cdot NH_3 + (x\text{-}1)NH_3 \; \rightleftharpoons \; I_2 \cdot 2NH_3 + (x\text{-}2)NH_3$$

brown   brown

$$(\div 2) \Big\|\, (\times 2) \qquad\qquad\qquad \Big\| \qquad\qquad (11)$$

$$[H_3N\text{---}I\text{---}NH_3]^+ + I_3^- \qquad\qquad [H_3N\text{---}I\text{---}NH_3]^+ + I^-$$

brown

The correctness of this hypothesis is supported by many experiments:

1. Many iodonium salts stabilized by N-bases are known (64, 56, 118, 125, 88). Among the factors determining their relatively high stability is, in our view, the absence of free electron pairs on the nitrogen.

2. Nitrogen triiodide-3-ammonia with little ammonium iodide and a large excess of ammonium nitrate dissolves to give a colorless solution. After pumping off the ammonia at room temperature, there is a stable colorless residue that gives iodine with aqueous sulfuric acid (79).

3. Nitrogen triiodide-3-ammonia when treated with a large excess of ammonium iodide in liquid ammonia solution at room temperature dissolves to give a stable brown solution in which iodine may be detected after removing the solvent [Eqs. (8) and (11)]. Conversely, treating of solid iodine with ammonia gas gives the solid brown compounds diiodine-1- and -2-ammonia, whereas dissolution of iodine in a large excess of liquid ammonia yields only iodonium cations and iodide anions as shown by the polarogram. When this solution is concentrated the solubility product of nitrogen triiodide-3-ammonia is exceeded and ammonium ions may be detected polarographically [Eqs. (8) and (11)] (79, 109).

4. Saturated solutions of nitrogen triiodide-3-ammonia, ammonium iodide, and ammonia in di-n-butyl ether at $-40°C$ contain a second conducting species in addition to ammonium iodide, as shown by conductivity measurements (122).

2. Reaction with Organic Compounds

a. Reaction with Primary and Secondary Amines. In accordance with Eq. (7), nitrogen triiodide-1-ammonia reacts with a large number of primary and secondary amines the more readily as the amine becomes more basic. The reactions with methylamine, dimethylamine, and ethylenediamine have already been discussed in Sections II, F and G. Recently this has been extended to the reaction in aqueous solution at 0°C with tri-, tetra-, penta-, and hexamethylenediamine, and the corresponding tetraiododiamines have been prepared. They are all deep red in color and decompose very rapidly at 0°C, their temperature sensitivity increasing with the chain length (*146*).

b. Reaction with Organic Compounds Containing Acidic CH *Groups.* In carrying out reactions, iodine was dissolved in excess of liquid ammonia at −33°C. According to Eqs. (8) and (11) of Section II, L, 1, the resulting suspension contains solid nitrogen triiodide-3-ammonia. In solution, ammonium and iodide ions as well as an equilibrium between iodonium cations and iodoamine-1-ammonia are to be expected. This equilibrium should be shifted almost completely in favor of iodonium cations because of the presence of ammonium ions. The suspension thus appears to contain several species of active iodine. The mixture remaining after reaction with the organic compound was worked up in different ways. Table XVI shows recent results (*32, 33, 39*). They are classified according to the final product, since only in a few cases (reactions 14–16) was it possible to isolate or stabilize the primary product because of secondary reactions in liquid ammonia.

The following points seem to be of interest:

1. As the primary products of reactions 14–15 (Table XVI) are formed via substitution of one or more protons by iodonium cations, it is concluded that this is the primary step in all reactions of organic compounds with acidic C—H groups.

2. Reactions 9 and 16 are exceptions. The primary product of reaction 16 is explained by addition of iodonium and iodide ions to the phosphorus atom. Reaction 9 could also be explained by a two-center reaction of iodamine with the double bond as the primary step. This view of the mechanism is supported by the observation that diethyl fumarate does not react at all. How far this phenomenon depends on the cis or trans configuration at the double bond requires further consideration. Steric factors alone are probably not decisive, otherwise cumarin and ethyl *cis*-cinnamate would likewise undergo a two-center reaction. In fact, the latter does not react at all and cumarin does so only at the aromatic ring (reactions 8 and 10).

3. In reacting organic compounds with the nitrogen triiodide-3-ammonia suspension, one often finds two or more reaction products (compare reactions 1 and 15; 5, 11, and 12; and 8 and 10). The kind of reaction product often depends on the way in which the reaction material is worked up (reaction 15).

4. The suspension is not only an iodinating but also an aminating reagent.

TABLE XVI

REACTION OF NITROGEN TRIIODIDE-3-AMMONIA WITH C—H ACIDIC ORGANIC COMPOUNDS IN LIQUID AMMONIA AT $-33°C$

Reaction No.	Starting material	Molar ratio, organic compound:I$^+$	Yield (%)	Final product
		Final product only iodinated		
1	Diethyl malonate	1:2, 1:3	∼20	Tri-, tetraiodomethane
2	Barbituric acid	1:1	80	Iodobarbituric acid
3	Methone	1:1	∼100	Ammonium salt of 2-iodo-methone
4	Acetone		good ⎫	
5	Ethyl acetoacetate	1:2	∼13 ⎬ Triiodomethane	
6	Mesityl oxide	1:1	∼25 ⎭	
7	Phenol	1:1	60	p-Iodophenol
			Very low	Triiodophenol
8	Cumarin	1:1	40	6-Iodocumarin
			Small	3,5-Diiodosalicylic aldehyde
		Final product iodinated and aminated		
9	Diethyl maleate	1:1	20	α-Amino, α'-iodomaleinimide
10	Cumarin	1:1	15	5-Iodocumaric amide
			Very low	3,5-Diiodocumaric amide
11	Ethyl acetoacetate	1:2	4	Diiodoacetamide
		Final product only aminated		
12	Ethyl acetoacetate	1:2	25 ⎫	
13	Diaminomalondi-amide	1:1	66 ⎭ Oxalamide	

Reaction No.	Starting material	Molar ratio	Primary product	Yield (%)	Final product
14	Ethyl- (methyl) phenylacetate	1:1	Ethyl- (methyl-) α-iodophenyl-acetate	60–70	Phenyglycin-amide
15	Diethylmalonate	1:2	Diethyldiiodo-malonate	30	Diaminomalon-diamide
				20	Ethyl carba-mate
16	Triphenylphosphine	1:1	Triphenylphos-phinediiodide	73	Triphenylphos-phineiimide

5. Similar organic compounds give similar reactions (compare the iodoform in reactions 1, 4, 5, and 6).

6. A comparison shows that iodination reactions with dimethyl-iodamine also take place with a suspension of nitrogen triiodide-3-ammonia in liquid ammonia. Conversely, however, not all of the iodination reactions that can be carried out with this suspension will

occur with dimethyliodamine; for example, maleic ester (reaction 9) and cumarin (reactions 8 and 10) do not react with dimethyliodamine. This may be ascribed to the positive inductive effect ($+I$ effect) of the methyl groups on the nitrogen, which makes cleavage of the N—I bond in iodonium and amide ions more difficult.

III. Nitrogen Trichloride

There is a fundamental difference between the behavior of nitrogen–chlorine and nitrogen–iodine compounds. The greater covalent bond energy of the nitrogen–chlorine bond and the much smaller tendency of the chlorine to interact with a second nitrogen by a 3 center–4 electron bond or an oxygen (12) or chlorine atom implies nitrogen–chlorine compounds in general to be not polymeric and to be more soluble than nitrogen–iodine compounds.

A. Structure of Dissolved and Gaseous Nitrogen Trichloride, NCl_3

The high solubility of nitrogen trichloride in many nonpolar solvents indicates a monomeric structure for this compound. Its molecular structure in solution has been elucidated simultaneously by several groups of workers using IR and Raman spectroscopy (8, 15, 22, 69, 129). The spectra (Table XVII) can be satisfactorily assigned on the assumption of a pyramidal structure with C_{3v} symmetry. (For force constant calculations, see Section III, C.) Electron diffraction investigations (13) and the microwave spectrum (17, 16) of the gaseous trichloride confirm the pyramidal structure ($\sphericalangle$Cl—N—Cl, 107°; r N—Cl, 1.76 Å).

B. Infrared Spectrum and Structure of Solid Nitrogen Trichloride

The IR spectrum of solid nitrogen trichloride (137) (Table XVII) shows about the same stretching frequencies (ν_1, ν_3) as the dissolved compound. Of the two bending frequencies, the asymmetric (ν_4) is displaced from 258 to 230 cm^{-1}; the symmetrical bending frequency (ν_2) cannot be observed but its position may be calculated as 389 cm^{-1} from the observed combination band $\nu_2 + \nu_4$ at 619 cm^{-1}. The deviation of the bending frequencies from those of dissolved nitrogen trichloride may be attributed to the change to the solid state, which has a greater influence on the bending vibrations than on the stretching vibrations.

TABLE XVII

INFRARED AND RAMAN SPECTRA OF NITROGEN TRICHLORIDE-^{14}N and -^{15}N IN THE RANGE OF FUNDAMENTAL VIBRATIONS[a]

IR spectra			Raman spectra		Assignment
$^{14}NCl_3(CCl_4)$ (cm^{-1})	$^{14}NCl_3$ (solid $-185°C$) (137) (cm^{-1})	$^{15}NCl_3(CCl_4)$ (cm^{-1})	$^{14}NCl_3(CCl_4)$ (cm^{-1})	$^{15}NCl_3(CCl_4)$ (cm^{-1})	
643 ss	642 ss	627 ss	643 w	628 w	$\nu_3\ \nu_{as}$ N—Cl
608 sh	619 sh				$\nu_2 + \nu_4$
520–540 w	542 m	510–530 w	541 s	528 s	$\nu_1\ \nu_s$ N—Cl
385 w		371 w			$\nu_3 - \nu_4$
349 w		349 w	349 ss	348 ss	$\nu_2\ \delta_s$ Cl—N—Cl
258 m	230 m	258 m	257 m	257 m	$\nu_4\ \delta_{as}$ Cl—N—Cl

[a] Intensities: ss, very strong; s, strong; sh, shoulder; m, medium; w, weak.

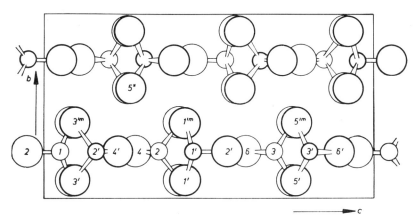

FIG. 10. Projection of the rhombic structure of nitrogen trichloride on the (100) plane. Molecules shown in heavy print lie at $x \sim 0.5$; the others are at $x \sim 0$ (*60*).

A further doubtful band observed at 155 cm^{-1} might be associated with weak intermolecular N···Cl interaction.

X-Ray structural analysis of solid nitrogen trichloride (*60*) (Figs. 10 and 11) shows a molecular lattice made up of nitrogen trichloride pyramids, which are stacked in layers at right angles to the *b*-axis (parallel to the *ac* plane). Pyramids within a layer are arranged with

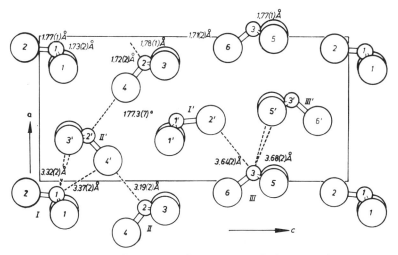

FIG. 11. Projection of the rhombic structure of nitrogen trichloride on the (010) plane. Only molecules with $y \sim 0.25$ are shown for the sake of clarity (molecules with $y \sim 0.75$ are omitted) (*60*).

the same orientation, whereas those in neighboring layers have an antiparallel orientation. The pyramids form undulating layers at right angles to the a-axis (parallel to the bc plane). Their projection on the bc plane shows that the chlorine atoms are approximately hexagonally close-packed.

The N—Cl bond distance and the Cl—N—Cl bond angle within a nitrogen trichloride pyramid vary appreciably, from 1.71 to 1.78 Å, with a weighted mean of 1.75 Å, and from 105.1 to 109.6° and, as a result of this, their ideal C_{3v} symmetry is lowered to m. The cause of this distortion is to be sought in generally very weak but varying intermolecular interactions. It is worth mentioning the distorted octahedral or trigonal prismatic coordination of the nitrogen atoms N1 and N3 (Fig. 11), although this will not be discussed further here since distances N1—Cl3′ (3.3), N1—Cl4′ (3.4), N3—Cl2′ (3.6), and N3—Cl5′ (3.7 Å) either reach or exceed the sum of the van der Waals radii of nitrogen and chlorine. Of greater significance in our context is the distance N2—Cl4′ (or N2′—Cl4) which, at 3.19 Å, is smaller than the sum of the van de Waals radii, and the angle N2—Cl4′—N2′ (or (N2′—Cl4—N2) which, at 177.4°, is practically 180°. As a consequence of this, part of the structure of nitrogen trichloride is made up of chains of NCl_4 tetrahedra parallel to the a-axis. These are formally similar to the chains of NI_4 tetrahedra discussed in Section II and with the linear twofold coordination of iodine which was stressed there. For nitrogen trichloride, the N—Cl $\cdots$ N grouping is, of course, unsymmetrical, and intermolecular nitrogen–chlorine interactions are very weak compared with those in nitrogen triiodide. Intermolecular chlorine–chlorine distances that are shorter than the sum of the van der Waals radii (3.60 Å) also occur. These are between the layers at right angles to the b-axis (parallel to the ac plane) (not shown in Fig. 10). These distances (3.37 and 3.42 Å) resemble the intermolecular distances found in solid chlorine (3.34 Å) (18), where they are interpreted as indicative of weak intermolecular interactions (155, 113). The structure shows very many similarities to that of chloroform, except that there intermolecular interactions play no part (35, 18, 94, 60).

C. Extent of Variations in the N—Cl Bond

In Table XVIII, results for nitrogen trichloride are compared with those for other nitrogen–chlorine compounds. Force constants cannot always be quoted without reservations since their calculation depends in part on estimated or extrapolated N—Cl distances.

The almost ideal tetrahedral angle in nitrogen trichloride and the

TABLE XVIII

BOND LENGTHS, FORCE CONSTANTS, AND IR WAVE NUMBERS
FOR THE N—Cl BOND

Compound [a]	$r_{\text{N—Cl}}$ (Å; mean value m)	$f_{\text{N—Cl}}$ (mdyn/Å)	ν N—Cl (cm^{-1})	Refs.[b]
With sp^3 hybridization at nitrogen				
$Ag[N(C_6H_{11})ClO_3] \cdot \frac{1}{2}H_2O$ (f)	1.63m		730	126, 6
F_2NCl (g)	1.73	3.13	697	
$HNCl_2$ (g)		2.93	687	
			666	
H_2NCl (g)		3.06	686	
CH_3NCl_2 (g)	1.74		667	
			656	
$[Pt(NH_3)_3(NCl_2)_2Cl]Cl$ (f)	1.75 m		664	156, 157
			644	
NCl_3 (g)	1.76		652	13, 17
(l)		2.72	643	
			541	
(f)	1.75 m		642	60, 137
			542	
$(CH_3)_2NCl$ (g)	1.77		600	
With sp^2 hybridization at nitrogen				
K_2NClO_3 (f)	1.41	8.81	1267	42
$KHNClO_3$ (f)	1.64		865	2
NCl (l)		4.0	824	
N_3Cl (g)	1.745			19
(l)			724	
$F_2S(O)NCl$ (g)	1.72		670	115
F_2SNCl (g)	1.72		642	54
$OCNCl$ (g)	1.70			
(f)		2.84	603	
$H_2C—CO$ $\diagdown$ NCl (f) $H_2C—CO$ $\diagup$	1.69		530	
$OSNCl$ (g)	1.70	2.63	526	114
CH_3CO $\diagdown$ NCl (f) CH_3 $\diagup$			483	
$[(CH_3)_3Si]_2NCl$ (f)			434	
O_2NCl (g)	1.83	2.46	370	
$ONCl$ (g)	1.95	1.27	332	

(continued)

Table XVIII—*Continued*

Compound[a]	r_{N-Cl} (Å; mean value m)	f_{N-Cl} (mdyn/Å)	ν N—Cl (cm^{-1})	Refs.[b]
	With *sp* hybridization at nitrogen			
Cl$_3$VNCl (g)	1.60			116
(f)	1.59		500	34
	Without hybridization at nitrogen			
H$_2$C | NCl (l) H$_2$C			563	

[a] Abbreviations: (g) gaseous; (fl) liquid; (l) in solution; (f) solid.
[b] References cited only where data are not given by Höhne *et al.* (*74*).

almost complete absence of intermolecular contacts in the solid state emphasize the correctness of the assumption that there is a relatively pure N—Cl single bond in this molecule. The experimental N—Cl stretching force constant of about 2.72 mdyn/Å accordingly represents that of a single N—Cl bond. In fact, comparable f_{N-Cl} values are found for other nitrogen–chlorine compounds (HNCl$_2$, OCNCl, OSNCl) in which a single N—Cl bond may also be assumed. When Siebert's equation (*140*) is used in calculating the single bond force constant, using the correction factor of Goubeau (*51*) which takes account of mutual repulsion between the free electron pairs on nitrogen and chlorine, a quite similar value of 2.4 mdyn/Å is obtained.

The results in Table XVIII show the strong influence of charge effects on the N—Cl stretching force constant. In nitrosyl and nitryl chloride, for example, f_{N-Cl} values are strikingly reduced, which can be explained by the participation of ionic structures. Moreover, other effects are also able to influence the N—Cl stretching force constant: The inductive effect of strongly electronegative substituents, for example, leads to an increase in f_{N-Cl} (F$_2$NCl). An increase can also be brought about particularly by the double-bond effect, which is favored by the large total electronegativity and the small electronegativity difference for nitrogen and chlorine; it is also influenced by substituents on both elements (NCl, K$_2$NClO$_3$) (*51*, *74*).

Changes in the N—Cl stretching force constant (*F*-matrix) that have been discussed will also change the N—Cl wave number (λ) according to

$$G \cdot F - E \cdot \lambda = 0 \tag{12}$$

where $G = G$-matrix (masses, valence angle, internuclear distance), $F = F$-matrix (force constants, internuclear distance), $E =$ unit matrix, and $\lambda =$ wavelength (wave number). Thus, for example, in nitrosyl or nitryl chloride, there is a decrease in ν_{N-Cl} as a result of a decrease in f_{N-Cl}, whereas an increase in f_{N-Cl}, as for example in chlorodifluoramine, nitrogen monochloride, or potassium amidoperchlorate causes an increase in ν_{N-Cl} compared with the value for nitrogen trichloride.

The N—Cl wave number (λ) changes not only with the force constant (F-matrix), but also with the mass of the ligands bonded to nitrogen and chlorine and the effective angle made by the nitrogen–chlorine bond and these masses (G-matrix). Thus, for example, N-chlorosuccinic acid imide and N-chloro-N-methylacetamide show a lower wave number than compound OCNCl because two masses instead of one are directly bonded to sp^2-hybridized nitrogen. The wave number for compounds with sp^2 hybridization at nitrogen is displaced relative to that for compounds of about the same weight but sp^3 hybridization, by about 100 cm^{-1} toward higher wavelengths because the angle with which the masses operate at the nitrogen–chlorine bond increases from 109° to 120°.

The bond lengths shown in Table XVIII are all determined experimentally. The mean value of r_{N-Cl} in nitrogen trichloride lies at 1.75 Å, which is close to $r_{N-Cl} \sim 1.7$ Å found for all compounds with an N—Cl single bond (OCNCl, OSNCl). The sum of the covalent radii (1.69 Å) (121) indicates too that a single bond requires a bond length of ca. 1.7 Å. Substantially shorter values of r_{N-Cl} correspond with multiple bonding, whereas longer distances are indicative of an ionic component in the bond.

Reference may be made again to the expected inverse relationship between force constants and bond length. Overall, there is an astonishing range of variation in the N—Cl bond.

D. STRUCTURE OF SOLID TRIAMMINE BISDICHLORAMIDOCHLORO-
PLATINUM(IV) CHLORIDE, [Pt(NH$_3$)$_3$(NCl$_2$)$_2$Cl]Cl

It is evident from Table XVIII that replacement of a chlorine atom in nitrogen trichloride by platinum, i.e., the use of NCl$_2$ ligands in platinum complexes, does not have a substantial influence on structural relationships of the NCl$_2$X group (X = Cl, Pt). The N—Cl distances and hybridization at nitrogen remain unchanged on average, and the N—Cl wave number is also not altered. Figure 12 shows the structure of the [Pt(NH$_3$)$_3$(NCl$_2$)$_2$Cl]$^+$ cation elucidated by X-rays (156, 157), whereas

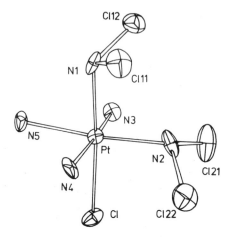

Fig. 12. Structure of the triammine bisdichloroamidochloroplatinum(IV) chloride cation (156).

Table XIX gives relevant bond distances and angles. The compound, which was originally prepared according to (101)

$$[\text{Pt}(\text{NH}_3)_5\text{OH}]\text{Cl}_3 + 4\text{Cl}_2 \longrightarrow [\text{Pt}(\text{NH}_3)_3(\text{NCl}_2)_2\text{Cl}]\text{Cl} + 5\text{HCl} + \text{H}_2\text{O} \quad (13)$$

has recently been obtained by the following reaction (117) as beautiful red crystals that are shock-sensitive:

$$[\text{Pt}(\text{NH}_3)_5\text{Cl}]\text{Cl}_3 + 4\text{Cl}_2 \longrightarrow [\text{Pt}(\text{NH}_3)_3(\text{NCl}_2)_2\text{Cl}]\text{Cl} + 6\text{HCl} \quad (14)$$

The reactions are interesting examples of the extensive current studies on the reactivity of ligands in transition metal complexes (143).

TABLE XIX

DISTANCES AND ANGLES IN CATION $[\text{Pt}(\text{NH}_3)_3(\text{NCl}_2)_2\text{Cl}]^+$

Pt—N1	2.07 Å	Pt—N1—Cl11	108.6°
Pt—N2	2.10 Å	Pt—N1—Cl12	111.0°
Pt—N3	2.04 Å	Cl11—N1—Cl12	102.6°
Pt—N4	2.02 Å	Pt—N2—Cl21	111.3°
Pt—N5	2.12 Å	Pt—N2—Cl22	114.0°
Pt—Cl	2.35 Å	Cl21—N2—Cl22	109.1°
N1—Cl11	1.82 Å		
N1—Cl12	1.77 Å		
N2—Cl21	1.75 Å		
N2—Cl22	1.66 Å		

E. N-PERCHLORYL COMPOUNDS

Table XVIII gives three examples showing the wide variation in N-perchloryl compounds. This first came to light after the recent preparation of a large number of organic N-perchloryl compounds by reaction of amines with the new perchlorylating reagent, dichlorine heptoxide (6, 10),

$$2R_1R_2NH + Cl_2O_7 \xrightarrow[25° \text{ to } -20°C]{CCl_4} R_1R_2NClO_3 + R_1R_2NH_2^+ClO_4^- \qquad (15)$$

where R_1 = organic group and R_2 = organic group or hydrogen. The resulting colorless or, at the most, faintly yellow compounds (oils) are not isolated as a rule because of their explosive nature but are identified in carbon tetrachloride solution by ratio analysis as well as by IR and ^{1}H NMR spectroscopy. When they still have hydrogen bonded to nitrogen, they are appreciably acidic:

$$pK_a \text{ for } (CH_3)_3CNHClO_3 \quad \text{or} \quad H_2C\underset{\underset{H_2\ H_2}{C-C}}{\overset{\overset{H_2\ H_2}{C-C}}{\diagup\diagdown}}CHNHClO_3 = 7.0 \text{ or } 6.9$$

compared with pK_a for $1(H_2NClO_3)$ or $2(H_2NClO_3)$ which is 5.5 or 11.9, respectively. More stable solid potassium and silver salts may be made which can be identified by element analysis and their IR spectra. Table XX shows the new N-perchloryl compounds with some of the older ones for comparison.

Table XX also shows the most important IR stretching frequencies of these compounds and their assignments so far as thay may be obtained by comparison with assignments for the spectra of the ions and compounds $NClO_3^{2-}$ (93, 52), $HNClO_3^-$ (93), $ROClO_3$, $HOClO_3$, $XOClO_3$ (X = halogen), and $O_3ClOClO_3$ (128, 135, 134). Splitting of the ν_{as} O—Cl frequencies in some N-perchloryl compounds can be attributed to lowering of the C_{3v} symmetry of the $NClO_3^-$ ion to C_s (119), whereas the partial doubling of the ν_s Cl—O frequencies in the cyclic N-perchloryl compounds may be explained by equatorial and axial positioning of the ClO_3 group in relation to the organic ring.

Once it is realized that, because of the mutual coupling of ν_{N-Cl} and ν_s O—Cl, identification of these frequencies is not straightforward and that they can also be influenced by differing mass effects, then it can be established that in the series in Table XX, going from top to bottom

$$(K_2NClO_3—Ag_2NClO_3—KRNClO_3—KHNClO_3—$$
$$AgRNClO_3—R_2NClO_3—RHNClO_3),$$

TABLE XX

THE MOST IMPORTANT STRETCHING FREQUENCIES OF THE NEWER N-PERCHLORYL COMPOUNDS (WITH OLDER VALUES FOR COMPARISON)

Compound	ν N—Cl (cm^{-1})	ν_s O—Cl (cm^{-1})	ν_{as} O—Cl (cm^{-1})	Refs.
K_2NClO_3	1264	815	890	26, 105, 93, 52
Ag_2NClO_3	1021 973	848	944	93
$K\left[\ \begin{array}{c}H_2C{-}CH_2\\ \ \ \ \ CH{-}NClO_3\\ H_2C{-}CH_2\end{array}\right]$ (cyclohexyl)	940	~1000	1120 1081	6
$K[(CH_3)_3CNClO_3]$	937	1020	1080	6
$KHNClO_3$	865	988	1121	93
$Ag\left[\ \begin{array}{c}H_2C{-}CH_2\\ \ \ \ \ CH{-}NClO_3\\ H_2C{-}CH_2\end{array}\right]\ \cdot\tfrac{1}{2}H_2O$ (cyclohexyl)	730	951	1136	6
$O\underset{H_2C{-}CH_2}{\overset{H_2C{-}CH_2}{\diagup\diagdown}}NClO_3$ (morpholine)	695	1005 974	1240 1202	6
$H_2C\underset{HC{-}CH_2,\ CH_3}{\overset{H_2C{-}C}{\diagup\diagdown}}NClO_3$ (methylpiperidine)	680	1000 955	1210 1180	6
$CH_3{-}CH\underset{H_2C{-}CH_2}{\overset{H_2C{-}C}{\diagup\diagdown}}NClO_3$ (methylpiperidine)	680	1010 960	1215 1185	6

(continued)

Table **XX**—*Continued*

Compound	ν N—Cl (cm^{-1})	ν_s O—Cl (cm^{-1})	ν_{as} O—Cl (cm^{-1})	Refs.
(cyclopentyl ring) $NClO_3$	680	1040 990	1225 1190	*37, 6, 10*
$(C_5H_{11})_2NClO_3$	680	1020	1240 1200	*10*
$C_2H_5—CH / H_2C$ ring $NClO_3$	680	1020	1245 1210	*10*
$(C_2H_5)_2NClO_3$	670	995	1220 1185	*6, 10*
$(C_3H_7)_2NClO_3$	665	1005	1230 1195	*6*
$(C_4H_9)_2NClO_3$	660	1000	1220 1185	*6*
$(CH_3)_2NClO_3$	655	1000	1220 1190	*6*
(cyclobutyl ring) $NClO_3$	640	1005	1215 1175	*6*
$CH_3—(CH_2)_5—NHClO_3$	665	1020	1240 1210	*10*
$(CH_3)_2CH—NHClO_3$	660	1020	1230 1205	*10*
$CH_3—(CH_2)_3—NHClO_3$	655	1010	1230 1200	*6, 10*
(cyclopentyl ring) $CH—NHClO_3$	654	1009	1204	*6*
$(CH_3)_3C—NHClO_3$	647	1012	1254 1198	*6, 10*
$CH_3—(CH_2)_2—NHClO_3$	645	1030	1250 1210	*10*

$\nu_{\text{N--Cl}}$ decreases from ~ 1260 to 650 cm^{-1} and ν_{as} O—Cl rises from ~ 900 to ~ 1250 cm^{-1}, corresponding to a decrease in the double-bond character of the N—Cl bond and to an increase for the O—Cl bond. The decrease in double-bond character for the N—Cl bond is to be associated with increase in the electronegativity of nitrogen with change in the atom attached to it: It attracts the π-bonding electron pair of the N—Cl double bond increasingly to nitrogen. The resulting increasing electron requirement of chlorine is then satisfied in increasing measure by π electrons from oxygen, which will explain the increase in the double-bond character of the O—Cl bond. A corresponding effect has already been observed and described in a comparison of the IR spectra of NClO$_3^{2-}$ and FClO$_3$ (52).

X-Ray structural analysis of potassium amidoperchlorate (42), potassium hydrogen amidoperchlorate (2), and silver cyclohexyl-amidoperchlorate·$\frac{1}{2}$H$_2$O (126) confirm this interpretation from interatomic distances (Table XXI). The small value of $r_{\text{N--Cl}}$ in the amidoperchlorate anion (in which there are three pairs of electrons available for π bonding and a low electronegativity at nitrogen) contrasts with the significantly greater value of $r_{\text{N--Cl}}$ (which is close to that for a single bond) in potassium hydrogen amidoperchlorate and silver cyclohexylamidoperchlorate (with no free electron pair, but with two silver ligands, distorted sp^3 hybridization, and a higher electronegativity at nitrogen). Conversely, Cl—O distances decrease progressively in the series K$_2$NClO$_3$—KHNClO$_3$—AgN(C$_6$H$_{11}$)ClO$_3$·$\frac{1}{2}$H$_2$O.

TABLE XXI

DISTANCES AND ANGLES IN POTASSIUM AMIDOPERCHLORATE (K$_2$NClO$_3$), POTASSIUM HYDROGEN AMIDOPERCHLORATE (KHNClO$_3$), AND SILVER CYCLOHEXYLAMIDOPERCHLORATE–WATER [AgN(C$_6$H$_{11}$)ClO$_3$·$\frac{1}{2}$H$_2$O]

Distance and angle measured	AgN(C$_6$H$_{11}$)ClO$_3$·$\frac{1}{2}$H$_2$O	KHNClO$_3$	K$_2$NClO$_3$
O—Cl	1.38–1.44 Å	1.43–1.46 Å	1.50–1.52 Å
N—Cl	1.61–1.65 Å	1.64 Å	1.41 Å
O—Cl—O N—Cl—O	104°–113°		109°
N—C	1.52–1.53 Å		
N—Ag	2.16–2.27 Å		
Cl—N—C	110°–112°		
Cl—N—Ag	107°–110°		
C—N—Ag	115°–121°		
Ag—N—Ag	87°–92°		

F. KINETICS OF FORMATION OF NITROGEN TRICHLORIDE IN SOLUTION

When ammonia is chlorinated with t-butyl hypochlorite the changes in concentration with time shown in Fig. 13 (*11*) may be observed by extinction coefficient measurements at the UV absorption maxima for the reagents (ammonia, t-butyl hypochlorite) and products (mono-chloramine, dichloramine, and nitrogen trichloride). A reaction scheme is then developed, on the basis of which concentration time curves may be calculated that are almost completely superposable on the experimental curves in Fig. 13. The scheme includes the basic steps

$$NH_3 + (CH_3)_3COCl \longrightarrow H_2NCl + (CH_3)_3COH \qquad (16)$$

$$H_2NCl + (CH_3)_3COCl \longrightarrow HNCl_2 + (CH_3)_3COH \qquad (17)$$

$$HNCl_2 + (CH_3)_3COCl \longrightarrow NCl_3 + (CH_3)_3COH \qquad (18)$$

as well as two indispensable reactions,

$$H_2NCl + NCl_3 \longrightarrow 2HNCl_2 \qquad (19)$$

$$NH_3 + HNCl_2 \longrightarrow 2H_2NCl \qquad (20)$$

Only by Eq. (19) is it possible to explain why the nitrogen trichloride concentration remains small and the dichloramine concentration

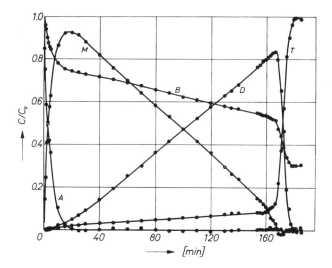

FIG. 13. Change in concentration with time for various substances in the reaction of 3.8 mmoles/liter ammonia with 15.4 mmoles/liter t-butyl hypochlorite in a mixed solvent containing cyclohexane and t-butanol (200 mmoles/liter) at 20°C. A = NH$_3$, B = (CH$_3$)$_3$COCl, M = H$_2$NCl, D = HNCl$_2$, and T = NCl$_3$. For A, M, D, and T the ratio C/C_0 denotes the ratio of the concentration at a given time to the initial concentration of A (or in the case of B, the initial concentration of B) (*11*).

increases simultaneously to over 80%, although the rate of formation of nitrogen trichloride is very large in comparison with the rate for dichloramine. The same is true for Eq. (20). The autocatalytic nature of the reaction leading to nitrogen trichloride when all the monochloramine has been used up is explained by

$$HNCl_2 + NCl_3 + (CH_3)_3COCl \longrightarrow 2NCl_3 + (CH_3)_3COH \tag{21}$$

Finally, the two reactions,

$$(CH_3)_3COCl \longrightarrow Cl\cdot + (CH_3)_3CO\cdot \tag{22}$$

$$NH_2Cl + 2Cl\cdot \longrightarrow NCl_3(d) + 2H\cdot \tag{23}$$

are necessary to explain the slow increase in nitrogen trichloride concentration in presence of monochloramine; it involves the mechanism for forming a deactivated nitrogen trichloride molecule $[NCl_3(d)]$ that can no longer react with monochloramine according to Eq. (19).

For the kinetics of the thermal decomposition of nitrogen trichloride in chlorobenzene, see Ref. 127.

IV. Nitrogen Tribromide

When the structures of solid polymeric nitrogen triiodide-1-ammonia and of solid nonpolymeric nitrogen trichloride are known, it is interesting to discover how solid nitrogen tribromide is constituted. Does the bromine of a nitrogen–bromine bond also form a contact with a second nitrogen (3 center–4 electron bond) or bromine atom? From all that is so far known about nitrogen–bromine compounds (78), it seems that the nitrogen–bromine bond occupies an intermediate position between an N—I bond, which readily forms such contacts, and an N—Cl bond, which does not do so.

It was first necessary to obtain a reproducible method for preparing nitrogen tribromide in solution and/or in the free state, for, unlike mono- and dibromamine, it is so far hardly known (78).

A. SYNTHETIC ROUTES TO NITROGEN TRIBROMIDE

Nitrogen tribromide was first observed as a product of the reaction of an acidic aqueous ammonia salt solution (pH = 4) with bromine and was characterized by determination of the nitrogen:bromine ratio and by UV spectroscopic methods (36, 144, 78). It was possible to isolate for the first time a solid containing nitrogen tribromide from these water or water–methanol solutions by extraction with ether and subsequent precipitation in Freon 12 at −110°C (144).

In the course of further attempts to prepare the pure tribromide (*96, 83*) several possible ways were found for brominating ammonia directly. In dichloromethane at $-87°C$, it is possible to use bromine, bromine chloride, *N*-bromsuccinimide, dibromisocyanuric acid (DBI) (*44*), acetyl hypobromite, and *t*-butyl hypobromite as brominating agents. Bromination experiments with a polymeric brominating agent [*N*-brominated divinyl benzene–styrene–vinyl pyrrolidone copolymer (*68*); see also Ref. *106*] were unsuccessful. Addition of 20% acetone is necessary in the case of DBI to bring the reagent into solution and thus enable it to exert its brominating effect. Solid nitrogen tribromide, which is precipitated, is contaminated with various secondary products. Dissolution of the tribromide from the orange-red mixture with dichloromethane gave pure orange-yellow solutions (nitrogen:bromine ratio, 1:3); they are, however, of little use for preparative or spectroscopic investigations since nitrogen tribromide concentrations are about 4×10^{-3} moles/liter.

Pure solid nitrogen tribromide may be obtained in a reaction analogous to that used for *N*-iodamines (*139, 146*) from bistrimethylsilylbromamine (*152*) and bromine chloride:

$$[(CH_3)_3Si]_2NBr + 2BrCl \xrightarrow[-87°C]{pentane} NBr_3 \downarrow + 2(CH_3)_3SiCl \qquad (24)$$

The deep red substance (nitrogen:bromine ratio, 1:3) is very temperature-sensitive and explodes even at $-100°C$ in a Nujol–pentane suspension (1:2.7) on the least mechanical disturbance. Nitrogen tribromide is soluble without decomposition at low temperatures ($< -80°C$) in polar solvents that do not undergo bromination or oxidation.

B. REACTIONS AND SPECTRA OF NITROGEN TRIBROMIDE; NITROGEN DIBROMIDE MONOIODIDE, NBr₂I

Nitrogen tribromide reacts instantly with ammonia to give dark violet monobromamine (*87*):

$$NBr_3 + 2NH_3 \xrightarrow[-87°C]{CH_2Cl_2} 3NH_2Br \downarrow \qquad (25)$$

With tertiary N bases such as pyridine and urotropine, the bromine adducts of the bases are formed directly. In this, nitrogen tribromide differs from nitrogen triiodide-1-ammonia; the latter with tertiary bases first gives the nitrogen triiodide adducts, and the iodine adducts are formed from these in a subsequent reaction (Section II, E).

Methylene chloride solutions of iodine react with nitrogen tribromide in the same solvent to give solid red-brown nitrogen dibromide iodide (nitrogen:bromine:iodine ratio, 1:2:1), which is stable up to $-20°C$:

$$NBr_3 + I_2 \xrightarrow[-87°C]{CH_2Cl_2} NBr_2I \downarrow + IBr \tag{26}$$

Solutions of nitrogen tribromide in the polar solvents studied are not suitable for IR and Raman spectroscopy because of numerous solvent bands. Since, on the other hand, pure solid nitrogen tribromide is difficult to handle (see above), the less explosive orange-red mixtures of solid nitrogen tribromide and ammonium bromide or acetate were finally examined (96, 83). The band-rich spectra—the IR spectrum described earlier (144, 78) requires some correction—rule out the assignment of C_{3v} symmetry to the monomeric NBr_3 molecule and bring up the question of intermolecular contacts. Further spectroscopic investigations should clarify which structural principle is operative in solid nitrogen tribromide. The ultraviolet spectra of nitrogen tribromide solutions have been known for some time (36, 144, 78).

C. INFRARED AND FURTHER STRUCTURAL STUDIES ON NITROGEN–BROMINE COMPOUNDS

Hitherto there has been only scanty information on the position of N—Br stretching frequencies. A whole range of N—Br compounds has been measured down to 250 cm^{-1} (Table XXII), but often no assignments are given or the assignment given is regarded as uncertain. The most reliable are those where assignments for the corresponding N—Cl and N—I compounds are also considered, so that there is a possibility of making comparisons. The N—Br frequencies are then always found to lie between those of the N—Cl and N—I compounds (73, 136). Some N—Br frequencies that have been assigned with some measure of certainty are 691 cm^{-1} [nitrogen monobromide in a matrix (108)], 560 and 470 cm^{-1} [solid nitrogen tribromide (96)], 540 cm^{-1} [dissolved monobromamine (73)], 451 cm^{-1} [liquid N-bromsulfinylimine (31)], and 265 cm^{-1} [gaseous nitrosyl bromide (14, 104, 102)].

The N—Br stretching force constants have been calculated on the basis of these assignments as 3.35 (nitrogen monobromide), 2.46 (N-bromsulfinylimine), and 1.13 mdyn/Å (nitrosyl bromide). As a rule they cannot be supported by measured N—Br distances.

The only N—Br distance measured by X-ray analysis is 1.82 Å in solid N-bromacetamide (24). This is close to the sum of the single covalent bond radii [$r_N = 0.70$, $r_{Br} = 1.14$ Å; $\Sigma r_{N+Br} = 1.84$ Å (121)]. Microwave spectra and electron diffraction investigations on gaseous

TABLE XXII

NITROGEN–BROMINE COMPOUNDS THE IR SPECTRUM OF WHICH HAVE BEEN MEASURED TO 250 CM^{-1}

Compound[a]	References	Compound[a]	References
With sp^3 hybridization at nitrogen			136
O KOSNBr$_2$ (f) O	136		
O KOSNHBr (f) O	136		73
CH$_3$NBr$_2$ (l)	47		
CH$_2$—NBr$_2$ | (f) CH$_2$—NBr$_2$	48		136
CH$_3$NHBr (l)	66		
(CH$_3$)$_3$CNBr$_2$ (l)	48		
NBr$_3$ (f)	96	F$_2$SNBr (fl)	40
H$_2$NBr (f)	30		
(l)	73		
(CH$_3$)$_2$NBr (g)	73		136
(l)	66		
With sp^2 hybridization at nitrogen			
NBr (g)	108	OSNBr (fl)	149, 31
N$_3$Br (g)	108	C$_2$H$_5$OCONBr$_2$ (fl)	136
	76		73
HCONBr$_2$ (f)	47	HCl$_2$CCONBr$_2$ (f)	47
CH$_3$CONBr$_2$ (f)	47		
OCNBr (f)	43, 46		47
CH$_3$CONHBr (f)	24, 73, 154		
CH$_3$CONBr$_2$ (f)	136	ONBr (g)	150, 95, 14, 104, 102
C$_2$H$_5$CONHBr (f)	136		
H$_2$C—CONBr$_2$ | (f) H$_2$C—CONBr$_2$	47	(f)	71, 102
C$_2$H$_5$CONBr$_2$ (f)	136	**Without hybridization at nitrogen**	
BrNSNBr (f)	138		132
CClH$_2$CONBr$_2$ (f)	47		
(C$_2$H$_5$O)$_2$CNBr (f)	136		

[a] Abbreviations: (g) gaseous; (fl) liquid; (l) solution; (f) solid.

nitrosyl bromide are in agreement in giving an N—Br distance of 2.14 Å (95, 150). This long distance, which is supported by a low stretching frequency and a small force constant, is explained in terms of participation of an ionic limiting structure.

Potential energy curves have been calculated for the $^1\Sigma^+$ and the $^3\Sigma^-$ states of nitrogen monobromide (75). For a calculation of the dissociation energy of the N—Br bond in nitrogen monobromide, see Ram et al. (130), and for measurements of the heat of decomposition of bromine azide, see Dupre et al. (25). Thermodynamic functions have been calculated from spectroscopic data for compound F_2SNBr (131).

D. Nitrogen–Bromine Compounds in Solution

1. Solutions of Bromine in Liquid Ammonia

Dilute solutions of bromine in liquid ammonia do not contain monobromamine at lower temperatures. This may be concluded from the absence of protons (ammonium ions) which would be formed in the reaction (70),

$$Br_2 + 2NH_3 \rightleftharpoons H_2NBr + NH_4^+ + Br^- \tag{27}$$

The solvated bromine cation clearly has a sufficiently high stability in these solutions. In concentrated solutions, on the other hand, mono-bromamine and ammonium bromide are formed according to Eq. (27) (86).

2. New Synthetic Routes to Organic Nitrogen–Bromine Compounds

Although new preparative methods have been developed in recent years for organic nitrogen–chlorine and nitrogen–iodine compounds (e.g., Sections II, J and II, L, 2, a), the main emphasis in recent synthetic work has been on nitrogen–bromine compounds. The sudden increasing use of nitrogen–bromine compounds, especially N-bromo-sulfono- and carboxylic acid amides, as selective brominating agents or cyclization reagents (aziridine) and, therefore, as adjuncts in metallurgy, photography, agriculture, the textile industry, pharmacy, and the perfumery industry makes this understandable.

Thus alkyl compounds with —CONBr$_2$ and $\diagdown$ CNBr$_2$ groups may $\diagup$

be obtained in good yields at room temperature from the corresponding amides or amines and dibromisocyanuric acid in aprotic solvents (*45, 47, 48*). The synthesis of a crystalline monobromamine, *N*-brom-*t*-butylamine, by this route was also successfully carried out. Mono-brominated alkyl compounds with the —CONHBr group may be obtained by the same method using less dibromisocyanuric acid (*50*). On the other hand, they were obtained in very good yield by reaction of the amide with sodium hypobromite and subsequent protolysis of —CONNaBr with dilute sulfuric acid in the presence of methylene chloride to extract the desired product (*3*).

REFERENCES

1. Andrews, M. V., Shaffer, J., and McCain, D., *J. Inorg. Nucl. Chem.* **33**, 3945 (1971).
2. Atovmyan, L. O., *Izv. Akad. Nauk SSSR, Ser. Khim.*, in preparation.
3. Bachand, C., Driguez, H., Patron, J. M., Touchard, D., and Lessard, J., *J. Org. Chem.* **39**, 3136 (1974).
4. Banks, D. F., *Chem. Rev.* **66**, 243 (1966).
5. Barber, H. J., and Henderson, M. A., *J. Chem. Soc., C* p. 862 (1970).
6. Baumgarten, D., Hiltl, E., Jander, J., and Meussdoerffer, J., *Z. Anorg. Allg. Chem.* **405**, 77 (1974).
7. Bayersdorfer, L., Ph.D. Thesis, Technische Hochschule München, 1966.
8. Bayersdorfer, L., Engelhardt, U., Fischer, J., Höhne, K., and Jander, J., *Z. Naturforsch. B* **23**, 1602 (1968); *Z. Anorg. Allg. Chem.* **366**, 169 (1969).
9. Bayersdorfer, L., Minkwitz, R., and Jander, J., *Z. Anorg. Allg. Chem.* **392**, 137 (1972).
10. Beard, C. D., and Baum, K., *J. Amer. Chem. Soc.* **96**, 3237 (1974).
11. Bekiaroglou, P., Drusas, A., and Schwab, G. M., *Z. Phys. Chem.* [N.F.] **77**, 43 (1972).
12. Brown, R. N., *Acta Crystallogr.* **14**, 711 (1961).
13. Bürgi, H. B., Stedman, D., and Bartell, L. S., *J. Mol. Struct.* **10**, 3138 (1971).
14. Burns, W. G., and Bernstein, H. J., *J. Chem. Phys.* **18**, 1669 (1950).
15. Carter, J. C., Bratton, R. F., and Jackowitz, J. F., *J. Chem. Phys.* **49**, 3751 (1968).
16. Cazzoli, G., *J. Mol. Spectrosc.* **53**, 37 (1974).
17. Cazzoli, G., Favero, P. G., and Dal Borgo, A., *J. Mol. Spectrosc.* **50**, 82 (1974).
18. Collin, R. L., *Acta Crystallogr.* **5**, 431 (1952).
19. Cook, R. L., and Gerry, M. C. L., *J. Chem. Phys.* **53**, 2525 (1970).
20. Cotton, F. A., and Wilkinson, G., "Advanced Inorganic Chemistry," p. 133. Wiley (Interscience), New York, 1972.
21. Dahl, T., and Hassel, O., *Acta Chem. Scand.* **24**, 377 (1970).
22. Delhaye, M., Durrieu-Mercier, N., and Migeon, M., *C. R. Acad. Sci., Ser. B.* **267**, 135 (1968).
23. Dregval, G. F., Rozvaga, R. I., and Petrunkin, V. E., *Ukr. Khim. Zhr.* **38**, 905 (1972); *Chem. Abstr.* **78**, 3910t (1973).
24. Dubey, R. J., *Acta Crystallogr.* **27**, 23 (1971).

25. Dupre, G., Paillard, C., and Combourien, J., *C. R. Acad. Sci.*, *Ser. C* **273**, 445 (1971).
26. Engelbrecht, A., and Atzwanger, H., *J. Inorg. Nucl. Chem.* **2**, 348 (1956).
27. Engelhardt, U., Ph.D. Thesis, University of Freiburg/Br., 1964.
28. Engelhardt, U., and Jander, J., *Fortschr. Chem. Forsch.* **5**, 663 (1966).
29. Erckel, R., Diploma Work, University of Heidelberg, 1971.
30. Erckel, R., Ph.D. Thesis, University of Heidelberg, 1974.
31. Eysel, H. H., *J. Mol. Struct.* **5**, 275 (1970).
32. Fenner, J., and Jander, J., *Liebigs Ann. Chem.* p. 1253 (1974).
33. Fenner, J., and Jander, J., *Z. Anorg. Allg. Chem.* **406**, 153 (1974).
34. Fernandez, V., and Dehnicke, K., *Naturwissenschaften* **62**, 181 (1975).
35. Fourme, R., and Renaud, M., *C. R. Acad. Sci.*, *Ser. B* **263**, 69 (1966).
36. Galal-Gorchev, H., and Morris, J. C., *Inorg. Chem.* **4**, 899 (1965).
37. Gardner, D. M., Helitzer, R., and Mackley, C. J., *J. Org. Chem.* **29**, 3738 (1964).
38. Gayles, J. N., *J. Chem. Phys.* **49**, 1841 (1968).
39. Geursen, R., Jander, J., Knuth, K., and Michelbrink, R., *Z. Anorg. Allg. Chem.* **414**, 10 (1975).
40. Glemser, O., Mews, R., and Roesky, H. W., *Chem. Ber.* **102**, 1523 (1969).
41. "Gmelins Handbuch der anorganischen Chemie," 8th ed., System No. 8. Verlag Chemie, Berlin, 1933.
42. Golovina, N. I., Klitskaya, G. A., and Atovmyan, L. O., *J. Struct. Chem. (USSR)* **9**, 919 (1968).
43. Gottardi, W., *Angew. Chem.* **83**, 445 (1971).
44. Gottardi, W., *Monatsh. Chem.* **100**, 42 (1969).
45. Gottardi, W., *Monatsh. Chem.* **103**, 878 (1972).
46. Gottardi, W., *Monatsh. Chem.* **103**, 1150 (1972).
47. Gottardi, W., *Monatsh. Chem.* **104**, 421 (1973).
48. Gottardi, W., *Monatsh. Chem.* **104**, 1681 (1973).
49. Gottardi, W., *Monatsh. Chem.* **105**, 611 (1974).
50. Gottardi, W., *Monatsh. Chem.* **106**, 611 (1975).
51. Goubeau, J., *Angew. Chem.* **78**, 565 (1966).
52. Goubeau, J., Kilcioglu, E., and Jacob, E., *Z. Anorg. Allg. Chem.* **357**, 190 (1968).
53. Gscheidmeyer, H. J., Ph.D. Thesis, University of Heidelberg, 1974.
54. Haase, J., Oberhammer, H., Zeil, W., Glemser, O., and Mews, R., *Z. Naturforsch. A* **25**, 153 (1970).
55. Hagedorn, R., Diploma Work, University of Heidelberg, 1974.
56. Haque, I., and Wood, J. L., *Spectrochim. Acta, Part A* **23**, 959 (1967).
57. Hartl, H., Private communication, Freie Universität Berlin, 1973.
58. Hartl, H., Bärnighausen, H., and Jander, J., *Z. Anorg. Allg. Chem.* **357**, 225 (1968).
59. Hartl, H., Pritzkow, H., and Jander, J., *Chem. Ber.* **103**, 652 (1970).
60. Hartl, H., Schöner, J., Jander, J., and Schulz, H., *Z. Anorg. Allg. Chem.* **413**, 61 (1975).
61. Hartl, H., and Ullrich, D., *Z. Anorg. Allg. Chem.* **409**, 228 (1974).
62. Hartl, H., and Ullrich, D., *Z. Naturforsch. B* **24**, 349 (1969).
63. Hassel, O., *Angew. Chem.* **82**, 821 (1970).
64. Hassel, O., and Hope, H., *Acta Chem. Scand.* **15**, 407 (1961).
65. Hassel, O., and Rømming, C., *Acta Chem. Scand.* **21**, 2659 (1967).

66. Heasley, V. L., Kovacic, P., and Lange, R. M., *J. Org. Chem.* **31**, 3050 (1966).
67. Hedvall, J. A., "Einführung in die Festkörperchemie." Vieweg-Verlag, Braunschweig, 1952.
68. Heitz, W., and Michels, R., *Makromol. Chem.* **148**, 9 (1971).
69. Hendra, P. J., and Mackenzie, J. R., *Chem. Commun.* p. 760 (1968).
70. Herlem, M., Thiébault, A., and Bobilliart, F., *J. Electroanal. Chem. Interfacial Electrochem.* **49**, 464 (1974).
71. Hisatsune, I. C., and Miller, P., *J. Chem. Phys.* **38**, 49 (1963).
72. Höhne, K., Diploma Work, Technische Hochschule Munich, 1966.
73. Höhne, K., Ph.D. Thesis, Technische Hochschule Munich, 1968.
74. Höhne, K., Jander, J., Knuth, K., and Schlegel, D., *Z. Anorg. Allg. Chem.* **386**, 316 (1971).
75. Itagi, S., and Shamkuwar, N. R., *Marathwada Univ. J. Sci., Sect. A* **11**, 29 (1972).
76. Jabay, O., Diploma Work, University of Heidelberg, 1974.
77. Jander, J., Bayersdorfer, L., and Höhne, K., *Z. Anorg. Allg. Chem.* **357**, 215 (1968).
78. Jander, J., and Engelhardt, U., *in* "Developments in Inorganic Nitrogen Chemistry" (C. W. Colburn, ed.), Vol. 2, pp. 70–203. Elsevier, Amsterdam, 1973.
79. Jander, J., and Engelhardt, U., *Z. Anorg. Allg. Chem.* **339**, 225 (1965).
80. Jander, J., and Engelhardt, U. *Z. Anorg. Allg. Chem.* **341**, 146 (1965).
81. Jander, J., Engelhardt, U., and Weber, G., *Angew. Chem.* **74**, 75 (1962).
82. Jander, J., Knackmuss, J., Knuth, K., and Pritzkow, H., *Z. Naturforsch. B* **27**, 1420 (1972).
83. Jander, J., Knackmuss, J., and Thiedemann, K. U., *Z. Naturforsch. B* **30**, 464 (1975).
84. Jander, J., Knuth, K., and Renz, W., *Z. Anorg. Allg. Chem.* **392**, 143 (1972).
85. Jander, J., Knuth, K., and Trommsdorff, K. U., *Z. Anorg. Allg. Chem.* **394**, 225 (1972).
86. Jander, J., and Kurzbach, E., *Z. Anorg. Allg. Chem.* **296**, 117 (1958).
87. Jander, J., and Lafrenz, C., *Z. Anorg. Allg. Chem.* **349**, 57 (1967).
88. Jander, J., and Maurer, A., *Z. Anorg. Allg. Chem.*, **416**, 251 (1975).
89. Jander, J., and Minkwitz, R., *Z. Anorg. Allg. Chem.* **405**, 250 (1974).
90. Jander, J., Pritzkow, H., and Trommsdorff, K. U., *Z. Naturforsch. B* **30**, 720 (1975).
91. Jander, J., and Schmid, E., *Angew. Chem.* **71**, 31 (1959); *Z. Anorg. Allg. Chem.* **304**, 307 (1960).
92. Jander, J., and Schmid, E., *Z. Anorg. Allg. Chem.* **292**, 178 (1957).
93. Karelin, A. I., Kharitonov, Y. Y., and Rosolovskii, V. Y., *in* "Kolebatel'nye Spektry neorg. Chim." [Vibration Spectra in Inorganic Chemistry] (Acad. Sci. USSR, Inst. Gen. Inorg. Chem. I. S. Kurnakov, ed.), pp. 182–218. Nauka Press, Moscow, 1970.
94. Kawaguchi, T., Takashina, K., Tanaka, T., and Watanabe, T., *Acta Crystallogr., Sect. B* **28**, 967 (1972).
95. Ketelaar, J. A., *Rec. Trav. Chim.* **62**, 289 (1943).
96. Knackmuss, J., Ph.D. Thesis, University of Heidelberg, 1975.
97. Knuth, K., Diploma Work, Freie Universität Berlin, 1968.
98. Knuth, K., Jander, J., and Engelhardt, U., *Z. Naturforsch. B* **24**, 1473 (1969); *Z. Anorg. Allg. Chem.* **392**, 279 (1972).

99. Knuth, K., Renz, W., and Jander, J., *Z. Anorg. Allg. Chem.* **400,** 67 (1973).
100. Krebs, H., *Z. Elektrochem.* **61,** 934 (1957).
101. Kukushkin, Y. N., *J. Inorg. Chem.* (*USSR*) **5,** 1943 (1960).
102. Laane, A. A., Jones, L. H., Ryan, R. R., and Asprey, L. B., *J. Mol. Spectrosc.* **30,** 485 (1969).
103. Ladell, J., and Post, B., *Acta Crystallogr.* **7,** 559 (1954).
104. Landau, L., and Fletscher, W. H., *J. Mol. Spectrosc.* **4,** 276 (1960).
105. Mandell, J., and Barth-Wehrenalp, G., *J. Inorg. Nucl. Chem.* **12,** 90 (1959).
106. Manecke, G., and Stärk, M., *Makromol. Chem.* **176,** 285 (1975).
107. Manley, T. R., and Williams, D. A., *Spectrochim. Acta* **21,** 1773 (1965).
108. Milligan, D. E., and Jacox, M. E., *J. Chem. Phys.* **40,** 2461 (1964).
109. Minet, J. J., Herlem, M., Thiébault, A., and Fave, G., *J. Electroanal. Chem.* **31,** 153 (1971).
110. Minkwitz, R., Diploma Work, Freie Universität Berlin, 1967.
111. Minkwitz, R., Ph.D. Thesis, Freie Universität Berlin, 1970.
112. Naae, D. G., and Gougoutas, J. Z., *J. Org. Chem.* **40,** 2129 (1975).
113. Nyburg, S. C., *J. Chem. Phys.* **48,** 4890 (1968).
114. Oberhammer, H., *Z. Naturforsch. A* **25,** 1497 (1970).
115. Oberhammer, O., Glemser, O., and Klüver, H., *Z. Naturforsch. A* **29,** 901 (1974).
116. Oberhammer, H., and Strähle, J., *Z. Naturforsch. A* **30,** 296 (1975).
117. Ohler, G., Diploma Work, University of Heidelberg, 1973.
118. Osborn, R. S., Ph.D. Thesis, University of London, 1972.
119. Paetzold, R., Dostal, K., and Ruzicka, A., *Z. Anorg. Allg. Chem.* **348,** 112 (1966).
120. Parrod, J., and Pornin, R., *C. R. Acad. Sci., Paris* **260,** 1438 (1965).
121. Pauling, L., "The Nature of the Chemical Bond." Oxford Univ. Press, London and New York, 1950.
122. Phelip, B., Ph.D. Thesis, University of Paris, 1962.
123. Pritzkow, H., *Monatsh. Chem.* **105,** 621 (1974).
124. Pritzkow, H., *Z. Anorg. Allg. Chem.* **409,** 237 (1974).
125. Pritzkow, H., *Acta Crystallogr., Sect. B* **31,** 1505 (1975).
126. Pritzkow, H., Private Communication, University of Heidelberg, 1975.
127. Radbil, B. A., and Kushnir, S. R., *Tr. Khim. Khim. Tekhnol.* p. 15 (1973).
128. Radell, J., Connolly, J. W., and Raymond, A. J., *J. Chem. Soc.* (*London*) **83,** 3958 (1961).
129. Rai, S. N., Rai, B., Nair, K. R., and Thakur, S. N., *Z. Naturforsch. A* **27,** 865 (1972).
130. Ram, R. S., Mishra, P. C., and Upadhya, K. N., *Spectr. Lett.* **6,** 541 (1973).
131. Randhawa, H. S., and Sharma, D. K., *J. Indian Chem. Soc.* **5,** 775 (1974).
132. Razumova, E., and Kostyanovskii, R., *Zh. Strukt. Khim.* **13,** 1080 (1972).
133. Renz, W., Ph.D. Thesis, Technische Hochschule Munich, 1965.
134. Savoie, R., and Giguere, P. A., *Can. J. Chem.* **40,** 991 (1962).
135. Schack, C. J., Christe, K. O., Philipovich, D., and Wilson, R. D., *Inorg. Chem.* [Washington] **10,** 1078 (1971).
136. Schlegel, D., Ph.D. Thesis, Freie Universität Berlin, 1970.
137. Schöner, J., Ph.D. Thesis, Freie Universität Berlin, 1972.
138. Seppelt, K., and Sundermeyer, W., *Angew. Chem.* **81,** 785 (1969).
139. Seppelt, K., and Sundermeyer, W., *Z. Naturforsch. B* **24,** 774 (1969).
140. Siebert, H., "Anwendungen der Schwingungsspektroskopie in der anorga-

nischen Chemie." Springer-Verlag, Berlin and New York, 1966; *Z. Anorg. Allg. Chem.* **273**, 170 (1953).
141. Stammreich, H., Forneris, R., and Tavares, Y., *J. Chem. Phys.* **25**, 580 (1956).
142. Strømme, K. O., *Acta Chem. Scand.* **13**, 268 (1959).
143. Suzuki, H., Itoh, K., Matsuda, I., and Ishii, Y., *Chem. Lett. (Tokyo)* p. 197 (1975).
144. Thiedemann, K. U., Ph.D. Thesis, University of Heidelberg, 1971.
145. Trommsdorff, K. U., Diploma Work, University of Heidelberg, 1972.
146. Trommsdorff, K. U., Ph.D. Thesis, University of Heidelberg, 1974.
147. Ullrich, D., Diploma Work, Freie Universität Berlin, 1969.
148. Ullrich, D., Ph.D. Thesis, Freie Universität Berlin, 1972.
149. Verbeek, W., and Sundermeyer, W., *Angew. Chem.* **81**, 331 (1969).
150. Weatherly, T. L., and Williams, Q., *J. Chem. Phys.* **25**, 717 (1956).
151. Wibenga, E. H., Havinga, E. E., and Boswijk, K. H., *Advan. Inorg. Chem. Radiochem.* **3**, 158 (1961).
152. Wiberg, N., and Raschig, F., *Angew. Chem.* **77**, 130 (1965.)
153. Wolf, W., and Steinberg, L., *Chem. Commun.* p. 449 (1965).
154. Wolfe, S., and Awang, D. V., *Can. J. Chem.* **49**, 1384 (1971).
155. Wong, P. T. T., and Whalley, E., *Can. J. Chem.* **50**, 1856 (1972).
156. Zipprich, M., Ph.D. Thesis, University of Heidelberg, 1975.
157. Zipprich, M., Pritzkow, H., and Jander, J., *Angew. Chem.* **88**, 225 (1976).

ASPECTS OF ORGANO-TRANSITION-METAL PHOTOCHEMISTRY AND THEIR BIOLOGICAL IMPLICATIONS

ERNST A. KOERNER VON GUSTORF,†

LUC H. G. LEENDERS,‡ INGRID FISCHLER, and ROBIN N. PERUTZ§

Institut für Strahlenchemie im Max-Planck-Institut für Kohlenforschung,
Mülheim/Ruhr, West Germany

I. Introduction

Two questions may demonstrate why the authors of this survey found it rather complicated to give a concise definition of its scope: What is

† Deceased.

‡ Present address: Bell Laboratories, Murray Hill, New Jersey.

§ Present address: Department of Chemistry, University of Edinburgh, Edinburgh, Scotland.

photochemistry? and What is organometallic chemistry? The answers we would like to suggest reflect our personal point of view.

Photochemistry is the study of chemical changes brought about by light. It encompasses all processes "participating in the formation and deactivation of electronically excited molecules" (*454*). Some of its aspects may be considered as photophysics, some as photobiology; however, any arbitrary demarcation would not be in accord with the interdisciplinary character of this field.

We consider *organometallic chemistry* to be the chemistry of compounds in which metals are bound to groups or molecules that are under the regime of organic chemistry. We do not make any distinction whether bonding occurs via carbon or another element, as for instance N, P, O, or S. In many transformations of organic material involving metals, the site of bonding and the primary reactive steps remain obscure anyhow.

Organometallic compounds are often made from "inorganic" starting materials, are applied in organic syntheses, and play a significant role in biochemistry. Too narrow a definition could only harm the possibilities of interdisciplinary cross-fertilization. Our view deviates somewhat from general practice. In a narrow sense, organometallic compounds should contain a direct carbon-to-metal bond. According to a recent IUPAC ruling, a differentiation is even made with respect to the nature of this bond. Whereas a σ-bonded metal alkyl is named an "organometallic compound," a π-alkylmetal compound is named a "coordination compound" (*273*).

After these programmatic remarks, it is obvious that a survey on organometallic photochemistry can only cover specifically selected areas of this broad field.

In this article we have tried to stress mechanistic and preparative aspects. We hope to address the inorganic and organometallic chemist who is interested in photochemistry as a method to be applied to his problems. The following points have, therefore, been emphasized: methods for the photochemical production and characterization of unstable and short-lived intermediates (discussed in a section on primary processes in organometallic photochemistry); important reaction principles; photosyntheses of new organometallic compounds (especially at low temperatures); photochemical transformations of organic compounds using metals as templates; mechanistic investigations using photochemical methods for the synthesis of, and the search for, postulated intermediates in organometallic reactions occurring in the dark; and application of organometallic photoreactions as models for biochemical studies.

The discussion is restricted to the photochemistry of organo-transition-metal compounds in low-oxidation states. It basically deals with the photochemistry of metal carbonyls and derivatives and metal compounds containing σ- and π-bonded organic systems. Sandwich compounds, such as ferrocene (68), have been omitted and will be discussed elsewhere (55).

Because an extensive survey of this field was published in 1969 from our laboratory (332), the coverage of literature in this present article is based on a search of *Chemical Abstracts* from 1969 to the middle of 1974. The volume of literature has been growing rapidly, and much of the information about photochemical work in organometallic chemistry is hidden and often hard to find. We, therefore, did not attempt a complete coverage of all work done in the field. We also felt unable to incorporate all the available information into one article of limited size. According to the plan of emphasizing preparative and mechanistic aspects, we have tried to present Sections III, D and IV as completely as possible. A more detailed discussion of the primary photoprocesses will be given elsewhere (55). The remarks on biological applications (Section V) are supposed to whet the reader's appetite rather than to give an extensive survey.†

We draw the reader's attention to two excellent books (1a, 27), some recent reviews on the photochemistry of coordination compounds (533, 388, 386, 26, 590), and two valuable surveys of metal carbonyl photochemistry (584, 609). Comprehensive annual reports on inorganic and organometallic photochemistry are found in the *Specialist Periodical Reports* (433). Finally, an annual highlights section on this subject is featured by *Molecular Photochemistry* (202).

II. Principles of Photochemistry

Quinkert (454) remarked that the chemistry of electronically excited states provides a new dimension in chemistry. This comment was founded on the different chemical reactivities of electronic ground and excited states. A photochemist must not only be concerned with the chemical consequences of photolysis but also with the ways in which the excited-state molecule can change its electronic state. Some of these processes are summarized briefly in Fig. 1 to familiarize the reader with the vocabulary of photochemistry (434). For a more detailed discussion the reader is referred to some of the excellent books in this field (568, 107, 27, 359, 143, 131, 534, 597).

† Much information on this subject is found in the numerous papers by R. J. P. Williams.

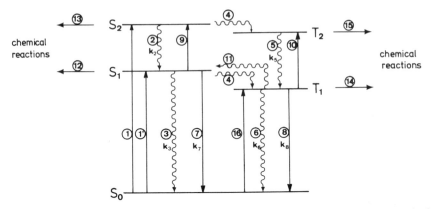

FIG. 1. Energy levels of excited states and transitions between them. Vertical straight lines represent radiative transitions; wavy lines nonradiative transitions. S = singlet, T = triplet, numbers are explained in text.

The state of a molecule produced on electronic excitation depends on the energy of the light quantum absorbed.† The excited states can dissipate their energy in different ways. Upper electronic states will usually be converted very fast to the lowest-excited states (process 2 in Fig. 1), from which further processes originate. The radiationless transition between states of the same multiplicity (e.g., processes 2, 3, and 5) is called internal conversion and that between states of different multiplicity (process 4), intersystem crossing. The radiative decay of excited states is termed fluorescence if it occurs between states of the same multiplicity (process 7) and phosphorescence if a change of multiplicity is involved (process 8). Emission after thermal back-population of S_1 from T_1 (process 11 followed by process 7) is called delayed fluorescence.

The probability of an electronic transition is controlled by symmetry, overlap, and spin requirements. The oscillator strengths of transitions are reduced by the following factors relative to a fully allowed transition: (a) about 10^{-5} if spin-forbidden (e.g., $S_0 \rightarrow T_1$); (b) 10^{-2} if overlap-forbidden (e.g., n-π^* transition of second-row heteroatoms); and (c) 10^{-1}–10^{-4} if symmetry or orbitally forbidden (e.g., parity or Laporte rule). The strictness of the spin rule is reduced by spin-orbit coupling, which plays an important role with the heavier elements. In some organometallic compounds, singlet–triplet absorptions (process 16 in Fig. 1) have been identified, but they are mostly

† Only light absorbed can bring about a photochemical change, and usually each quantum absorbed activates only one molecule in the primary state.

restricted to compounds of third-row transition metals [e.g., $Mo(CO)_5$-$(NHEt_2)$, no singlet–triplet absorption; $W(CO)_5(NHEt_2)$, $^1A_1 \rightarrow {}^3E$ at 438 nm, $\varepsilon = 730$) (*609, 619, 620*)].

Chemical reactions (processes 12–15 in Fig. 1) can arise from any excited state, but most photoconversions originate from the lowest-excited states (S_1, T_1), since they have a longer lifetime than the upper states. This difference can be attributed to the larger energy gap between the ground state and first excited state. In reactions of organometallic compounds, reactions from upper-excited states are observed quite frequently (see Section IV).

III. Primary Processes in Organometallic Photochemistry

This section is devoted to the methods used for studying primary photochemical processes (the production and decay of electronically excited states) (*359*) and their application to organometallic compounds. The following topics will be discussed: electronic absorption spectroscopy (UV/vis), luminescence, flash photolysis, photolysis in rigid media, and quenching and sensitization processes.

A. ELECTRONIC ABSORPTION SPECTROSCOPY

An indication of the number of excited states of a molecule accessible by direct absorption can often be derived from the absorption spectra (e.g., processes 1, 1', and 16 in Fig. 1). The resolution of the spectra is of critical importance and can be increased by (a) measuring the spectra in the gas phase, (b) by measuring at low temperatures in a matrix or single crystal, (c) by using solvents of different polarity or hydrogen-bonding ability, (d) by using magnetic circular dichroism (*541*), or (e) by using circular dichroism if the compounds are optically active (*252*, 392) [possibly by introduction of ligands containing optically active groups (*335*)]. The assignment of the absorption bands is a major problem often requiring detailed theoretical treatment of the system under investigation.

The important kinds of electronic transitions in an organometallic compound can be described schematically using the simplified case of an olefin–transition metal complex. According to the Chatt-Duncanson picture (*111*), bonding in such a system can be described as being composed of two major components.

An oversimplified MO scheme can be derived from such a picture (Fig. 2) by considering the frontier orbitals of the metal and olefin (e.g.,

for the hypothetical Fe$\cdots$ $\rangle\langle$; Fig. 3). Some of the resulting MO's

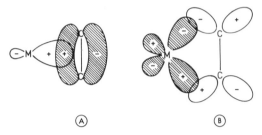

FIG. 2. Bonding between a metal and an olefin. (A) σ Donor bond, olefin $\rightarrow$ metal by overlap of an occupied olefin π-orbital with an unoccupied metal σ-orbital. (B) π-Backbonding, metal $\rightarrow$ olefin, by overlap of the unoccupied olefin π^*-orbital with an occupied metal d_π-orbital.

will have major contributions from the metal orbitals ($\pi_{1'}$, σ_2), others from the ligand orbitals (σ_1, π_2). Three important types of transitions may then occur.

    i. *d-d Transitions.* These involve transitions between two orbitals of predominantly metal d character (transition A in Fig. 3). The result of such an excitation is a redistribution of electron density at the metal atom without any major change of charge. In the case of transition A in Fig. 3, the depopulation of a π-bonding orbital is accompanied by the population of the antibonding orbital σ_2. Such

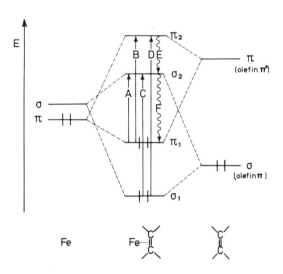

FIG. 3. Simplified MO scheme of an iron olefin species. Transitions A–F explained in text.

transitions often cause the dissociation of metal–ligand (M—L) bonds. This reaction is one of the most important phenomena in this branch of photochemistry.

ii. *Charge-transfer transitions.* In processes B and C, electronic transitions occur from orbitals of mainly metallic to orbitals of mainly ligand character or vice versa. Thus, transition B results in a charge-transfer from metal to ligands (CTML transition), whereas transition C causes a movement of charge in the opposite direction (CTLM transition). Either nucleophilic or electrophilic attack at the metal should be facilitated in the CT-excited states according to the direction of charge flow. Although d-d transitions are usually weak, the large transition moment of CT processes often results in strong absorption bands (for d-d processes $\varepsilon \simeq 1$–150 for spin-allowed transitions in octahedral complexes, $\varepsilon \simeq 1000$ for noncentrosymmetric complexes; for CT processes $\varepsilon \simeq 10^4$).

iii. *Intraligand transitions.* The third kind of transition relates to processes involving orbitals of mainly ligand character (e.g. transition D in Fig. 3.) It may also be possible to observe intraligand transitions of groups that are not involved in bonding to the metal. The frequencies of such transitions are often not influenced by coordination to the metal. In such cases they can be used to study the effects of the neighboring metal on the lifetime of the excited states and to check for intramolecular energy transfer, etc. (*604, 618, 615, 627, 628*).

A clear distinction among these different kinds of transitions is seldom possible in organometallic compounds. Although the excited states usually have mixed character, these approximations are still useful. The differences between the reactivity of the different types of excited states will be discussed in Section IV. However, most chemical reactions can be ascribed to the decay of d-d excited states.

Much work has recently been devoted to the development of theoretical models of excited state properties (*615, 629, 630, 631*). There have also been attempts to draw conclusions about the nature of the excited states from the observed chemical reactivity (*236, 615*).

Dependence of the photobehavior on the wavelength of irradiation is much more common among organometallic compounds than among organic compounds, for which the number of cases known is limited. It appears that fast chemical reactions can compete with internal conversion of excited states. As a result, small changes in the ligands often have a dramatic effect on the ratio of the corresponding rate constants.

Most of the available absorption data on metal carbonyl compounds have recently been reviewed in an excellent survey by Wrighton (*609*).

We have therefore restricted ourselves to the discussion of some examples from organoiron chemistry that illustrate the general points outlined above.

The UV spectra of $Fe(CO)_5$ shown in Fig. 4 demonstrate the effect of low-temperature measurements on the quality of the spectrum. Two major absorption bands can be observed at $77°K$ at $35,450$ cm^{-1} ($\varepsilon = 3,900$) and at $41,670$ cm^{-1} ($\varepsilon = 15,700$). The spectrum has been assigned (Table I) on the basis of detailed MO calculations (Fig. 5) (147).

Although the longest-wavelength transition occurs between two orbitals with predominantly d character, it gains intensity by mixing with ligand orbitals. The higher-energy absorption at $\sim 41,500$ cm^{-1} is assigned to a superposition of transitions to two orbitals with 77 and 99% π^* (CO) character, respectively.

The situation is even more complicated in the substituted carbonyls derived from $Fe(CO)_5$. Table II lists the UV spectra of a variety of olefin–tetracarbonyliron complexes (236).

The example of the fumaric acid–tetracarbonyliron complex (Fig. 6) demonstrates that circular dichroism (CD) spectra reveal more bands and are better resolved than the corresponding "normal" absorption spectra (392). Instead of using the tedious procedure of optical resolution, CD spectra of metal complexes can also be obtained by the introduction of optically active ligands. For instance, di-l-menthyl fumarate tetracarbonyliron shows extrema at [kK($\Delta\varepsilon$)] 27.8(-1.7), 32.9($+11.1$),

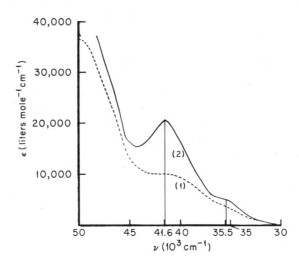

FIG. 4. Ultraviolet spectrum of $Fe(CO)_5$ in isopentane methylcyclohexane at $300°K$ (1) and at $77°K$ (2). [From Dartiguenave et al. (147).]

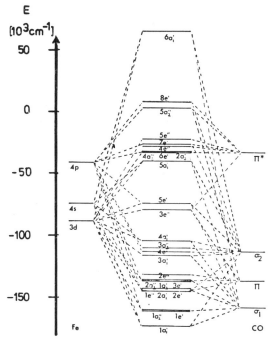

Fig. 5. Molecular orbital scheme of $Fe(CO)_5$. [From Dartiguenave et al. (147).]

TABLE I
ASSIGNMENT OF TRANSITIONS IN $Fe(CO)_5{}^a$

Absorptions observed in cm^{-1} (ε in liters $mole^{-1}$ cm^{-1})			Calculated absorptions (cm^{-1})
300°K	77°K	Assignment	
35,500 (3,800)	35,450 (3,900)	$5e' \rightarrow 5a_1'$ allowed	34,000
		$3e'' \rightarrow 5a_1'$ forbidden	39,200
41,500 (10,200)	41,670 (15,700)	$5e' \rightarrow 6e'$ allowed	41,300
		$5e' \rightarrow 2a_2'$ allowed	41,400
		$5e' \rightarrow 4a_2''$ forbidden	41,000
50,000 (37,000)	—	$5e' \rightarrow 4e''$ allowed	45,800
		$3e'' \rightarrow 6e'$ allowed	46,500
		$3e'' \rightarrow 4a_2''$ allowed	46,200
		$3e'' \rightarrow 2a_2'$ forbidden	46,600
		$5e' \rightarrow 7e'$ allowed	48,500
		$3e'' \rightarrow 4e''$ allowed	51,000
		$5e' \rightarrow 5e''$ allowed	52,600
		$3e'' \rightarrow 7e'$ allowed	53,700

^{a} Data from Dartiguenave et al. (147).

TABLE II

ULTRAVIOLET DATA OF SOME TETRACARBONYL–OLEFINIRON COMPLEXES IN n-HEXANE[a]

L in L—Fe(CO)$_4$	$cm^{-1} \times 10^{-3}$ (kK)[b]			
cis-1,2-Dibromoethylene	~28.5 (WS) (470)	36 (S) (5,400)	~43 (WS) (14,600)	46[c] (17,000)
trans-1,2-Dibromoethylene	~28 (WS) (490)	34.7 (M) (5,680)	~40.5 (WS) (10,200)	~46 (17,000)
cis-1-Bromo-2-fluoroethylene[d]	~29.5 (WS) (700)	~37 (WS) (5,100)	—	—
trans-1-Bromo-2-fluoroethylene	~29 (WS) (420)	~36.5 (WS) (4,500)	~40.5 (WS) (8,500)	46 (17,500)
cis-1,2-Dichloroethylene	~29 (WS) (550)	37 (S) (4,900)	—	~46 (18,300)
trans-1,2-Dichloroethylene	~29 (WS) (290)	36.5 (S) (4,550)	~43 (WS) (15,000)	46 (17,500)
Dimethylmaleate	~29 (WS) (630)	38.5 (S) (7,700)	—	46 (20,400)
Dimethylfumarate	~28.5 (WS) (710)	36.0 (M) (9,100)	—	46 (21,600)
Methylacrylate	~29 (WS) (700)	38.5 (S) (7,600)	—	46 (20,000)
Ethylene tetracarboxylic acid–tetramethylester	~30 (WS) (1,080)	37.3 (M) (9,430)	—	47 (19,000)
CO[e]	—	35.5 (3,800)	41.5 (10,200)	50 (37,000)

[a] From Grevels and Koerner von Gustorf (236).
[b] (WS) weak shoulder; (S) shoulder; (M) maximum.
[c] Measurement limit, possibly a shoulder or a maximum.
[d] Spectrum measured up to 41,000 cm^{-1}.
[e] From Dartiguenave et al. (147).

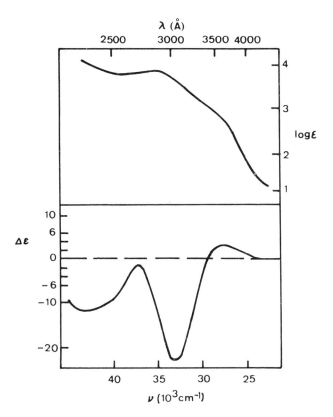

FIG. 6. Electronic and circular dichroism spectra of (−)-fumaric acid Fe(CO)$_4$ in methanol. [From Musco *et al.* (*392*).]

40(+ 4.5)(shoulder), 43.8(+ 6.2) (*335*). These positions compare quite well with those of fumaric acid–tetracarbonyliron shown in Fig. 6. The sign and size of $\Delta\varepsilon$ indicates a considerable enrichment of the *S,S*-dimentyl fumarate–iron tetracarbonyl complex.

Since the low symmetry (C_{2v}) of these complexes allows extensive mixing of Fe, CO, and olefin orbitals, the resulting MO's are of rather complex composition [for a detailed discussion, see Grevels and Koerner von Gustorf (*236*)]. It is not surprising, therefore, that photolysis results in the loss of both CO and olefin. However, the ratio of these two processes is wavelength-dependent (*93*).

The complexity of attempts to correlate observed photoreactions with the nature of the corresponding excited state is demonstrated by the study of the photoreactivity of η-(1,2-dihaloethylene)–tetracarbonyliron complexes (*236*). Irradiation in the 34–37,000 cm^{-1} region, led to (*3*).

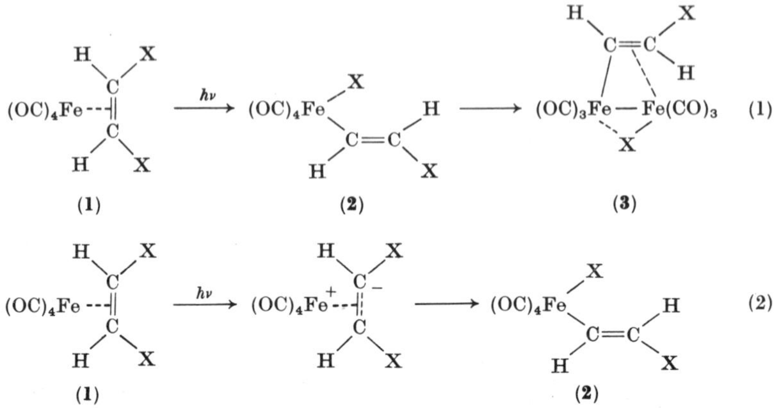

(1)

(2)

Compound **2** was detected by IR spectroscopy and transforms into **3** as the final product. Although compound **2** might be expected to originate from a CTML state, as shown in Eq. (2), a detailed study revealed that the primary process is the loss of CO, which is followed by oxidative addition of a C—X bond and readdition of CO. The reactive state was, therefore, assigned to a *d-d* excited state (leading to cleavage of a M—CO bond). Whether this reactive state is reached by direct excitation or by internal conversion remains an open question.

In Table III we have compiled a selection of absorption data for diene–metal complexes, the dienes being butadiene, substituted butadienes, and aza and oxa derivatives.

A comparison shows that absorption is observed in the same range ($\bar{\nu} > 23,000$ cm^{-1}) so long as the diene system functions as a 4π-electron donor system (**4**). Diazabutadiene complexes that are differently bonded (**5**) show a dramatically different spectrum with a strong transition in the 20,000 cm^{-1} area (*165*). This band shows remarkable solvent dependence indicating considerable CT character.

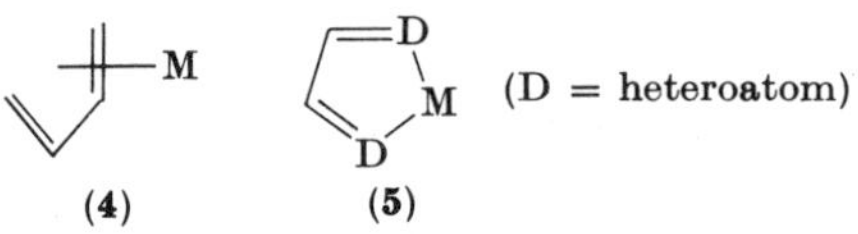

(4)   (5)   (D = heteroatom)

The absorption data given for **9** and **16** (Table III) are examples of transitions between ligand orbitals (*74*) that are little perturbed by the metal.

The assignment of the transitions in diene–tricarbonyliron complexes is difficult. The CD spectra of *trans,trans*-2,4-hexadienoic acid

TABLE III

ABSORPTION DATA FOR DIENE–METAL CARBONYLS

Complex	Absorptions^a $\bar{\nu}$ in kK (ϵ in liters mole^{-1} cm^{-1})	Solvent	Ref.
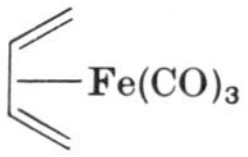 (6) —Fe(CO)$_3$	36 (2,500) (S at RT) (M at LT) 36 (S) (700) 49 (13,200)	n-Hexane Methylcyclohexane–isopentane Gas phase	275
(7) L = CO	24.7 (180) 32 (S) (950) 36.5 (S) (2,400) 45 (20,000)	n-Hexane	92
(8) L = PF$_3$	25.2 (180) 34 (S) (1,000) 38 (S) (2,300) 45 (17,000)	Cyclohexane	92
(9) Ph—...—CHO Fe(CO)$_3$ $n = 0$ $n = 1$ $n = 2$ $n = 3$	37.9 (22,700) 35 (22,300) 32 (24,500) 29.9 (23,000)	Dichloromethane	75

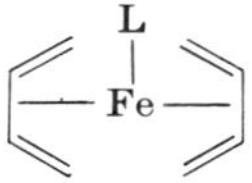

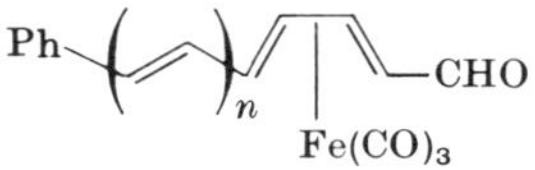

(continued)

TABLE III (Continued)

Complex	Absorptions^a $\bar{\nu}$ in kK (ε in liters mole^{-1} cm^{-1})	Solvent	Ref.
(10)	29.0 (S) (1,600) 41.0 (S) (12,000)	Cyclohexane	338
(11)	24.0 (S) (520) 30.7 (2,400)	n-Hexane	338
(12)	23.5 (S) (550) 29.7 (1,980) 42.2 (11,900)	Diethyl ether	333
(13)	29.5 (S) (5,500) 32.5 (S) (8,600) 38.8 (22,400)	n-Hexane	333
	29.0 (S) (4,400) 33.2 (S) (13,200) 40.0 (22,800)	n-Hexane	333

Complex		Solvent	Ref.
(15) PPh₃–Mo(CO)₃ camphor-imine complex (CH₃ on N)	15.1[b] 14.9[b]	Acetone Benzene	465 465
(16) p-R¹–C₆H₄–N=C₆H₄–R²-p, Fe(CO)₃ R¹ R² H H Cl Cl OMe H	35.6 (21,700) 32.8 (17,200) 35 (25,300) 32.6 (S) (22,100) 35 (20,900) 32.6 (S) (20,500)	Cyclohexane	74
(17) p-R¹–C₆H₄–N=C₆H₄–R²-p R¹ R² H H Cl Cl OMe H	33.8 (26,700) 30.3 (S) (14,200) 32.9 (29,900) 29.8 (S) (20,300) 32.2 (S) (23,200) 30.9 (24,200) 29.4 (S) (21,100)	Cyclohexane	74
(18) imidazole–Mo(CO)₄ (N–CH₃) R = H R = CH₃	20[b] 18.7[b] 21.7[b] 19.8[b]	Dimethylformamide Benzene Dimethylformamide Benzene	165

[a] (S) shoulder; (M) maximum, (RT) room temperature, (LT) low temperature (89°K).
[b] Longest-wavelength absorption.

iron tricarbonyl reveal at least five different transitions in the 25–45 kK area (*392*).

According to SCCC calculations comprising charge and configuration iteration of the metal as well as of the ligand atoms, we can expect ten transitions in the 24,000–50,000 cm^{-1} area for **6**, and thirteen transitions in this region for **7** (*537, 599*). These calculations also show that the MO's of photochemical interest have varying contributions from the diene, CO, and metal orbitals. However, for some of the electronically excited states, preferential cleavage of the Fe—CO or the Fe—diene bond is predicted in agreement with the experimental observations (see Section IV).

Valuable information about electronic spectroscopy of inorganic and main group organometallic compounds is found in Refs. *368* and *457*.

B. LUMINESCENCE

Luminescence appears to be a scarce phenomenon among organometallic compounds and can often only be studied in compounds isolated in glasses at temperatures of 77°K or below. Emission from complexes of the Group VIB metals (Cr, Mo, W) has received most attention. Although emission has been observed for most of the tungsten complexes studied, it is much less common among the lighter members of the group.

Luminescence has been reported for the following groups of metal carbonyl derivatives: (i) $W(CO)_5L$ (L = NR_3, NHR_2, NH_2R, pyridine, ROH, ROR, ketones) (*618, 619, 621*); (ii) $M(CO)_4LL$ (LL = 2,2'-bipyridyl, 1,10-phenanthroline, M = Mo, W) (*609, 621*); (iii) $W(CO)_4L_2$ (L = piperidine, pyridine, L_2 = ethylenediamine) (*609*); (iv) $Cr(CO)_4LL$ (LL = 2,2'-bipyridyl) (*291*); and (v) $ClRe(CO)_3LL$ (LL = 1,10-phenanthroline, 2,2'-bipyridyl) (*623*). The emission of the $W(CO)_5L$ complexes consists of a broad band centred at about 533 nm. Although the band maximum varies little, the emission lifetime is more sensitive to the nature of the ligand (0.65–25.5 μsec). The lack of emission from the corresponding $Cr(CO)_5L$ and $Mo(CO)_5L$ complexes led the authors to assign the bands to a *d-d* triplet–singlet transition (*621*) ($^3E \rightarrow \, ^1A_1$).†
The emission of the $M(CO)_4LL$ complexes (LL = bidentate heterocycle) can be observed at room temperature in the solid and was assigned to a

---

† Compare Section II: Only the $W(CO)_5L$ complexes show a singlet–triplet absorption. Such transitions are more intense in the W complexes because of spin-orbit coupling.

M → L(π*) transition (609, 291). The quantum yield for emission varies from 0.02 to 0.09. On the other hand, the luminescence of $W(CO)_4L_2$ (L = aliphatic amine, pyridine), which was at much longer wavelength, was again assigned to a d-d triplet–singlet transition.

Kaizu et al. (291) have associated luminescence with complexes that are photostable. Since the lowest-energy excited state of $Mo(CO)_4$-bipy has Mo → bipyridyl CT character, this complex fulfills the requirement particularly well. Wrighton has noted that a number of complexes that dissociate (618) rapidly in solution at room temperature, luminesce at 77°K, but no longer dissociate. However, considerably more evidence is needed before a detailed theory of the luminescence of these compounds can be formulated.

The series of Re complexes $ClRe(CO)_3LL$ have much higher emission quantum yields (at 77°K, $\Phi \simeq 0.3$) than the Group VIB complexes and are the first known examples of transition metal carbonyls that luminesce in fluid solution (609). The emission is assigned to a triplet–singlet M → L(π*) CT transition. Complexes that emit in solution should be useful for studies of energy transfer and nonradiative decay.

C. FLASH PHOTOLYSIS

Flash photolysis is one of the most important techniques for investigating short-lived electronically excited molecules as well as reactive species. With a light flash of about 10^{-6} (conventional flash lamps) (413, 449) or 10^{-9} sec (laser flash) (414) duration, a relative high concentration of transients is produced. A detailed description of the technique can be found in the literature (51, 448). Ultraviolet absorption spectroscopy is the normal method of characterizing the transients. Ultraviolet spectra of organometallics are often poorly resolved and assignment of bands of stable molecules is already difficult; flash spectra are correspondingly more problematical to interpret. Infrared spectra of organometallics carry much more structural information and are more readily assigned. The development of flash apparatus with IR detection is, therefore, an important goal; it should be attained in the not too distant future. That such measurements are possible in principle has been shown by Pimentel et al. (557). The application to organometallics (e.g., metal carbonyls), which should be extremely fruitful, is hampered by the lack of a suitably transparent solvent.

Flash photolysis of metal carbonyls has been restricted mainly to chromium and iron. Photolysis of $Cr(CO)_6$ in cyclohexane yielded a highly reactive species (λ_{max}: 503 nm), assigned to $Cr(CO)_5$, which reacts rapidly with CO to re-form $Cr(CO)_6$ (*300*):

$$Cr(CO)_5 + CO \longrightarrow Cr(CO)_6$$
$$[k = (3 \pm 1) \times 10^6 \text{ liters mole}^{-1} \text{ sec}^{-1}]$$

The observation that this reaction proceeds much faster ($k \sim 3.5 \times 10^8$ liters mole^{-1} sec^{-1}) in perfluoromethylcyclohexane (*301*) points toward a considerable interaction between $Cr(CO)_5$ and cyclohexane. Such effects were previously observed in matrix studies (*229*).

$$Cr(CO)_5 + \text{solvent} \underset{k_2}{\overset{k_1}{\rightleftharpoons}} Cr(CO)_5 \cdot (\text{solvent})$$

By laser flash photolysis the interaction of a variety of organic solvents ($\sim 10^{-2}$ M in cyclohexane) with $Cr(CO)_5$ could be studied directly (*298*). The rate constants for the formation and the stability constants of $Cr(CO)_5L$ have been deduced (see Table IV).

Constant k_1 appears to be a direct measure of the ability of a potential ligand to occupy a "relatively free" coordination site. Rate constants such as k_1 may be of some use as parameters describing solvent effects in organometallic reactions involving free coordination sites (*298*).

Earlier results, which were interpreted in terms of the formation of a $Cr(CO)_5$ unit with D_{3h} symmetry (trigonal bipyramid) from the initial C_{4v} $Cr(CO)_5$ (square pyramid) (*396*), now appear to be due to traces of impurities in the solvents. An alternative suggestion that the primary species observed in solution is an isocarbonyl $(OC)_5Cr \cdot OC$ (*609*) cannot be reconciled with the observed kinetics.

TABLE IV

RATE CONSTANTS AND STABILITY CONSTANTS FOR THE REACTION OF $Cr(CO)_5$ WITH L[a]

L	k_1 (liters mole^{-1} sec^{-1})	k_2 (sec^{-1})	K[b] (liters mole^{-1})	λ_{max}(nm) for $Cr(CO)_5L$[c]
Benzene	7.0×10^6	2.1×10^5	3.4×10^1	475
Diethyl ether	1.1×10^7	1.4×10^2	7.6×10^4	455
Methanol	3.1×10^7	—	—	460
Ethyl acetate	9.2×10^7	8.1×10^1	1.1×10^6	445
Acetone	1.3×10^8	1.4×10^1	9.3×10^6	438
Acetonitrile	1.6×10^8	$<1 \times 10^{-2}$	$<1 \times 10^{10}$	391

[a] Data from Kelly *et al.* (*298*).
[b] K = equilibrium constant (k_1/k_2).
[c] Measurements ± 5 nm.

A recent study concerned with the flash photolysis of $Fe(CO)_5$ (*299*), shows that in this case, too, the first step is probably the elimination of CO yielding $Fe(CO)_4$. Using cyclohexane as solvent, substantial amounts of $Fe_3(CO)_{12}$ were formed. The differences in the behavior of $Fe(CO)_4$ and $Cr(CO)_5$ may be due to different ground-state multiplicities. Although $Cr(CO)_5$ should have a singlet ground state (*96b*), a triplet ground state appears to be most probable for $Fe(CO)_4$ (C_{2v} symmetry) (*445*). The symmetry-allowed combination of triplet–$Fe(CO)_4$ units to $Fe_2(CO)_8$ [which has already been identified in the matrix (*444*)] and the insertion of $Fe(CO)_4$ into $Fe_2(CO)_8$ provide a reasonable pathway for the rapid formation of $Fe_3(CO)_{12}$.

Photodecomposition of $Ni(CO)_4$ in the gas phase, leading to the successive loss of CO groups, was studied by flash photolysis as early as 1961 (*102*). In the presence of oxygen, decomposition of $Ni(CO)_4$ and $Fe(CO)_5$ is initiated by the flash, followed by slow heterogeneous oxidation (*103*). The same authors reported spectra of excited atoms (Fe and Ni) and excited ions, produced by isothermal flash photolysis of $Ni(CO)_4$ and $Fe(CO)_5$ (*104, 105*).

In 1967, a Canadian group (*569*) described the flash photolysis of cyclobutadiene tricarbonyliron in the gas phase, using a fast-response mass spectrometer for detection of short-lived transients. In this way they observed the fragments ($C_4H_4{}^+$) and [$Fe(CO)_3{}^+$] and proposed the elimination of C_4H_4 as the major primary photolytic step (*204*):

$$(C_4H_4)Fe(CO)_3 \xrightarrow{\ h\nu\ } C_4H_4 + Fe(CO)_3$$

The same group studied photolysis of cyclobutadiene tricarbonyliron in a matrix and observed an intense ESR signal, attributed to $Fe(CO)_3$. These authors suggested that the elimination of C_4H_4 was also the primary step in the matrix, a result that was recently doubted by Chapman *et al.* (*110*) (see Section III, D).

The method of using a kinetic mass spectrometer as a detector for flash photolysis was also employed for the photoreaction of $Fe(CO)_5$ in presence of NO (*544*). In addition to the known stable compound $Fe(CO)_2(NO)_2$, the unstable neutral species $Fe(CO)(NO)_3$ was observed. The latter compound, which does not follow the noble gas rule, decays with a half-life of ~ 75 msec.

D. Photolysis in Rigid Media

The aim of the matrix isolation technique is to embed individual molecules of a compound in a large excess of host material. This matrix, which is usually produced by cocondensation of host and reactant, is typically solid argon; temperatures are in the range of 4° to 20°K. The isolated compounds are examined by spectroscopic methods, chiefly by UV and IR absorption spectroscopy. Unstable species may be produced by photolysis of the molecules deposited in the matrix, thus allowing the direct observation of many primary products of photochemical reactions which cannot be detected by other means. For detailed information on the method, the interested reader is referred to the literature (*247, 172, 382*).

1. Matrix Photochemistry of Binary Metal Carbonyls

Metal carbonyls are especially suitable for matrix isolation and have been investigated extensively. The high extinction coefficient of the CO-stretching vibration in M(CO) moieties facilitates the observation of reactions. According to current evidence the first observable effect on photolysis of an unsubstituted metal carbonyl is the loss of a carbon monoxide molecule. Typical examples investigated are $M(CO)_6$ [M = Cr, Mo, W], $Fe(CO)_5$, $Ni(CO)_4$, and $Fe_2(CO)_9$ (*230, 231, 442, 444, 445, 469, 471*). The primary species observed are compounds **19–23**.

By comparing the spectra of ^{13}CO-enriched fragments with calculated spectra, the structures can be determined much more reliably than from the ^{12}CO spectra alone (*430*). This technique has been applied, for instance, to $Fe(CO)_4$ (*445*). Further photolysis can produce secondary products such as $M(CO)_4$ (M = Cr, Mo, W) and $M(CO)_3$ (M = Fe, Cr, Mo, W) (*492, 441*).

a. *Group VIB Carbonyls.* The data for $M(CO)_5$ are collected in Table V. The reader may ask why $Cr(CO)_5$ does not react back immediately to $Cr(CO)_6$. In fact this back reaction needs further activation

TABLE V

INFRARED FREQUENCIES FOR $M(CO)_5$ (M = Cr, Mo, W) IN DIFFERENT MATRICES[a]

Cr(230) Ar (cm^{-1})	Cr(428) Xe (cm^{-1})	Cr(429) Ar + 2% Xe[b]		Mo(230) Ar (cm^{-1})	W(230) Ar (cm^{-1})	Assignment
		Xe (cm^{-1})	Ar (cm^{-1})			
2093	2086	2089	2092	2098	2097	A_1
1936	1929	1939	1936	1933	1932	A_1
1966	1956	1963	1965	1973	1963	E

[a] References are given in parentheses in column heads.

[b] Columns labeled Xe and Ar give frequencies for $Xe \cdots Cr(CO)_5$ and $Ar \cdots Cr-(CO)_5$, respectively, in an Ar + 2% Xe matrix.

(230) and occurs rapidly upon illuminating the matrix with wavelengths corresponding to the visible absorption of $Cr(CO)_5$. The back reaction is also observed when Ar matrices are doped with 1% CO and warmed to 45°K. Two explanations have been offered for the effect of visible light: (a) a local annealing of the matrix upon radiationless dissipation of the energy absorbed and (b) a specific photoreaction. We will return to this point later.

Although the matrix material has no major effect on the infrared spectra of $Cr(CO)_5$, a remarkable dependence of the long-wavelength UV absorption on the matrix was observed (229, 429) (Table VI). In

TABLE VI

VISIBLE ABSORPTION BANDS OF $Cr(CO)_5$ IN DIFFERENT MATRICES[a]

Matrix	Band maxima (nm)	Band maxima ($10^3 cm^{-1}$)
Ne	624	16
SF_6	560	17.1
CF_4	547	18.3
Ar	533	18.8
Kr	518	19.3
Xe	492	20.3
CH_4	489	20.4
CO	462	21.6
Ne/Xe	628	15.9
	487	20.5

[a] Data from R. N. Perutz and J. J. Turner (429).

two-component matrices (x and y) both species $x \cdots Cr(CO)_5$ and $y \cdots Cr(CO)_5$ can be observed, thus demonstrating a specific interaction between the fragment and the matrix species, which is significant even for noble gas atoms. The interacting matrix species can be exchanged by selective photolysis:

$$x \cdots Cr(CO)_5 \xrightarrow[h\nu\,(2)]{h\nu\,(1)} y \cdots Cr(CO)_5$$

By considering a simple d-orbital scheme for C_{4v} $Cr(CO)_5$ (615) (Fig. 7), it can be predicted that ΔE in $Cr(CO)_5$ will be increased by a ligand L approaching the system in the z direction. This will result in a shift of the longest-wavelength d-d transition to shorter wavelength (128). The change in the visible spectrum has been attributed to interaction of this type and a change in the axial-radial bond angle of $Cr(CO)_5$. (The IR spectra of species $x \cdots Cr(CO)_5$ and $y \cdots Cr(CO)_5$ also differ appreciably. The bond angles are calculated from relative intensities of IR bands (429).)

The significant change in the $Cr(CO)_5$ absorption between CF_4 and CH_4 matrices suggests that the latter interacts much more strongly. Occupation of a vacant coordination site by a hydrocarbon is a matter of current interest (126). Attempts to correlate the structural parameters with the degree of interaction may help toward an understanding of C—H activation by transition metal compounds.

Returning to the effect of light on the recombination of CO with $Cr(CO)_5$, Turner et al. (96a, 428) report experiments that clarify this behavior. By using polarized light, they have been able to produce oriented $Cr(CO)_5$ in CH_4 matrices. They showed that $Cr(CO)_5$ could undergo photochemically induced reorientation in a matrix. These results may be explained by a D_{3h} structure of the excited $Cr(CO)_5$ that can return to the C_{4v} ground state in several orientations by undergoing a Berry twist (393), so producing an apparent rotation. Recent calculations support this suggestion (96, 96b). This mechanism (Fig. 8) also

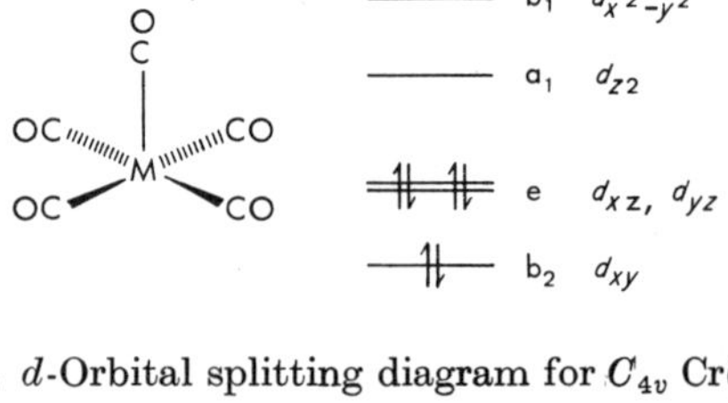

FIG. 7. d-Orbital splitting diagram for C_{4v} $Cr(CO)_5$.

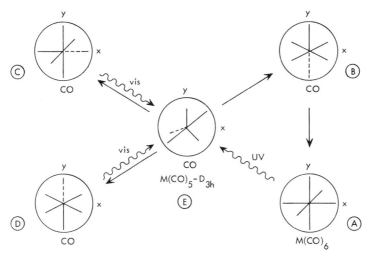

FIG. 8. Proposed mechanism of reactions of C_{4v} Cr(CO)$_5$. Central D_{3h} species represents postulated first-excited state of M(CO)$_5$. Symbols x and y indicate different matrix atoms. Diagram shows generation of M(CO)$_5$ from M(CO)$_6$ (A → E → C and A → E → D), exchange of matrix atoms in vacant site of M(CO)$_5$ (C → E → D), rotation of M(CO)$_5$ (C → E → D), and recombination of M(CO)$_5$ with CO (C → E → B → A and D → E → B →A).

explains the photochemical recombination of Cr(CO)$_5$ with CO and the photochemical exchange of the interacting matrix species.

The intermediacy of D_{3h} Cr(CO)$_5$ also allows the rationalization of the escape of the released CO on initial photolysis of Cr(CO)$_6$:

$$Cr(CO)_6 \xrightarrow{h\nu} Cr(CO)_6^* \xrightarrow{-CO} \underset{C_{4v}}{Cr(CO)_5^*} \longrightarrow \underset{D_{3h}}{Cr(CO)_5^*} \longrightarrow \underset{\text{reoriented } C_{4v}}{Cr(CO)_5}$$

However, in some cases local annealing of the matrix may be a valid explanation.

Another interesting problem is posed by the formation of Cr(CO)$_5$ in a pure CO matrix. The absorption maximum at 462 nm in CO matrices demonstrates that this species is different from all others observed (see Table VI) (430). There has been considerable discussion as to whether the CO interacts with the free coordination site of Cr(CO)$_5$ via oxygen (24) or sideways (25) (96c, 354, 430, 609).

We would like to introduce an additional suggestion. This is the formation of species (26), which may result either from the addition of CO to an excited $Cr(CO)_5$ molecule or from "incomplete bond rupture" in $Cr(CO)_6$. A fast thermal reversal of (26) to $Cr(CO)_6$ may be inhibited by the interaction of this species with CO in the matrix.

There have been several reports in the literature of the existence of a D_{3h} form of $M(CO)_5$; however, all these observations have been questioned. Sheline was the first to report a D_{3h} form of $M(CO)_5$ (542, 543) claiming that it resulted from warming up glasses containing C_{4v} $Mo(CO)_5$. Braterman et al. (67) as well as Turner et al. (229) showed that formation of the "Sheline D_{3h}" was dependent on the concentration of $M(CO)_6$ and was probably the result of a species of the composition $M(CO)_6-M(CO)_5$. More recently, Ozin (354) claimed to have observed a very stable D_{3h} $Cr(CO)_5$, resulting from the cocondensation of chromium atoms and CO in CO/Ar matrices. Turner et al. (96c) have suggested that the spectra observed by Ozin can be reassigned to a mixture of two C_{4v} species [$(OC)_5Cr\cdots Ar$ and $(CO)_5Cr\cdots CO$, see above) together with $Cr(CO)_4$.

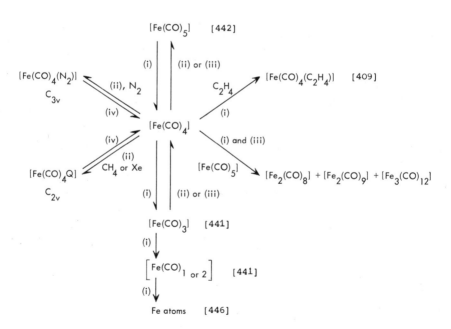

Fig. 9. Principal reactions of matrix-isolated $Fe(CO)_4$. (i) Ultraviolet photolysis; (ii) "Nernst-glower" photolysis ($\lambda > 320$ nm); (iii) annealing matrix; (iv) $\lambda > 375$ nm. [From Poliakoff and Turner (445).]

b. $Fe(CO)_5$ *and* $Fe_2(CO)_9$. Extensive work on $Fe(CO)_5$ (*442, 443, 441, 446, 445*) has led to the final conclusion that the primary product on photolysis is C_{2v} $Fe(CO)_4$. ^{13}CO Exchange experiments support this result. On the basis of MO calculations (*96*), a triplet ground state is suggested for $Fe(CO)_4$. (For an extensive report on the photochemistry of $Fe(CO)_5$, see Ref. *331*.)

The principal reactions of matrix-isolated $Fe(CO)_4$ are summarized in Fig. 9 (*445*).

Ogilvie *et al.* (*409*) provided the first evidence for the formation of an ethylene complex and also proposed a 1-butene-3-yneiron complex:

$$Fe(CO)_5 + 2HC{\equiv}CH \xrightarrow[\text{312 nm} < \lambda < \text{417 nm}]{\text{Ar/17°K}} H_2C{=}CH{-}C{\equiv}CH + CO$$
$$\underset{Fe(CO)_4}{|}$$

Photolysis of $Fe_2(CO)_9$ in matrices at 20°K yielded $Fe_2(CO)_8$ in two isomeric forms, probably **22** and **23** [see page 84 (*444*)]. This finding is of special interest for the understanding of the behavior of $Fe(CO)_4$ in solution. Compound **22** may be the key intermediate in the rapid formation of $Fe_3(CO)_{12}$ (see Section III, C) as well as in many other reactions (*201*).

2. *Matrix Photochemistry of Substituted Metal Carbonyls*

As discussed in Section IV, other photoprocesses have to be considered for substituted metal carbonyls apart from the elimination of CO. Unfortunately, as Braterman (*49, 50*) has explained, only a few of these reactions can be observed directly in the matrix.

a. Group VIB Metal Carbonyl Compounds. Although the photolysis of bis(triisopropylphosphine) tetracarbonyltungsten in solution under a CO atmosphere gives rise to the dissociation of the M—P bond,

$$W(CO)_4P_2 \xrightarrow{h\nu/CO} W(CO)_5P$$

the corresponding fragment is observed neither in a hydrocarbon nor in a methyl tetrahydrofuran glass (*49*). Dissociation of the metal–phosphine bond in the matrix may escape detection because of rapid recombination in the cage. Diffusion of such large species is extremely limited in the matrix. In the light of the observations by Turner *et al.* (*96a*), it remains an open question whether steric and/or electronic factors control the reorientation of an intermediate to a position where the free coordination site is removed from the expelled ligand. The largest ligand that has been observed to dissociate from a metal carbonyl molecule in a hydrocarbon glass is PH_3 from $Mo(CO)_5PH_3$ (*49*). Irradiation of $Mo(CO)_5P(C_6H_{11})_3$ at 77°K with $\lambda > 305$ nm in a hydrocarbon glass yields two isomers, whereas in methyl tetrahydrofuran only the cis isomer is formed (*50*). The reaction is described in Fig. 10. Such isomerization has also been observed for the species $Cr(CO)_4CS$ generated from $Cr(CO)_5CS$ in an Ar matrix. In this case both cis $\rightarrow$ trans and trans $\rightarrow$ cis isomerization have been observed (*440*).

b. Manganese Compounds. Photolysis of $HMn(CO)_5$ in Ar at 15°K results in loss of CO and formation of $HMn(CO)_4$. Upon irradiation with $\lambda > 285$ nm, the reaction is reversed (*470*):

$$HMn(CO)_5 \underset{h\nu/\lambda > 285 \text{ nm}}{\overset{h\nu/\lambda = 229 \text{ nm}}{\rightleftharpoons}} HMn(CO)_4$$

The photochemical reactions of $XMn(CO)_5$ (X = CH_3, CH_3CO, CF_3) were investigated by Ogilvie (*417*) and Rest (*468*). The results of Rest are summarized as follows:

$$CH_3COMn(CO)_5 \xrightarrow[300 \text{ nm} < \lambda < 600 \text{ nm}]{h\nu/-CO} CH_3COMn(CO)_4 \xrightarrow[\lambda > 320 \text{ nm}]{h\nu} CH_3Mn(CO)_5 \quad (3)$$

$$\underset{\substack{230 \text{ nm} < \lambda \\ < 280 \text{ nm}}}{h\nu} \Big\updownarrow \underset{\substack{h\nu/\lambda > \\ 280 \text{ nm}}}{}$$

$$CH_3Mn(CO)_4 + CO$$

The primary process in $CH_3COMn(CO)_5$ is the loss of a terminal CO (probably equatorial) and corresponds well with solution photochemistry (*410*).

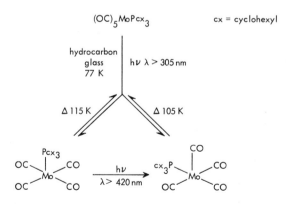

FIG. 10. Photochemical formation of cis and trans isomers of $(CO)_4MoPcx_3$ from $(CO)_5MoPcx_3$. [From Black and Braterman (*50*).]

The cyclopentadienyl carbonylmanganese complexes lose CO on photolysis and are regenerated upon softening the glass (69)†:

$$(\pi\text{-}RC_5H_4)Mn(CO)_3 \underset{\Delta}{\overset{\underset{\text{methylcyclohexane 80 K}}{h\nu}}{\rightleftharpoons}} (\pi\text{-}RC_5H_4)Mn(CO)_2 + CO$$

(R = H, CH_3)

Carbon monoxide is also eliminated (467) on photolysis of $Mn(CO)_4NO$:

$$Mn(CO)_4NO \underset{h\nu\ (\lambda > 550\ \text{nm})}{\overset{\underset{\text{Ar, 15° K}/h\nu\ (280\ \text{nm} < \lambda < 360\ \text{nm})}{h\nu}}{\rightleftharpoons}} Mn(CO)_3NO + CO$$

$$\underset{\substack{(230\ \text{nm} < \lambda \\ < 280\ \text{nm})}}{\overset{h\nu}{\Big\updownarrow}} \Delta\ (30°\text{K})$$

$$Mn(CO)_2NO + CO$$

c. *Iron Carbonyl Compounds.* Some controversy exists about the matrix photochemistry of cyclobutadiene tricarbonyliron. Gunning *et al.* (569) claimed on the basis of ESR evidence that $Fe(CO)_3$ was formed on photolysis ($\lambda > 220$ nm) at 77°K, both in the solid compound and in matrices of C_3F_8 and neopentane. However, Chapman (110) has shown in IR experiments that the only process at 8°K is the loss of CO. Chapman's observations are supported by the behavior of $(C_4H_4)Fe(CO)_3$ in solution. Photosubstitution of one CO by olefin is possible (594), and one or two CO groups have been substituted by phosphine (201). It was shown that the quantum yield for CO loss is at least 10 times larger than the quantum yield for the loss of cyclobutadiene (201).

3. Photochemistry in Reactive Matrices

When transition metal compounds are photolyzed in matrices consisting of material that can react to form primary products, two types of compounds may be formed. (a) Species normally not observed because of rapid recombination are detectable by reaction with the matrix; (b) unstable compounds not otherwise accessible can be produced.

In pure CO matrices, photolysis of carbonyl nitrosyls leads to substitution of NO by CO, thus demonstrating the cleavage of a metal–NO bond that is otherwise not observed (132, 134) [in Ar or CH_4 the loss of CO is the only detectable process (132, 134)]. The substitution of π-bonded ligands has also been achieved (134):

$$Fe(CO)_2(NO)_2 \xrightarrow[\text{CO matrix}]{h\nu} Fe(CO)_5 + NO$$

† Some interesting observations in an Apiezon N-matrix are found in Ref. 49.

$$Co(CO)_3(NO) \xrightarrow[\text{CO matrix}]{h\nu} Co(CO)_4 + NO$$

$$(\pi\text{-}C_5H_5)NiNO \xrightarrow[\text{CO matrix}]{h\nu} (\pi\text{-}C_5H_5)Ni(CO)_2 \xrightarrow{h\nu} Ni(CO)_4$$

$$Mn(CO)_4NO \xrightarrow[\text{CO matrix}]{h\nu} Mn(CO)_x + NO \quad (x = 5 \text{ or } 6)$$

Metal carbonyl nitrosyls can be produced by photolysis of carbonyls in NO-doped Ar matrices (*132, 134*):

$$Fe(CO)_5 \xrightarrow[\text{Ar/NO matrix}]{h\nu} Fe(CO)_2(NO)_2$$

$$Os(CO)_5 \xrightarrow[\text{Ar/NO matrix}]{h\nu} Os(CO)_2(NO)_2$$

Irradiation of $Fe(CO)_5$ and $Ni(CO)_4$ in nitrogen matrices results in CO substitution by N_2 (*443, 445*), yielding, for instance, $Ni(CO)_3N_2$ (*466*).

In carbonyl nitrosyl compounds, the substitution of CO by N_2 can also be observed (*134*):

$$Co(CO)_3NO \xrightarrow[\text{N}_2 \text{ matrix}]{h\nu} Co(CO)_2(NO)N_2 + CO$$

4. Ionic Carbonyls in Matrices

Heterolytic cleavage of metal CO bonds can be achieved by vacuum UV photolysis (H_2 or Ar arc) (*95, 72a*). In this way anions $Ni(CO)_3{}^-$, $Cr(CO)_5{}^-$, and $Fe(CO)_4{}^-$ have been detected. In the case of $Ni(CO)_4$, CO^+ has been observed, but the other cations have not yet been positively identified. Vacuum UV photolysis of $V(CO)_6$ results in formation of $V(CO)_6{}^-$. The mechanism of these reactions is still under investigation. Heterolytic cleavage according to

$$(\pi\text{-}C_5H_5)NiNO \xrightarrow{h\nu} (\pi\text{-}C_5H_5)Ni^+ + NO^- \tag{4}$$

has been claimed for $(\pi\text{-}C_5H_5)NiNO$ on UV photolysis (*133, 134*). The anionic species $Cr(CO)_5{}^-$, $Co(CO)_3NO^-$, and $Co(CO)_4{}^-$ can also be produced by incorporation of alkali metal atoms in matrices and subsequent photolysis (*72*) or by electron bombardment during deposition (*72a*):

$$Cr(CO)_6 \xrightarrow[h\nu]{\text{Na/Ar}} Cr(CO)_5{}^- + Na^+$$

The anion $V(CO)_6{}^-$ has also been photolyzed in a MeTHF glass at 77°K and formation of $[V(CO)_5MeTHF]^-$ has been observed (*67*).

In recent years, matrix photochemistry has turned out to be the most powerful means of elucidating structures of intermediates in metal carbonyl photochemistry. The structures are less symmetrical than had been anticipated (227, 303) and have required the development of new theoretical approaches (96, 96b). The application of matrix isolation to compounds less favorable than organometallics has been little explored and should profit from the advent of Fourier-transform IR spectroscopy. Processes that are observed in the matrix probably also occur in solution. However, processes not detectable in the matrix should also be considered in solution. The reviewers feel that a comparative study of systems by matrix and flash photolysis will be a rewarding approach.

E. Quenching and Sensitization Processes in Organo-Transition-Metal Photochemistry

In this section we will briefly survey quenching processes in which organometallic compounds are involved. We will consider quenching processes as those bimolecular processes in which an electronically excited compound A* interacts with a ground-state molecule B. We will distinguish three cases:

(a) A* = organic; B = organometallic
(b) A* = organometallic; B = organic
(c) A* = organometallic; B = organometallic

The major modes of quenching are as follows:

1. Electronic energy transfer, e.g., according to the general scheme (358),

$$A^* + B \longrightarrow A + B^*$$

2. Formation of excited complexes (exciplex),

$$A^* + B \longrightarrow (AB)^*$$

3. External heavy atom effect,† whereby the interaction of heavy atoms with the excited molecule A* may enhance radiationless deactivation via spin-forbidden processes by inducing spin-orbit coupling externally.

$$^1A^* \xrightarrow{M_{(heavy)}} {}^3A^*$$

or

$$^3A^* \xrightarrow{M_{(heavy)}} {}^1A$$

4. Electron transfer processes, e.g.,

$$\ddot{A}^* + B \longrightarrow A^{\cdot +} + B^{\cdot -}$$

5. Quenching by chemical processes, e.g.,

$$(ML_n)^* + L' \longrightarrow [L'ML_n] \longrightarrow ML'L_{(n-1)} + L$$

† Another bulk property of metal-containing systems that is probably of some importance in quenching processes is the so-called spin-catalyzed deactivation. For detailed information see Refs. 51 and 55.

The example given would be typical of quenching by a bimolecular chemical process. It should be noted that electron transfer can be considered as a special case of chemical quenching.

A very critical and detailed discussion of quenching and sensitization processes of *coordination compounds* can be found in the literature (*28*). A more extensive discussion of quenching processes involving organometallics than is possible in the present review will be given elsewhere (*55*). We will restrict ourselves here to the discussion of a few examples illustrating the principles outlined above.

1. Electronic Energy Transfer

One of the most commonly used sensitizers in organic photochemistry is triplet-excited benzophenone. It is therefore, not surprising that the first attempts to sensitize photochemical reactions of transition metal carbonyls were carried out with benzophenone. Vogler (*586*) irradiated $Cr(CO)_6$ and benzophenone and then added pyridine. He concluded that triplet energy transfer had occurred from benzophenone to $Cr(CO)_6$, leading to $Cr(CO)_5$. The limiting quantum yield of the process was determined to be unity.

Diffusion-controlled quenching of triplet fluorenone ($k_q = 3\text{–}8 \times 10^9$ liters mole^{-1} sec^{-1}) by iron carbonyl complexes with olefins, dienes, and α,β-unsaturated ketones and by some sandwich complexes has been attributed to a triplet-triplet energy transfer (*226*).

The quenching of a variety of triplet-excited organic molecules (benzophenone, triphenylene, phenanthrene, 2-acetonaphthone, chrysene, pyrene, 1,2-benzanthracene, anthracene) by Ni(II) and Pd(II) chelates of the general structure **27** has been studied by Wilkinson *et al.* (*8, 1, 606*).

R	R′
CH_3	OH
$C_{11}H_{23}$	OH
CH_3	OBu
CH_3	Bu

(**27**)

Slightly different rates for the quenching of high- and of low-energy triplet donor molecules by these complexes have been explained by two different energy-transfer processes. The high-energy donors are quenched by energy transfer to intraligand $\pi\text{-}\pi^*$ states [benzophenone–NiL_2 (R = CH_3, R′ = OH): $k_q = 5.9 \times 10^9$ liters mole^{-1} sec^{-1}], whereas ligand field states of the acceptors are initially produced in the case of low-energy triplet donors [anthracene–NiL$_2$ (R = CH$_3$, R' = OH): k_q = 3.6 × 10^9 liters mole^{-1} sec^{-1}]. Although for the first transfer, orbital overlap is not critical, the latter process requires good overlap of the metal d orbitals with the donor molecule. Here the energy-transfer efficiency can be affected by changing the geometry of the complexes, e.g., by variation of the ligands in the nickel(II) chelates.

These examples demonstrate energy transfer from excited organic molecules to an organometallic system; the following reactions are examples for the reverse reaction.

Electronically excited [Ru(bipy)$_3$]$^{2+}$ is quenched by *trans*-stilbene, *trans*-2-styrylpyridine, *trans*-4-styrylpyridine, and anthracene (*610a*) (Φ *trans* → *cis*-styrylpyridine = 0.4; Φ *trans* → *cis*-stilbene ~ 0.5). Large amounts of cis isomers (> 90%) are found in the photostationary state achieved by this energy-transfer process, both in the case of stilbene and styrylpyridines. This result is consistent with a donor triplet energy below 57 kcal/mole (E_T *cis*-olefin = 57 kcal/mole; E_T *trans*-olefin = 49–50 kcal/mole); *cis*-pentadiene (E_T = 57 kcal/mole) does not show any quenching activity.

An interesting example for energy transfer from an organometallic to an organic substrate and vice versa is the intramolecular energy transfer observed with several metalloporphyrin complexes (M = Mg, Zn, Co), containing azastilbene ligands (L) (*604*). The remarkably efficient cis-trans isomerizations of stilbazole coordinated to Zn etioporphyrin (Φ = 0.4), Mg etioporphyrin (Φ = 0.17), and especially of 1-(1-naphthyl)-2-(4-pyridyl)ethylene (NPE) coordinated to Zn etioporphyrin (Φ = 6.6) and Mg etioporphyrin (Φ = 3) have been interpreted by the following mechanism:

$$\text{MP—L} \longrightarrow {}^1\text{MP*—L} \longrightarrow {}^3\text{MP*—L} \tag{5}$$

$$^3\text{MP*—L} \longrightarrow \text{MP—}^3\text{L*} \longrightarrow \text{MP—}^3\text{L'*} \tag{6}$$

$$\text{MP—}^3\text{L'*} \longrightarrow \alpha\text{MP—L'} + (1 - \alpha)\text{MP—L} \tag{7}$$

$$\text{MP—}^3\text{L'*} \longrightarrow {}^3\text{MP*—L'} \tag{8}$$

$$^3\text{MP*—L'} + \text{L} \longrightarrow {}^3\text{MP*—L} + \text{L'} \tag{9}$$

$$^3\text{MP*—L} \longrightarrow \text{MP—L} \tag{10}$$

$$^3\text{MP*—L'} \longrightarrow \text{MP—L'} \tag{11}$$

(L = cis isomer, L' = trans isomer)

In the primary photoprocess, a porphyrin-localized triplet state is produced [Eq. (5)], which undergoes intramolecular energy transfer, leading to an olefin-localized triplet state [Eq. (6)]. This species is expected to be able to undergo isomerization. The resulting isomerized molecule, which is still in a ligand-localized excited state, can either decay to a mixture of the ground-state molecules [Eq. (7)] or undergo reversible intramolecular energy transfer [Eq. (8)], leading back to a porphyrin-localized triplet state, now containing a *trans*-olefin ligand. The authors postulate that this species should be capable of ligand exchange during its lifetime [Eq. (9)]. (The reader can only admire the patience of the porphyrin molecule!) The regenerated but still excited starting complex must now play this game again. In the case of NPE, it has to pass through the sequence at least 6 times. It should be noted that the authors took considerable care to rule out trivial mechanisms, as for instance a chain mechanism via radical ions. The interesting point about Eq. (9) is that it describes the rare process of a bimolecular reaction of an electronically excited transition metal complex (see below).

A special case of two organometallic species, working as donor and acceptor in triplet energy transfer, is the intramolecular energy transfer from triplet-excited Zn porphyrin to a Cu porphyrin, which are linked by a —CO—NH—C_2H_4—NH—CO— chain (*522*).

Other important reports on this subject include Refs. *612, 618, 615,* and *291*.

2. Exciplex Formation

Quenching processes involving the intermediate formation of excited complexes can best be demonstrated for cases in which emission from the excited complexes is observed. For example (*25*),

$$cis\text{-Ir(phen)}_2\text{Cl}_2{}^+ \xrightarrow{\ h\nu\ } (^1d\pi^*)\ cis\text{-Ir(phen)}_2\text{Cl}_2{}^+ \qquad (12)$$

$$(^1d\pi^*)\ cis\text{-Ir(phen)}_2\text{Cl}_2{}^+ \longrightarrow (^3d\pi^*)\ cis\text{-Ir(phen)}_2\text{Cl}_2{}^+ \cdot \qquad (13)$$

$$(^3d\pi^*)\ cis\text{-Ir(phen)}_2\text{Cl}_2{}^+ + \text{naphthalene} \longrightarrow [cis\text{-Ir(phen)}_2\text{Cl}_2{}^+ \cdot \text{naphthalene}] \quad (14)$$

The formation of the exciplex [Eq. (14)] was indicated by the appearance of a new emission centered around 560 nm. Other examples have also been observed (*475, 476, 106*); they will be discussed in full elsewhere (*55*).

3. Heavy Atom Effects

Although quenching of excited states of transition metal complexes via heavy atom effects does not appear to be unequivocally established (*28*), it has been quite well investigated in the quenching of excited anthracene with $(CH_3)_2Hg$. The singlet-excited anthracene is quenched by $Hg(CH_3)_2$ at a diffusion-controlled rate by enhanced conversion to triplet anthracene, which is radiationlessly deactivated. The effect of $Hg(CH_3)_2$ on the latter process is small, the quenching rate of the triplet state is $k_q = 1.2 \times 10^3\ M^{-1}\ \text{sec}^{-1}$. A more detailed discussion is given in Ref. *55*.

4. Electron-Transfer Processes

Many processes involving transfer of electronic energy from an excited donor D to an acceptor A occur with rates close to those for diffusion control, provided that the energy of the accepting state is below that of the donor state. Usually a sharp decrease of quenching rate constants is observed when the electronic energy transfer becomes endothermic (358). The situation can be quite different with electron transfer, which may be exothermic, but still much slower than corresponds to diffusion control.

The principle of electron-transfer reactions can be explained using a very simple MO scheme (Fig. 11). We will return to this point later on in connection with photosynthesis (Section V, B).

Two special cases of electron-transfer quenching processes are illustrated in Fig. 11—(a) reversible electron transfer and (b) secondary electron transfer. Figure 11 may explain how electronically excited chlorophyll could be involved in electron transport phenomena (from X to A) in photosynthesis. One of the best established examples of electron transfer from an electronically excited transition metal compound to other transition metal complexes, as well as to organic

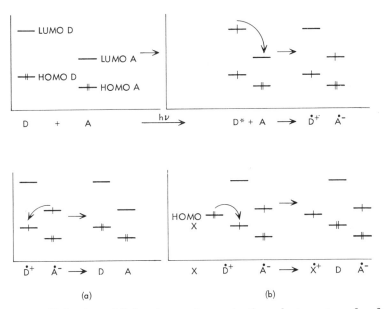

FIG. 11. Molecular-orbital scheme demonstrating electron transfer from donor molecule D to an acceptor A (upper half), reversible electron transfer (a), and secondary electron transfer from X to A (b). HOMO = highest occupied molecular orbital; LUMO = lowest unoccupied molecular orbital.

substrates, is that investigated by Bock and his collaborators (53). Flash spectroscopic investigation allowed the direct observation of the following processes:

$$[Ru(bipy)_3]^{2+*} + Fe(H_2O)_6^{3+} \xrightarrow{k_q} [Ru(bipy)_3]^{3+} + Fe(H_2O)_6^{2+} \qquad (15)$$

$$[Ru(bipy)_3]^{3+} + Fe(H_2O)_6^{2+} \xrightarrow{k_1} [Ru(bipy)_3]^{2+} + Fe(H_2O)_6^{3+} \qquad (16)$$

$$(k_q \sim 3 \times 10^9 \ M^{-1} \ sec^{-1})$$

Similar quenching constants were observed with $[Ru(NH_3)_6]^{3+}$, trans-1,2-bis(N-methyl-4-pyridyl)ethylene^{2+}, and 1,1'-dimethyl-4,4'-bipyridine^{2+}. The back reaction [Eq. (16)] occurs at a rate of approximately $10^6 \ M^{-1} \ sec^{-1}$, although it is exothermic ($\Delta H = -12$ kcal), but the back reaction in the case of $[Ru(NH_3)_6]^{2+}$ is even faster than the corresponding forward reaction ($k_q = 2 \times 10^9 \ M^{-1} \ sec^{-1}$, $k_1 = 3.68 \times 10^9 \ M^{-1} \ sec^{-1}$).

Other examples of this process are found in Refs. 522, 52, and 362.

5. Chemical Quenching

Photochemical ligand substitution is a very important process in organo-transition-metal chemistry, as discussed in detail in Section IV. However, quenching reactions according to the general scheme,

$$L^* + ML_n' \longrightarrow LML_{n-1}' + L'$$

$$[ML_n']^* + L \longrightarrow LML_{n-1}' + L'$$

are not supported by well-documented examples. The S_N2 type of reaction of PPh_3 with $Mn(CO)_4NO$ (297), discussed in Section IV, and the ligand-exchange reaction of the excited stilbene–porphyrin complexes (see page 95) can be considered as examples.

IV. Main Reaction Pathways in the Photochemistry of Organo-Transition-Metal Complexes

In this section the main reaction paths and mechanisms in organo-transition-metal photochemistry will be considered in more detail. In the first part we will try to survey the important types of reactions; in the second part a more explicit discussion of some additional examples is given.

A. SURVEY OF IMPORTANT REACTIONS IN ORGANO-TRANSITION-METAL PHOTOCHEMISTRY

As discussed in Section III, most photoreactions of transition metal compounds result from the direct or indirect electronic excitation of d-d transitions. Reactions of CT-excited molecules are the exception rather than the rule. As pointed out in Section III, the result of d-d excitation is very often the dissociation of the metal–ligand bond. This process has been widely used for the elimination of CO, dienes, phosphines, and other ligands. The primary species is usually a coordinatively unsaturated compound in which the metal has not undergone a change of its oxidation state [Eq. (17)]. This free coordination site can be occupied either by (a) addition of a new ligand with retention of the oxidation state [(Eq. (18)] or (b) addition with increase of the oxidation state of the metal. This reaction is termed oxidative addition [Eq. (19)]. Although the mechanism of process a appears to be well understood, the details of mechanism b are a matter of current discussion (*324, 345, 249, 424*). In the oxidative addition reaction a formal transfer of two electrons from the metal to the ligand occurs, often resulting in the cleavage of an intraligand bond. Many reactions involving oxidative addition are, therefore, combined with transformation of ligands. The reverse of an oxidative addition reaction is called reductive elimination [Eq. (20)].

$$M(0)L_n \xrightarrow{h\nu} M(0)L_{n-1} + L \qquad (17)$$

$$M(0)L_{n-1} + L' \longrightarrow M(0)L_{n-1}L' \qquad (18)$$

$$M(0)L_{n-1} + XY \longrightarrow L_{n-1}M(II){\overset{\displaystyle X}{\underset{\displaystyle Y}{<}}} \qquad (19)$$

$$L_{n-1}M(II){\overset{\displaystyle X}{\underset{\displaystyle Y}{<}}} \longrightarrow L_{n-1}M(0) + XY \qquad (20)$$

1. Substitution of CO in Simple Metal Carbonyls

The principal process on irradiation of metal carbonyls is the substitution of CO by other ligands such as n donors (phosphines, amines) or π donors (olefins, dienes). Most of these reactions occur by dissociative activation (S_N1): CO is eliminated from the excited molecule and the ligand L adds to the coordinatively unsaturated species (*332*):

$$M(CO)_n \xrightarrow{h\nu} [M(CO)_n]^* \longrightarrow M(CO)_{n-1} + CO$$

$$M(CO)_{n-1} + L \longrightarrow M(CO)_{n-1}L$$

The quantum yield of the dissociative process for $M(CO)_6$ (M = Cr, Mo, W) was found by Strohmeier *et al.* (*545*) to be one independent of the solvent or the wavelength of excitation. The unit efficiency of this process has been questioned (*318*) because inner light-filter action of the products may have limited the accuracy of the measurement. This is a general problem for quantum yield determinations whenever the degree of conversion is so high that the products absorb a significant amount of light. The interested reader is referred to a recent paper containing a rate expression for consecutive first-order (photochemical) reactions

$$A \xrightarrow{h\nu} B \xrightarrow{h\nu} C$$

which takes account of the absorptions of B and C (*375*).

Substitution of CO in metal carbonyl compounds by associative (S_N2) activation seems to be an unusual occurrence. A detailed study of the reaction,

$$MnNO(CO)_4 + L \xrightarrow{h\nu} MnNO(CO)_3L + CO \qquad (21)$$

revealed that the quantum yield is dependent on the concentration of L if L = PPh_3, but not when L = $AsPh_3$ (*297*) [Eq. (21)]. It is suggested that the reaction with the weaker nucleophile $AsPh_3$ proceeds via an S_N1 mechanism, whereas in the case of PPh_3 an associative activation also occurs.

2. *Wavelength-Dependent Substitution Reactions of Metal Carbonyl Derivatives*

Photolysis of substituted metal carbonyls causes substitution not only of CO but also of other ligands:

$$M(CO)_{n-1}L \xrightarrow{h\nu/L'} M(CO)_{n-2}LL' + CO$$

$$M(CO)_{n-1}L \xrightarrow{h\nu/L'} M(CO)_{n-1}L' + L$$

In several cases the ratio of the quantum yields of these two processes changes with the wavelength of excitation (Table VII). A well-studied example of such wavelength-dependent photosubstitution is the reaction of diene–iron carbonyl complexes with phosphines (P) (*315, 316*):

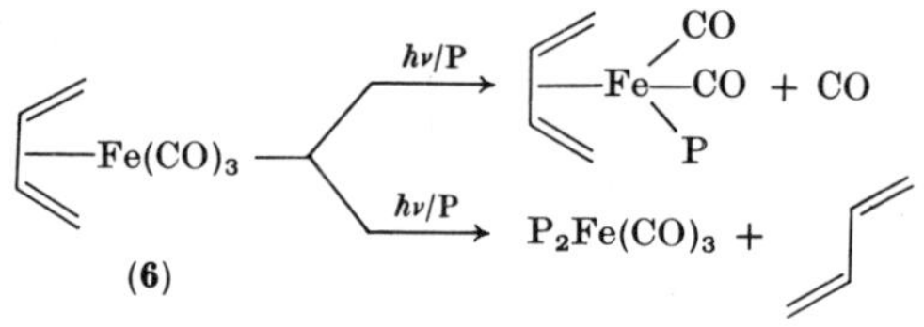

(6)

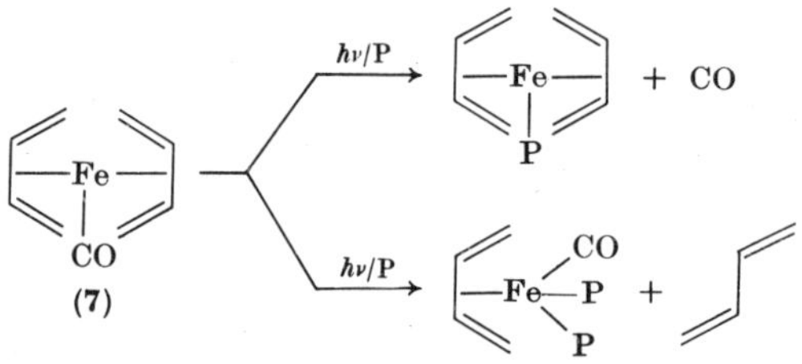

(7)

Although CO substitution predominates upon irradiation at short wavelengths, the diene substitution becomes more important at longer wavelengths (Table VII).

The possibility of influencing the course of reaction by varying the excitational energy offers interesting preparative applications, for instance in the synthesis of bis(butadiene)trimethylphosphiteiron. This complex is only obtained in a satisfactory way using 254-nm irradiation.

There have been several attempts at explaining such wavelength-dependent reactions (see Section III, A) (*615, 629, 630, 631, 599*). Of the models developed to predict the course of reaction, we will discuss that of Wrighton, which is based on simple ligand field considerations (*610*). The pattern of d-orbital splitting of a $M(CO)_5L$ complex with C_{4v} structure (e.g., M = Mo, W; L = NH_3, piperidine) is shown in Fig. 7 (Section III, D). The six d electrons occupy the d_{xy}, d_{xz}, and d_{yz} orbitals. From this scheme two types of one-electron excitation can be derived. The lowest-energy excitation results in population of the d_{z^2} orbital causing labilization along the z axis (OC—M—L), leading to preferential cleavage of the M—L bond. Labilization along the x and y axes, due to population of the $d_{x^2-y^2}$ orbital is achieved by higher-energy excitation and is expected to cause loss of equatorial CO. Photosubstitution experiments on $M(CO)_5L$ (M = Mo, W; L = NH_3, n-propylamine, piperidine) are consistent with this model (*610*). Irradiation in the long-wavelength region yields efficient substitution of L, whereas at shorter wavelength CO substitution increases. The additional L substitution occurring upon high-energy excitation is explained by competitive internal conversion to the lower-excited state (compare Fig. 1, Section II).

Analogous behavior was shown by $W(CO)_5L$ (L = pyridine, 2-styrylpiperidine, 4-styrylpiperidine), but in the case of styrylpiperidine isomerization of the ligand was also observed (Φ trans $\rightarrow$ cis = 0.49 at 436 nm) (*618*). The authors conclude that the lowest-excited ligand field state (labilization in direction of the z axes) leads either to dissociation of the W—L bond or to isomerization by electronic energy migration. Substitution of L is reduced when the isomerization begins to appear. The decrease in isomerization at short wavelengths parallels that for the photosubstitution of L.

3. Photochemical Isomerization of Organo-Transition-Metal Complexes

a. Cis–Trans Isomerization of Ligands. Irradiation of olefins or dienes in the presence of transition metal carbonyl compounds can

TABLE VII

WAVELENGTH-DEPENDENT QUANTUM YIELDS FOR THE SUBSTITUTION OF CO (Φ^{CO}) AND DIENES (Φ^D) IN DIENE–CARBONYLIRON COMPLEXES[a]

Complex	[complex] (moles/liter)	[P(OCH₃)₃] (moles/liter)	254 nm		405 nm	
			Φ^{CO}	Φ^D	Φ^{CO}	Φ^D
⌐Fe(CO)₃	1.1×10^{-2}	0.1	0.28	0.08	0.05	0.01
CO ⌐Fe	5.7×10^{-3}	0.2	0.14	0.03	0.01	0.01
CH₃ ⌐Fe(CO)₃	9.6×10^{-3} 9.5×10^{-3}	0.1 0.1	0.46	0.064	0.05	0.088
CH₃ ⌐Fe(CO)₃	10.4×10^{-3} 10.2×10^{-3}	0.1 0.1	0.23	0.016	0.03	0.009

[a] Data from Kirsch et al. (316).

result in their cis–trans isomerization. Three different mechanisms have to be considered for this reaction.

i. Isomerization is brought about by a sequence of addition–elimination reactions, according to the so-called π-allyl-hydride mechanism (*374*, *332*), which is often accompanied by a migration of the double bond via a 1,3-hydrogen shift (Scheme 1).

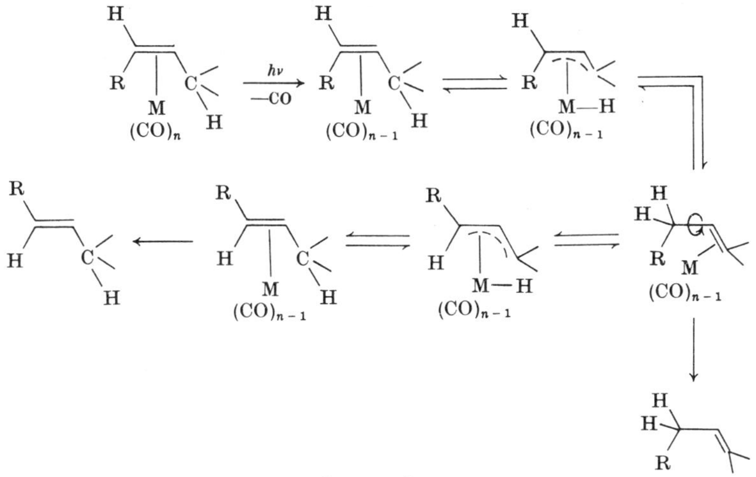

SCHEME 1

This mechanism appears to be operative in the mutual isomerization of *cis*- and *trans*-2-pentene (*622*),

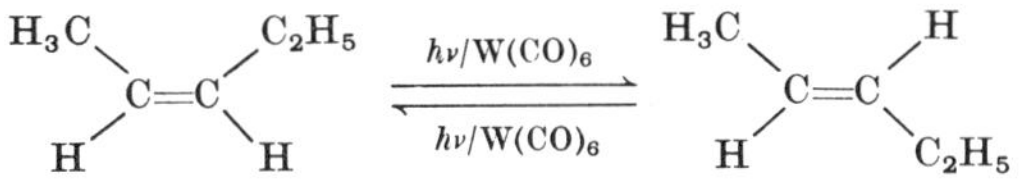

$$(22)$$

as well as in that of 1,3-pentadiene,

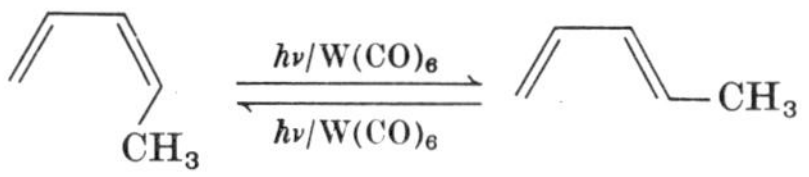

$$(23)$$

and 2,4-hexadiene (*617*, *622*). The action of light in this case is solely restricted to the formation of a free coordination site by photochemical cleavage of a metal–CO bond.

ii. In those cases where no readily abstractable β-hydrogen is available, an intermediate resulting from the electronic excitation of the metal–olefin system appears to be responsible for cis-trans isomerization. Proposals for the isomerizing species include **28** (*332*) and **29** (*622*):

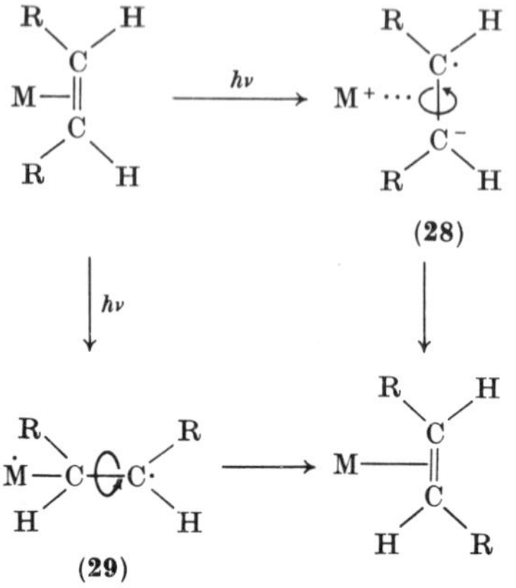

In neither species is the metal–olefin bond completely dissociated; however, in both cases facile rotation around the C—C bond is possible. Examples for this type of mechanism are

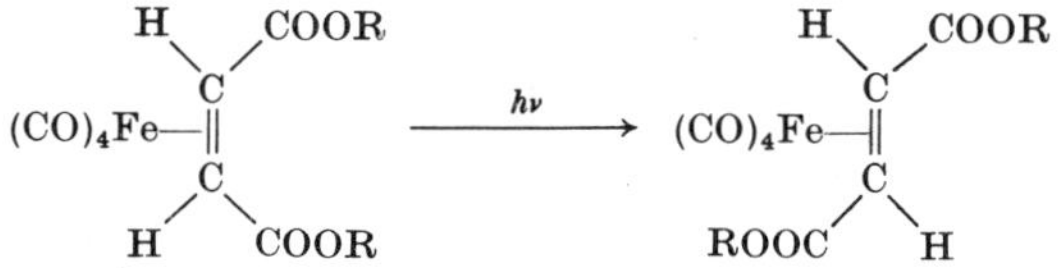

Only in the first case have the complexes of the *cis*- and *trans*-olefin been isolated (*492*).

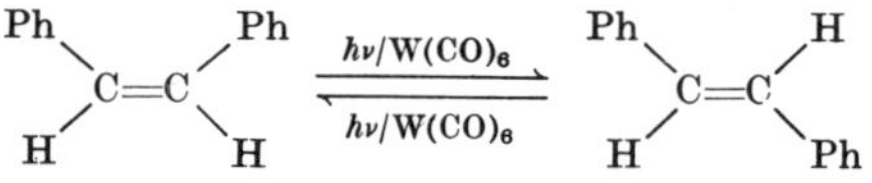

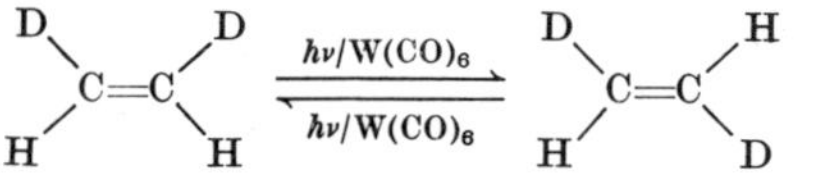

iii. Cis-trans isomerization of a ligand double bond not directly coordinated to the metal may result from excitation of a charge-transfer transition (e.g., CTML), by excitation of an intraligand transition, or by energy transfer (see Section III, E for the latter case).

The best-studied example for ligand isomerization by CTML excitation is the cis-trans isomerization of the stilbazole ligand in [Ru(2,2'-bipy)$_2$(*cis*-4-stilbazole)$_2$]$^{2+}$ and [Ru(2,2'-bipy)$_2$(*cis*-4-stilba-zole)Cl]$^+$ (*627, 628*):

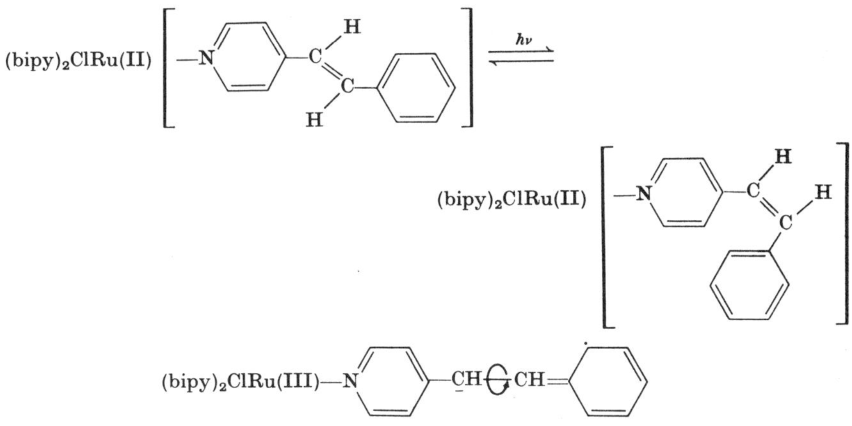

(30)

Investigation of the wavelength dependence of the isomerization pointed to at least two different types of excited states. Irradiation in the long-wavelength region converted the cis complexes efficiently to the trans forms, whereas the trans complexes gave only small yields of the cis isomers. By contrast, irradiation at shorter wavelengths of the trans complexes resulted in efficient isomerization to the cis forms. The authors suggest that long-wavelength excitation results in a CTML transition (d-π^*) to a state that can be described as a metal-oxidized radical anion of the ligand **(30)** that preferentially undergoes cis-to-trans isomerization. Irradiation at shorter wavelengths produces predominantly intraligand excited states of the olefinic system (π-π^*) which decay to cis and trans isomers with nearly equal probability.

b. Positional Isomerization of Ligands. Photolysis of the bis(carbene) complexes **(31)** (M = Cr, Mo, W) results in isomerization, yielding the corresponding trans complexes **(32)** (*415, 416*):

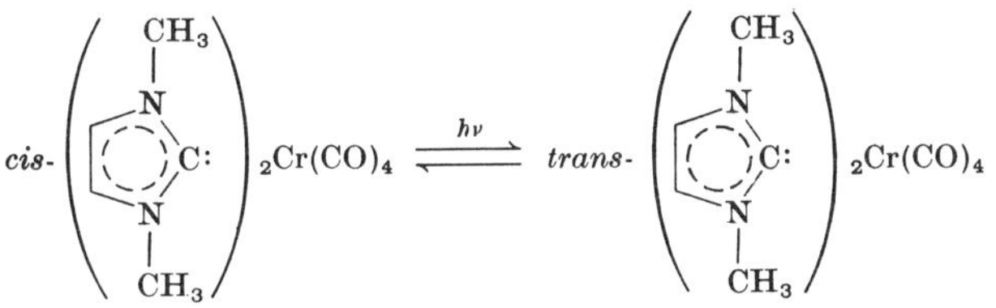

Irradiation of species **31** is expected to lead to the elimination of one CO group. In the resulting coordinatively unsaturated intermediate, the large heterocyclic ligand can change to the sterically preferred trans position. The species is then stabilized by the recoordination of CO.

Linkage isomerism has also been observed (167, 168):

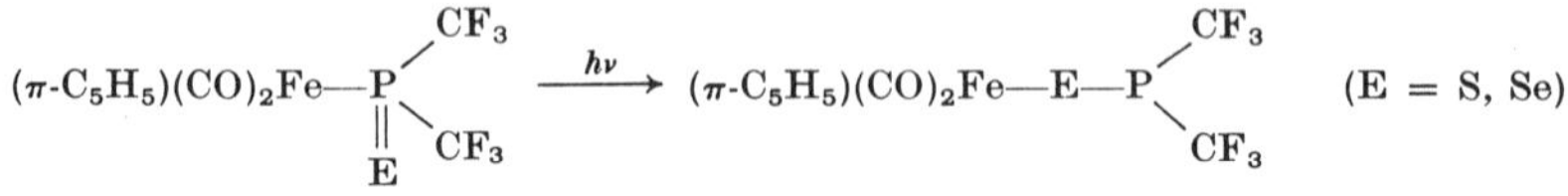

$$(\pi\text{-}C_5H_5)(CO)_2Fe\text{—}P\underset{\underset{E}{\overset{\parallel}{\diagdown}}CF_3}{\overset{\diagup CF_3}{}} \xrightarrow{h\nu} (\pi\text{-}C_5H_5)(CO)_2Fe\text{—}E\text{—}P\underset{CF_3}{\overset{\diagup CF_3}{}} \qquad (E = S, Se)$$

4. Photodissociation of Metal–Metal Bonds

The structure of the dimeric metal carbonyls $M_2(CO)_{10}$ (M = Mn, Re) can be viewed as the combination of two $M(CO)_5$ units, both having C_{4v} structures and a d^7-electron configuration (Fig. 12). The metal–metal bond is formed by interaction of the d_{z^2} orbitals of the $M(CO)_5$ moieties, which are occupied by the unpaired electrons. In Fig. 12 a simple MO scheme demonstrates this bonding situation (609).

The UV spectra of the $M_2(CO)_{10}$ complexes exhibit strong absorption bands [e.g., $Re_2(CO)_{10}$: $\lambda_{max} = 310$ nm, $\varepsilon = 17,000$], which were assigned to the σ-σ^* transition on the basis of their polarization (367). Excitation in this wavelength region can result in cleavage of the metal–metal bond, caused by the population of the antibonding σ^* orbital.

Kinetic studies of polymerization reactions, which were photo-initiated by $M_2(CO)_{10}$, appeared to indicate an unsymmetric cleavage of the metal carbonyl (36, 37):

$$M_2(CO)_{10} \xrightarrow{h\nu} M(CO)_4 + M(CO)_6$$

By contrast, results of other authors (248) pointed toward a symmetric cleavage. Irradiation of $Mn_2(CO)_{10}$ ($\lambda \geq 350$ nm) in THF produced paramagnetic species, which were suggested to be $\cdot Mn(CO)_5$ on the basis of their reactions. However, in a recent study it was found

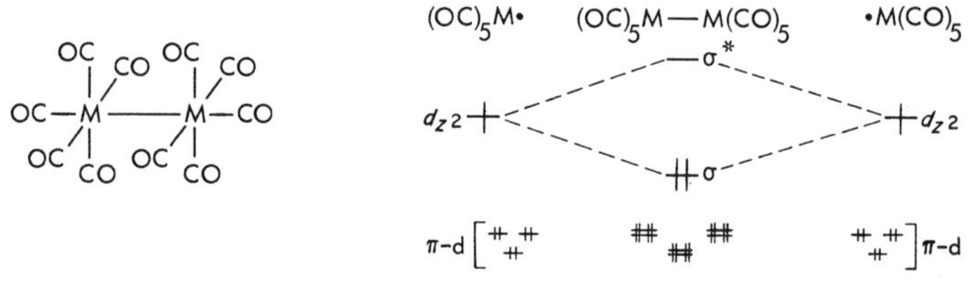

FIG. 12. Molecular orbital scheme for the metal–metal bond in $M_2(CO)_{10}$. [From Wrighton (609).]

that this radical species is generated only in the presence of traces of oxygen (187). The authors, therefore, assigned it to the peroxocomplex $\cdot O_2Mn(CO)_5$.

Storage of the irradiated solution led to quantitative recombination to $Mn_2(CO)_{10}$, whereas addition of iodine resulted in the formation of $Mn(CO)_5I$. No re-formation of $Mn_2(CO)_{10}$ could be observed when N_2 was bubbled through the irradiation solution, but the radical species eliminated CO (248).

Study of the photochemistry of $Re_2(CO)_{10}$ (611) gave further evidence for a symmetrical splitting. Irradiation in CCl_4 yielded $Re(CO)_5Cl$. The quantum yield for the appearance of $Re(CO)_5Cl$ is about twice the disappearance yield of $Re_2(CO)_{10}$. The value of 0.6 for the latter indicates that M—M bond cleavage is the major photoreaction in $Re_2(CO)_{10}$.

A detailed quantitative investigation of the photochemistry of the complexes **33–37** [$Mn_2(CO)_{10}$ (**33**); $Mn_2(CO)_9PPh_3$ (**34**); $Mn_2(CO)_8(PPh_3)_2$ (**35**); $Re_2(CO)_{10}$ (**36**); $MnRe(CO)_{10}$ (**37**)] was recently reported (613). Some of the results are briefly summarized.

Photolysis of **33–37** in CCl_4 ($\lambda \geqslant 366$ nm) yielded the corresponding mononuclear metal carbonyl chlorides ($\Phi = 0.5$). Irradiation of **33**, **36**, or **37** in the presence of I_2 resulted in nearly quantitative formation of $M(CO)_5I$ species. A mixture of **33** and **36** yielded **37** on irradiation. Flash photolysis of **37** and **34** occurred according to

$$MnRe(CO)_{10} \xrightarrow{h\nu} Mn_2(CO)_{10} + Re_2(CO)_{10} \quad (\sim 1{:}1) \qquad (24)$$

$$Mn_2(CO)_9PPh_3 \xrightarrow{h\nu} Mn_2(CO)_{10} + Mn_2(CO)_8(PPh_3)_2 \quad (\sim 1{:}1) \qquad (25)$$

These reactions are interesting "preparative" applications of high-intensity photochemistry, using flash techniques. All the observed photochemistry can be interpreted as arising from homolytic metal–metal bond cleavage.

The mechanism of the formation of products resulting from CO substitution (420) (e.g., **34** from **33**) is still a matter of discussion (609). In contrast to the direct photochemical cleavage of a M—CO bond in $M_2(CO)_{10}$ (420), a recent proposal by Wrighton suggests that metal–metal bond cleavage precedes the loss of CO:

$$M_2(CO)_{10} \underset{}{\overset{h\nu}{\rightleftharpoons}} 2M(CO)_5 \xrightarrow{L} M(CO)_4L + CO$$

$$\downarrow M(CO)_5$$

$$(CO)_5M\text{—}M(CO)_4L$$

$$5. \; L_{n-1}M + X—Y \rightarrow L_{n-1}M\underset{Y}{\overset{X}{<}}$$

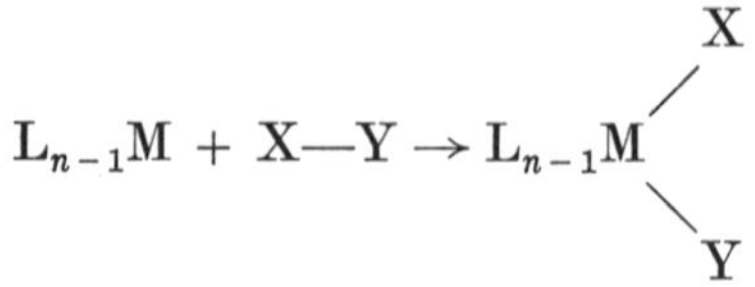

The coordinatively unsaturated species $L_{n-1}M$, which is produced by photodissociation of ML_n, can add to covalent molecules X—Y by cleavage of the X—Y bond (oxidative addition).

a. X = H.   Reaction of $Fe(CO)_4$ with HBr or HI yielded the hitherto unknown dimeric halogen iron tricarbonyls $[Fe(CO)_3Y]_2$ (Y = Br, I) (337):

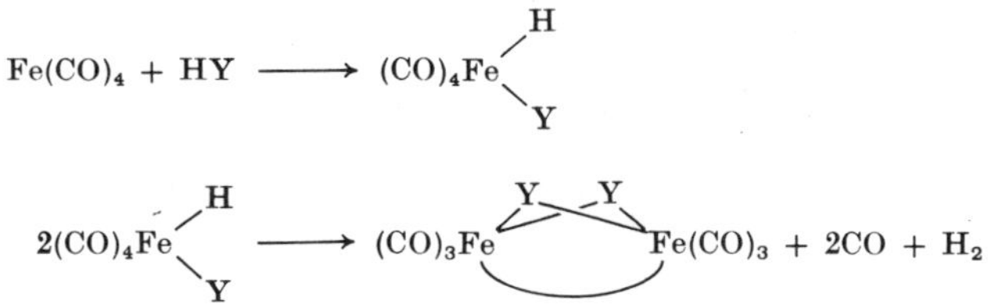

$$Fe(CO)_4 + HY \longrightarrow (CO)_4Fe\underset{Y}{\overset{H}{<}}$$

$$2(CO)_4Fe\underset{Y}{\overset{H}{<}} \longrightarrow (CO)_3Fe\overset{Y}{\underset{}{\bigwedge}}\overset{Y}{}Fe(CO)_3 + 2CO + H_2$$

b. X—Y = RSSR.

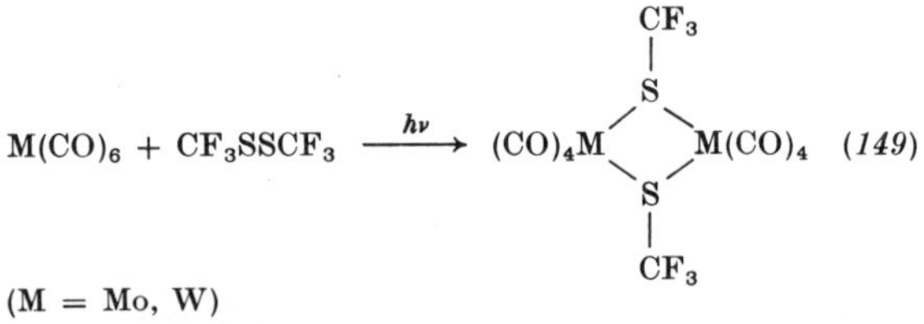

$$M(CO)_6 + CF_3SSCF_3 \xrightarrow{h\nu} (CO)_4M\begin{smallmatrix}CF_3 \\ | \\ S \\ \diagup \diagdown \\ \diagdown \diagup \\ S \\ | \\ CF_3\end{smallmatrix}M(CO)_4 \quad (149) \qquad (26)$$

(M = Mo, W)

$$[(\pi\text{-cp})M(CO)_3]_2 + CF_3SSCF_3 \xrightarrow{h\nu} [(\pi\text{-cp})M(CO)_3SCF_3] \; (150) \qquad (27)$$

(M = Mo, W)

The classification of reactions (26) and (27) as oxidative addition reactions is not clear-cut. However, the reviewers surmise that the first step in Eq. (26) is the oxidative addition of RSSR to $M(CO)_5$, giving

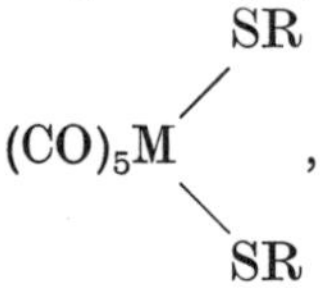

$$(CO)_5M\underset{SR}{\overset{SR}{<}}$$

, from which the final product originates.

c. X—Y = R—$Halogen$.   An example for this reaction, the photolysis of 1,2-dihaloethylene tetracarbonyliron complexes, has already been discussed in Section III, A (*236*). For X—Y = R—pseudo-halogen, similar reactions occur (*22, 560*):

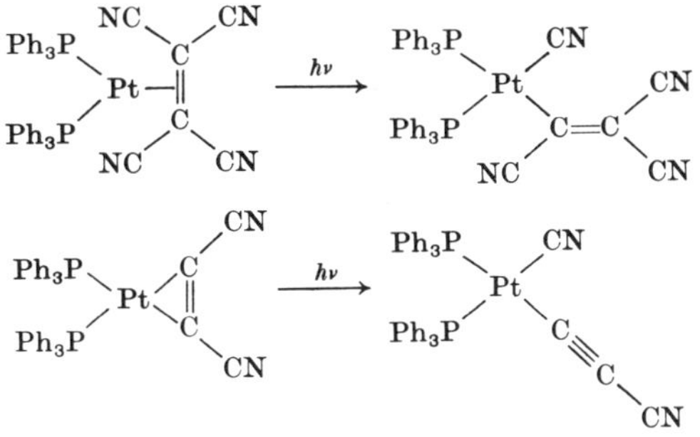

d. X—Y = M—M'. Irradiation of Ph_4As—$M'Cl_3$ (M' = Ge, Sn) in the presence of $Cr(CO)_6$, $Mo(CO)_6$, $W(CO)_6$, or $Fe(CO)_5$ yielded **38** (*479*).

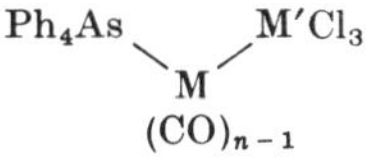

(M = Cr, Mo, W, Fe)

(**38**)

The formation of **38** can be considered as the oxidative addition of an As—M' bond to $M(CO)_{n-1}$, as well as insertion of $M(CO)_{n-1}$ into the As—M' bond. Because the transition metal is the subject of this article, the former classification is relevant.

6. Insertion and Elimination Reactions

Wojcicki (*607*) suggested the following definition of insertion reactions of organometallic compounds:

$$M—X + Y \longrightarrow M—Y—X$$

where M is a metal, and X and Y are monoatomic or polyatomic species.

This type of reaction may be considered as a special case of addition of M—X to the substrate Y, involving cleavage of the M—X bond. Very often this reaction is reversible; the extrusion of Y is called elimination.

Insertion reactions may be classified in two ways (*607*): (a) according to the nature of the entering ligand (e.g., CO, SO_2, SO_3, S, organic isocyanides) or (b) according to the nature of the M—X linkage, which is engaged in the reaction. We will follow the last suggestion.

a. M—X = M—M′. The behavior of various fluoroalkenes and alkynes toward $(CH_3)_3M$—$Mn(CO)_5$ was studied by Clark *et al.* (*115*).

$$(CH_3)_3Si-Mn(CO)_5 + CF_2{=}CF_2 \xrightarrow{h\nu} (CH_3)_3Si-CF_2-CF_2-Mn(CO)_5$$

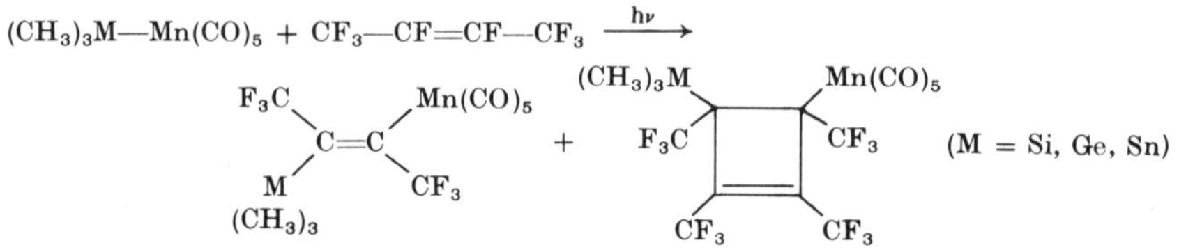

$$(CH_3)_3M-Mn(CO)_5 + CF_3-CF{=}CF-CF_3 \xrightarrow{h\nu}$$

(M = Si, Ge, Sn)

(39)

Besides the products of insertion into the M—M′ bond, they also observed secondary products, e.g. **39**.

b. M—X = M—C. Photolysis of **40** in the presence of diphenyl-diazomethane yielded complex **41** (*23*). The formation of **41** can result from insertion of diphenylcarbene into the Fe—C_1 bond, rupture of the newly formed Fe—C bond, rotation about C_1—C_2, and subsequent coordination of the oxygen to iron.

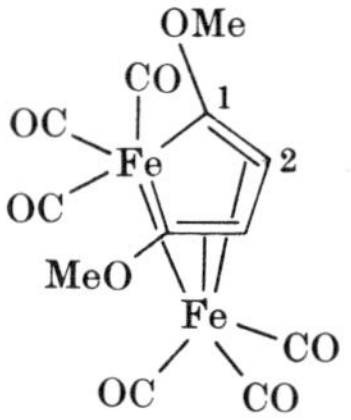

(40)

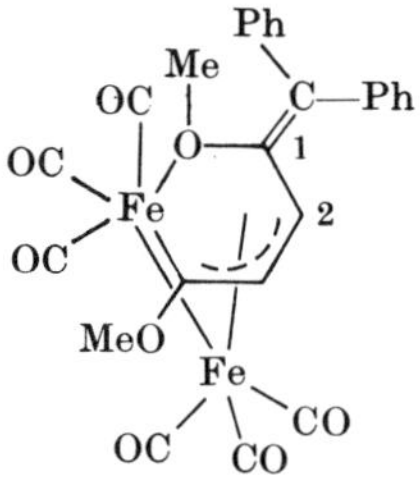

(41)

In recent years Giannotti *et al.* (*216, 221, 205*) have studied the photochemical insertion of O_2 into the Co–carbon bond of a series of cobaloximes (**42**), giving the stable peroxo compounds (**43**):

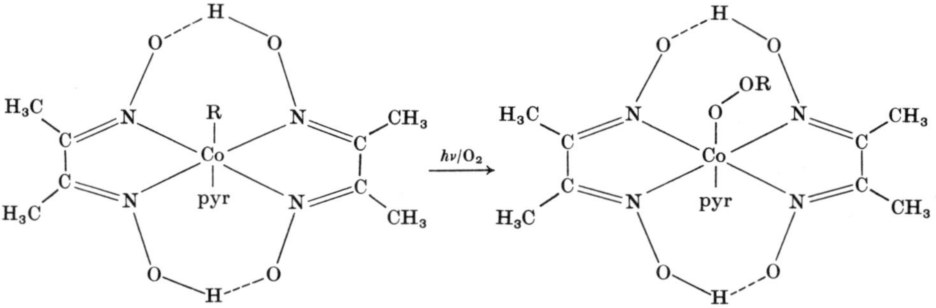

(42)

(43)

This reaction is discussed in more detail in Section V, E.

Several examples for photochemical insertion of CO in metal–carbon bonds (carbonylation reaction) have been reported (607), for instance (402):

$$(C_5H_5)(CO)_3WC_6H_5 \underset{-CO}{\overset{h\nu/CO}{\rightleftarrows}} (C_5H_5)(CO)_3W{-}\overset{\overset{\displaystyle O}{\|}}{C}{-}C_6H_5$$

Carbonylation reactions often occur with concomitant skeletal rearrangement (see following). Several examples of decarbonylation will be discussed in detail in Section IV, B, which also contains some remarks on the mechanism of this reaction. For a survey of carbonylation and decarbonylation reactions, see Ref. 607.

c. M—X = M—H.

$$(CH_3)_3SiOs(CO)_4H \xrightarrow{h\nu/C_2F_4} (CH_3)_3SiOs(CO)_4(C_2F_4H) \quad (321)$$

7. Skeletal Rearrangements

Light-induced reaction of the iron–tricarbonyl complexes (**44** and **45**), formed from cyclooctatetraene dimers (463), afforded **46** as shown by X-ray analysis (591). In the conversion of **44** to **46**, a vinyl-substituted six-membered ring has been formed from an eight-membered ring.

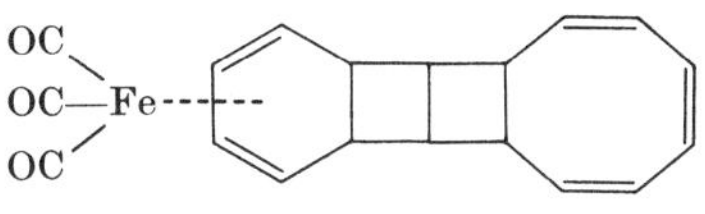

(44) 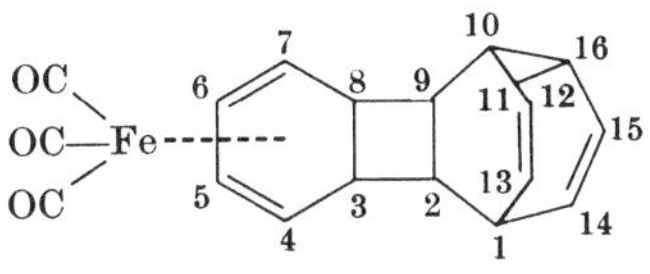   (45)

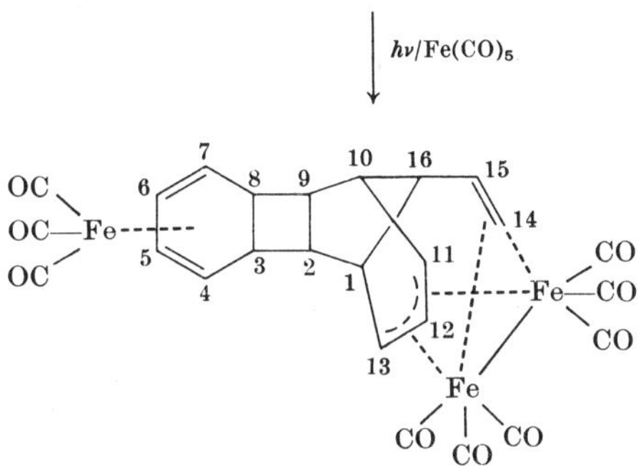

$h\nu/Fe(CO)_5$

(46)

Photoinduced insertion of CO into metal–carbon bonds can accompany skeletal rearrangements. For example the photoreaction of vinylcyclopropane with $Fe(CO)_5$ yielded **47** and **48** (*20*):

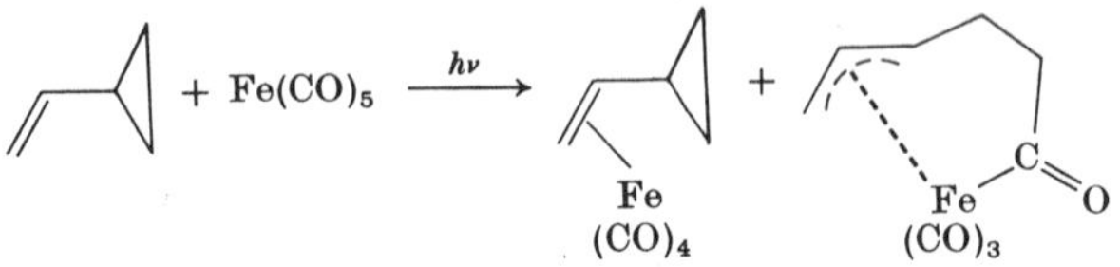

(47)   (48)

8. C—C Bond Formation

A detailed study of photochemically induced C—C bond formation between two olefins with subsequent CO insertion, yielding cyclopentanone derivatives has been published (*373, 372*):

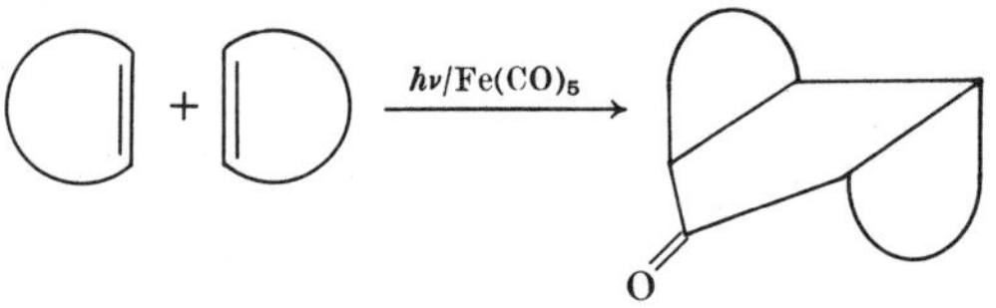

Insight into the mechanism of this reaction has been obtained by irradiation of $Fe(CO)_4L$ in the presence of L (L = methylacrylate). At room temperature, a ferracyclopentane (**50**) is formed whose reaction with CO yields the cyclopentanone derivative (**51**) (*237*):

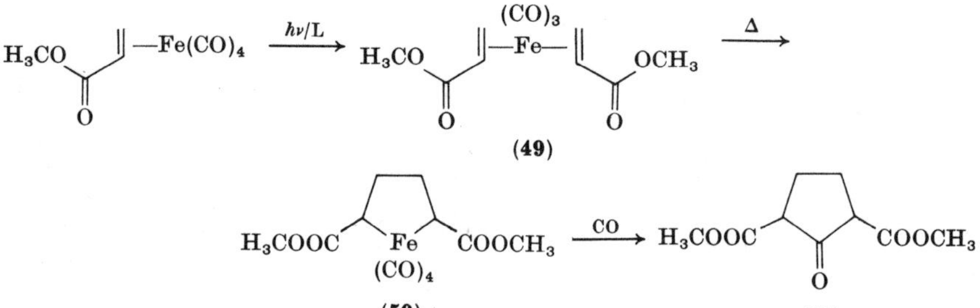

(49)

(50)   (51)

(L = CH_2=CH—$COOCH_3$)

By performing the reaction at low temperature ($-30°C$), the hitherto unknown bis(olefin)tricarbonyliron (**49**) has been isolated and shown to be an intermediate in this reaction.

Evidence is accumulating that the transition metal-catalyzed formation of cyclobutanes from 2 olefin molecules occurs via metallacyclopentanes as intermediates (*253*):

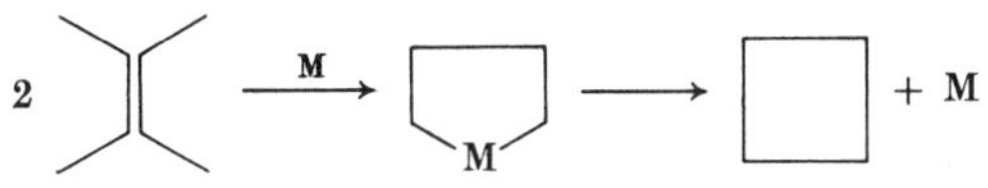

(M = transition metal)

Much mechanistic work has, therefore, been directed toward the synthesis and reactions of metallacyclopentanes. Compounds of this type have also been discussed as intermediates in metathesis reactions (239).

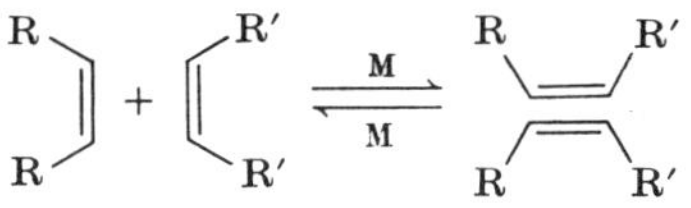

9. M—C Bond Cleavage

The involvement of metallacyclopentanes as intermediates in the formation of cyclobutanes and in metathesis reactions, as discussed above, makes their photochemistry interesting.

An investigation of the photoreactions of **50** gave the following results (238):

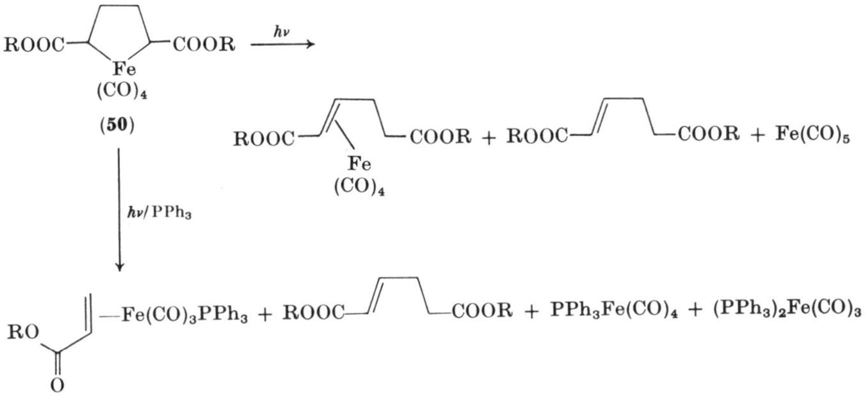

The understanding of the mechanistic details of this reaction will require more knowledge about the photochemical cleavage of transition metal–carbon single bonds. The photochemical formation of radicals from nontransition organometallic compounds, e.g., mercury alkyls, is a well-known reaction (44, 119):

$$R—Hg—R \xrightarrow{h\nu} R—Hg\cdot + R\cdot$$

The corresponding reaction with transition metal alkyls is restricted to a few examples, especially to alkylcobalt complexes (see Sections V, D and E).

Irradiation of aqueous $[C_6H_5-CH_2-Co(CN)_5]^{3-}$ in the CTLM (benzyl → Co) transition resulted in a redox reaction yielding $Co(CN)_5^{3-}$ and dibenzyl (from benzyl radicals). If this reaction was carried out in the presence of oxygen, no $Co(CN)_5^{3-}$ was observed and the benzyl radicals were oxidized to benzaldehyde (585):

$$[C_6H_5-CH_2-Co(CN)_5]^{3-} \begin{cases} \xrightarrow[H_2O]{h\nu} & [Co(CN)_5]^{3-} + \cdot CH_2-C_6H_5 \\ & 2\ \cdot CH_2-C_6H_5 \longrightarrow C_6H_5-CH_2-CH_2-C_6H_5 \\ \\ \xrightarrow[H_2O/O_2]{h\nu} & [Co(CN)_5H_2O]^{2-} + C_6H_5CHO + OH^- \end{cases}$$

Recently, the photochemical reaction of dimethylderivatives of titanocene, zirconocene, and hafnocene have been reported (11). Irradiation resulted in a homolytic cleavage of the methyl–metal bond, by which the metal is reduced to a lower oxidation state:

$$(C_5H_5)_2M(CH_3)_2 \xrightarrow{h\nu} (C_5H_5)M + 2\ \cdot CH_3$$
$$(M = Ti, Zr, Hf)$$

The methyl radicals undergo further reactions (e.g., methane formation).

When photolysis is carried out in the presence of acetylenes, metallacycles are the major products:

$$cp_2M(CH_3)_2 + Ph-C{\equiv}C-Ph \xrightarrow{h\nu} cp_2M \begin{matrix} Ph \\ | \\ \end{matrix}$$

(M = Ti, Zr, Hf)

10. Photocatalysis

It is difficult to give a concise definition of the term photocatalysis. In an organometallic catalytic photoreaction, a catalyst, which is formed and (or) activated photochemically *in situ*, brings about the thermal conversion of more than 1 substrate molecule (per catalyst molecule). Depending on the quantum yield of catalyst formation (or activation), the total turnover (T) of substrate can be less than ($\Phi_T \leqslant 1$) or more than ($\Phi_T > 1$) 1 molecule per photon absorbed.

Wrighton (614) has suggested that the term photocatalysis be reserved for the case of $\Phi_T > 1$. He also writes: "We find, empirically, that continuous photolysis accelerates many photocatalytic reactions, and in many cases we find no measurable substrate transformation without continuous photolysis. The term *photoassisted* is reserved for substrate transformations that appear to proceed according to

$$\text{metal complex} \xrightarrow{h\nu} [\text{catalyst}] \tag{28}$$

$$\text{metal complex + substrate} \xrightarrow{\Delta} \text{no reaction} \tag{29}$$

$$[\text{catalyst}] + \text{substrate} \xrightarrow{\Delta} \text{transformed substrate} + [\text{catalyst}] \tag{30}$$

$$[\text{catalyst}] \xrightarrow{\Delta} \text{poisoning} \tag{31}$$

but which require continuous photolysis" (*614*).

The reviewers would like to suggest the term *photoassisted catalytic reactions* for the general case of $\Phi_T \leqslant 1$. A determination of the quantum yield of transformed substrate provides an experimental criterion to distinguish the two kinds of catalytic reactions.

In many photoassisted catalytic reactions, light is not only required for the production of a catalyst but also for its activation and/or reactivation according to the modified (*614*) Michaelis-Menten mechanism (*232*):

$$[\text{catalyst}] + \text{substrate} \xrightleftharpoons{\Delta\text{ or }h\nu} [\text{catalyst}\cdots\text{substrate}] \tag{32}$$

$$[\text{catalyst}\cdots\text{substrate}] \xrightarrow{\Delta\text{ or }h\nu} [\text{catalyst}\cdots\text{transformed substrate}] \tag{33}$$

$$[\text{catalyst}\cdots\text{transformed substrate}] \xrightarrow{\Delta\text{ or }h\nu} [\text{catalyst}] + \text{transformed substrate} \tag{34}$$

In such cases $\Phi_T < 1$ can be easily rationalized.

The possibility of preparing catalysts photochemically (especially at low temperatures) outside of the system undergoing the catalytic reaction, may indicate which of the steps of Eqs. (28)–(34) require light. A few examples should illustrate these principles.

a. Photochemical Production of a Catalyst (Outside of the System To Be Catalyzed). Irradiation of $Fe(CO)_5$ in the presence of excess butadiene leads, by substitution of four CO groups, to bis(butadiene)monocarbonyliron (7) (*327, 340*) in excellent yield:

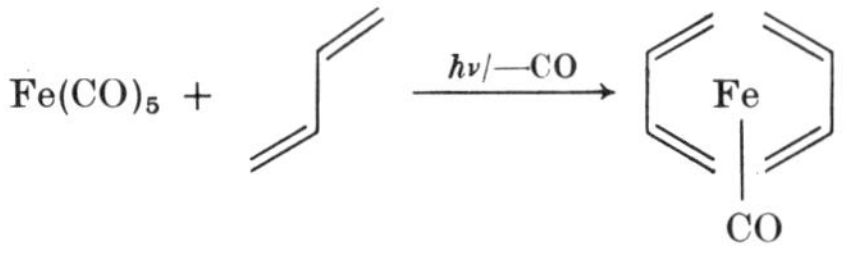

In the presence of PPh_3, 7 acts in the dark as a catalyst for the oligomerization of butadiene to give 4-vinylcyclohexene and 1,5-cyclooctadiene. 7 also brings about the mixed oligomerization of acrylate and dienes to give heptadienoic esters (*92*) (see Section IV, B). In these two catalytic reactions, only Eq. (28) requires light.

The cluster compound (53), which is formed either thermally by reaction of the corresponding carbonyl-bridged complex (52) (*341*) with

HCl, or by its photochemical reaction in the presence of H_2, serves as a catalyst for the photohydrogenation of olefins or dienes (*330*):

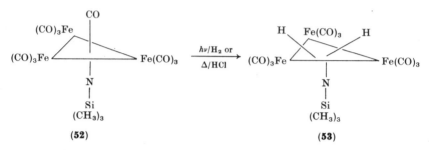

(52)   (53)

b. Photocatalytic Reactions. The hydrogenation of methylacrylate, cyclobutene, cyclobutene-3,4-dimethylycarboxylate, and 2,3-dimethyl-butadiene with H_2 and **53** as a catalyst, require light for Eq. (32). Actually, one CO has to be eliminated photochemically to allow the formation of the catalyst–substrate complex. This was demonstrated by the formation of the corresponding (stable) $P(OCH_3)_3$ complex, when the photoreaction was carried out in the presence of $P(OCH_3)_3$ (*330*).

After the photochemical activation [Eq. (32)] of **53**, catalytic hydrogenation of the unsaturated substrates occurs in the dark, giving direct evidence for $\Phi_T > 1$.

According to recent evidence, the photohydrogenation of dienes with $Cr(CO)_6/H_2$ (*625, 614, 395, 507, 435, 436, 439, 437, 624*) also belongs to the group of truly photocatalytic reactions. 1,3-Diene–tetracarbonylchromium complexes, which are catalysts of this reaction, have recently been prepared (e.g., by low-temperature photolysis) and tested (*328, 339*).

Norbornadiene–$Cr(CO)_4$ is also an efficient catalyst in the photohydrogenation of dienes (*435, 436, 439, 437*). Also, $IrCl(CO)(PPh_3)_2$ has been employed as catalyst in the photohydrogenation of olefins (*547*).

Another example in which light is needed for the formation of a catalytically active species is the 1,4-hydrosilation of 1,3-dienes in the presence of $Cr(CO)_6$ (*626*):

$$\text{/\!\!/}\quad\text{\\} \quad + \text{ HSi(CH}_3)_3 \xrightarrow{h\nu/\text{Cr(CO)}_6} \text{/}\!\!=\!\!\text{\\}_{\text{Si(CH}_3)_3} \qquad (35)$$

In some cases the resulting allylsilanes are obtained in almost quantitative yields.

c. Photoassisted Catalytic Reactions. Many catalytic photoisomerization and -dimerization reactions appear to belong in this group; however, detailed quantum yield studies are missing in most cases. Table VIII provides a survey of such reactions.

d. *Photoinitiation of Free-Radical Polymerization by Transition Metal Derivatives.* A special case of organometallic photocatalysis is the initiation of free-radical polymerization (*29*). Many of these reactions are based on metal carbonyl photolysis. Three initiation mechanisms are commonly considered (*29, 35*).

i. Formation of radicals from organic halides (*36, 33*):

$$M(CO)_n \xrightarrow{h\nu} M(CO)_{n-1} + CO$$

$$M(CO)_{n-1} + RX \xrightarrow[\text{addition}]{\text{oxidative}} R\cdot + [XM(CO)_{n-1}]$$

ii. Formation of radicals by hydrogen abstraction from C—H bonds (*34*):

$$M(CO)_n \xrightarrow{h\nu} M(CO)_{n-1}$$

$$M(CO)_{n-1} + R'—H \xrightarrow[\text{addition}]{\text{oxidative}} R'\cdot + [HM(CO)_{n-1}]$$

iii. Formation of radicals by metal–metal bond cleavage and addition of $CF_2{=}CF_2$ (*35*):

$$M_2(CO)_{2n} \xrightarrow{h\nu} 2 \cdot M(CO)_n$$

$$\cdot M(CO)_n + CF_2{=}CF_2 \longrightarrow (CO)_nM—CF_2—CF_2\cdot$$

The radicals formed according to processes i–iii can polymerize a variety of vinyl monomers (*36, 32, 29*) and olefinic substrates (*17, 30, 35*). Much work has been done to elucidate the primary processes in these systems (*31, 32, 37, 33, 36*); some of the complications with $Mn_2(CO)_{10}$ have already been discussed (Section IV, A 4). The detailed mechanism of radical formation by oxidative addition is still under discussion (see Section IV, A). In some cases formation of X—M—R may precede radical production. It has been suggested, that addition of a vinyl monomer to X—M—R could initiate a (non-free-radical) Ziegler-type of polymerization (*332*).

The polymerization of ethylene oxide with $Mn_2(CO)_{10}$ as a starter (*549, 548*) and of styrene with tetrabenzylzirconium (*24*) are further examples of this application of organometallic photochemistry.

B. Discussion of Some Selected Examples

In the following section we will discuss some examples selected to give some idea of the scope and versatility of preparative applications

TABLE VIII

PHOTOASSISTED CATALYTIC REACTIONS

Reaction	Substrate	Catalyst	Ref.
Isomerization	$C_2H_2D_2$	$W(CO)_6$	622
	Pentenes	$W(CO)_6$ or $Mo(CO)_6$	622
	Hexenes	$W(CO)_6$	622
	Pentadienes	$W(CO)_6$ or $Mo(CO)_6$	617
	Hexadienes	$W(CO)_6$	617
	Stilbenes	$W(CO)_6$ or $Mo(CO)_6$	616
	N-Allylamide	$Fe(CO)_5$	270
	Allyl ethers	$Fe(CO)_5$	268
	N-Allylcarbamate	$Fe(CO)_5$	269
	O-Phenylcarbamate	$Fe(CO)_5$	269
	N-Propenylcarbamate	$Fe(CO)_5$	269
	1-Heptene	$IrClCO(PPh_3)_2$	546
		$RhHCO(PPh_3)_3$	546
Dimerization	Norbornadiene	$Cr(CO)_6$	281, 263
		$(C_6H_6)Cr(CO)_3$	263
		$Ni(CO)_4$	582, 263
		$(PPh_3)_2Ni(CO)_2$,	263
		$Co(CO)_3NO$, $(Ph_3P)Co(CO)_2NO$,	
		$(Ph_3P)CoCONO$, $Fe(CO)_2(NO)_2$,	
		$(Ph_3P)FeCO(NO)_2$, $(Ph_3P)_2Fe(NO)_2$,	
		1,4-diphenyl-1,3-butadiene-$Fe(CO)_3$	

of organometallic photochemistry. The ordering of these reactions reflects principles discussed in the preceding Section IV, A.

Substitution of CO in metal carbonyls plays a key role in the synthesis of transition metal complexes. Irradiation of $Cr(CO)_6$ in the presence of NO yielded $Cr(NO)_4$, the final member of the so-called pseudocarbonyl-nickel series, $Ni(CO)_4$, $Co(CO)_3(NO)$, $Fe(CO)_2(NO)_2$, $Mn(CO)(NO)_3$, $Cr(NO)_4$ (*259, 553*).

Photosubstitution of CO in $Cr(CO)_6$ and $(\pi\text{-cp})Mn(CO)_3$ by the relatively weak donor THF and subsequent addition of P_2H_4 gave the first complexes containing diphosphine as a ligand (*525*):

$$2Cr(CO)_6 \xrightarrow[-2CO]{h\nu/THF} 2Cr(CO)_5THF \xrightarrow{-78°, \ P_2H_4} P_2H_4[Cr(CO)_5]_2$$

$$2(\pi\text{-cp})Mn(CO)_3 \xrightarrow{h\nu/THF} 2(\pi\text{-cp})Mn(CO)_2THF \xrightarrow{-78°, \ P_2H_4} P_2H_4[(\pi\text{-cp})Mn(CO)_2]_2$$

Carbene complexes of the general structure $(CO)_5MC(OR')R''$ (M = Cr, W; R, R'' = alkyl, aryl; R' = alkyl) add PR_3 at $-20°C$ to form $(CO)_5M—C(OR')R''PR_3$ by nucleophilic addition of PR_3 to the carbene carbon. Photolysis of this addition product (M = Cr) at $-15°C$ gave $cis\text{-}(CO)_4(R_3P)CrC(OCH_3)CH_3$, whereas at room temperature a mixture of the *cis*- and the *trans*-tetracarbonylchromium complex was obtained (*200*).

An interesting example for CO substitution by a π-system is reported by Herberhold *et al.* (*254*). They prepared the tetramethoxyethylene (TME) complexes $Fe(CO)_4TME$ and $(\pi\text{-cp})Mn(CO)_2TME$. The CO frequencies of these complexes indicate considerably more M → TME backbonding than was expected for this "electron-rich" olefin.

The reader's attention should be drawn to the substitution of four CO groups in $Fe(CO)_5$ by dienes, leading to bis(diene)monocarbonyliron (*326, 327, 340*) (see Table X in Section IV, C); in a related reaction of dienyl tricarbonyliron salts with dienes, the (dienyl)(diene)mono-carbonyliron salts are obtained, which can be reduced with hydrides to bis(diene)monocarbonyliron compounds (*16, 288*) (Table X).

Complete photochemical substitution of CO by cycloheptatriene in cyclopentadienyltricarbonylmanganese has been reported (*423*):

(**54**)

The resulting complex (54) can easily be converted by hydride abstraction to the tropylium salt.

Photolysis of $Fe(CO)_5$ in the presence of styrene yielded complex (55) (334). Further irradiation leads to carbonyliron derivatives (56) and (57) where 4 or 8 electrons of the π-system are involved in coordination (581, 577, 580, 578):

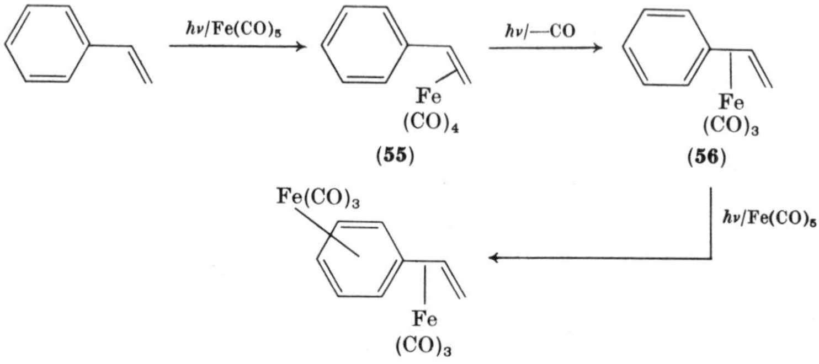

From reactions with meta-substituted styrenes, positional isomers of bistricarbonyliron complexes, corresponding to the two trapped Kékulé structures were isolated (577).

An additional example of diene-type reactivity shown by an originally aromatic system is the reaction of benzocyclobutadienetricarbonyliron with pentacarbonyliron (575). Besides the expected compound 58, products 59–62, which result from cyclobutadiene ring opening and insertion of an $Fe(CO)_3$ unit, are obtained.

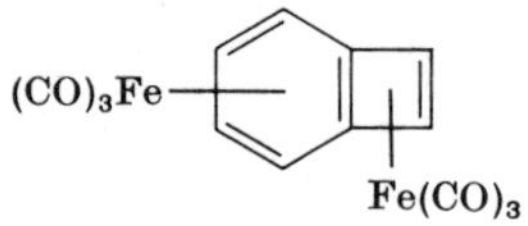

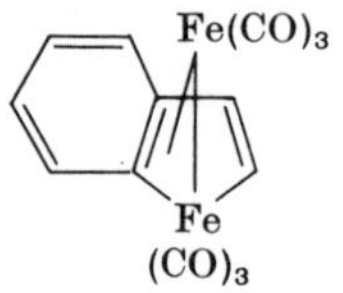

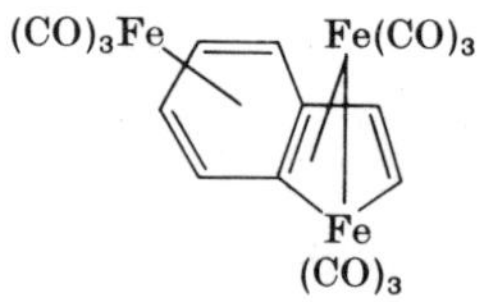

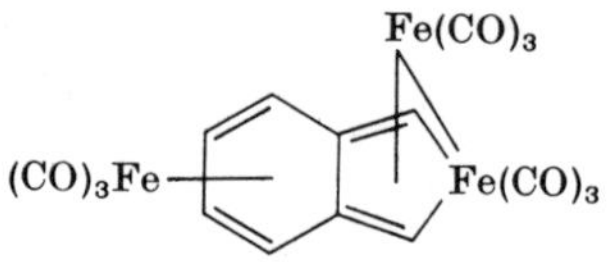

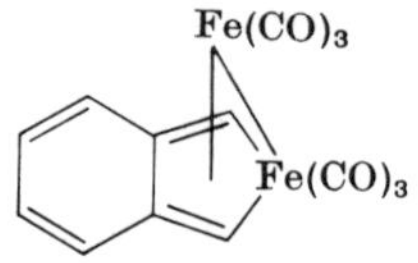

Photochemical synthesis of pinocarvone–, pulegone–, and benzalacetophenone–$Fe(CO)_3$ complexes have been reported (333). In all cases bonding to the iron involves only the π-system of the 1-oxadiene systems, and not the nonbonding electrons of the keto group. The same bonding feature was found in the $Fe(CO)_4$ (63) and $Fe(CO)_3$ (64) complexes of isobutylidene Meldrum's acid, which were prepared photochemically (*338*).

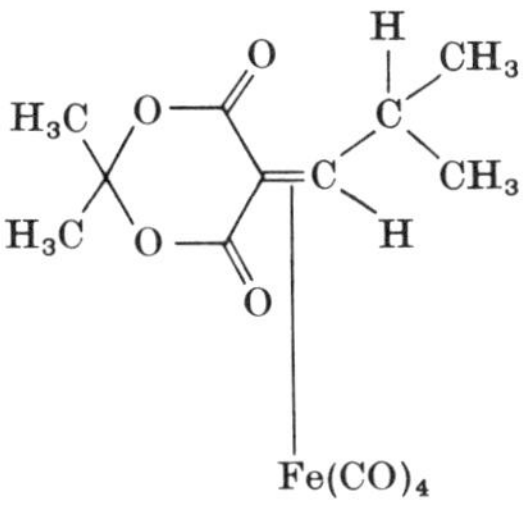

(63) 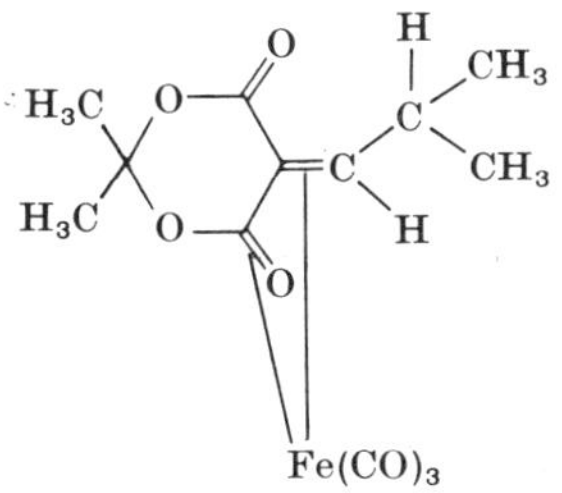   (64)

A reversible intramolecular CO substitution was suggested as the first step in the preparation of **66** via the intermediate complex **65** starting from **65a** (*240*) (Scheme 2).

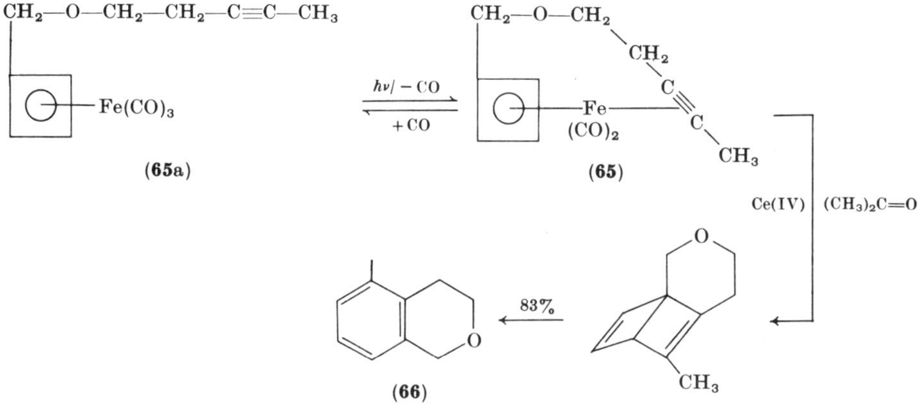

SCHEME 2

The mechanism and preparative application of arene–tricarbonyl-chromium complexes has recently attracted considerable attention. Although in thermal substitution reactions the aromatic ligand is always replaced, it is exclusively CO, which is substituted photochemically. The only exception was the observation by Strohmeier (*552*) that the arene can be replaced photochemically by labeled benzene. Re-investigation, using hexadeuterobenzene and ^{13}CO revealed that this process is suppressed by CO (*225*). This suggests that arene–dicarbonyl-chromium is also the precursor in the exchange of the aromatic system.

A series of maleic anhydride (MA) complexes of the type $ArCr(CO)_2MA$ (Ar = substituted benzene) has been prepared photo-chemically (*255*). A comparison of IR and NMR data of $ArCr(CO)_3$ and

ArCr(CO)$_2$MA indicated that π-bonded MA is a better acceptor ligand than carbon monoxide.

The stepwise substitution of CO by different ligands in ArCr(CO)$_3$ yielded chiral complexes due to the pseudotetrahedral configuration around the metal (*277, 279, 276, 278*). The relatively easy dissociation of the Cr—P bond in complex **67** interfered with a detailed investigation; even **68** racemized in acetone.

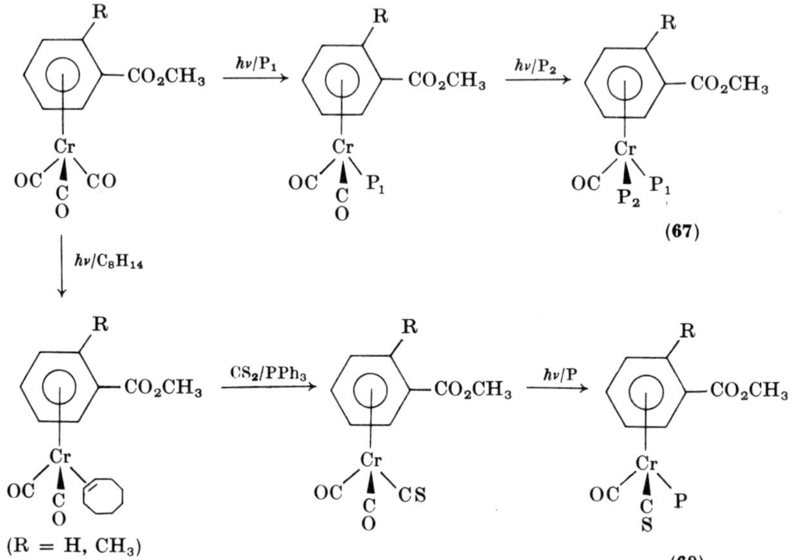

(67)

(68)

(R = H, CH$_3$)

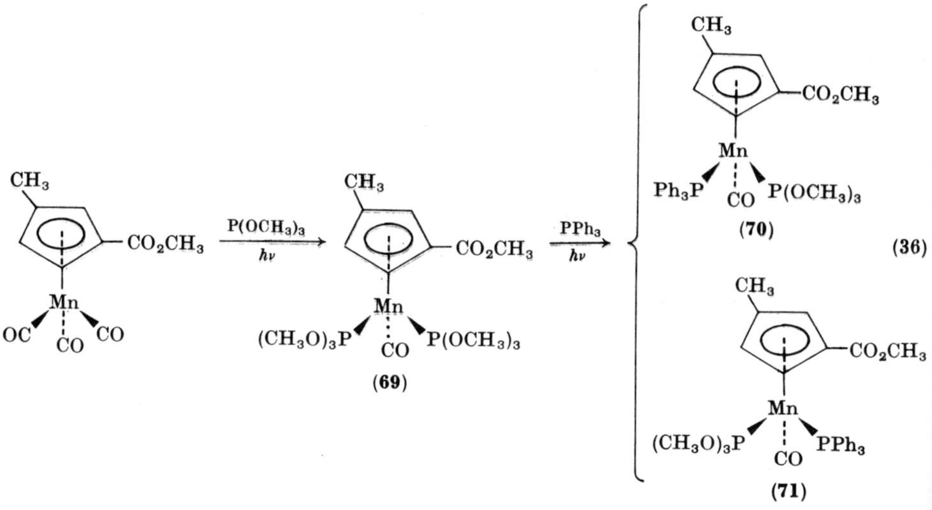

(69)

(70)

(71)

(36)

Other chiral systems have also been synthesized by photochemical substitution reactions [Eq. (36)]. The formation of the two diastereomers (70) and (71) occurs to the same extent (366). It is interesting to note that phosphite and not CO is substituted in the intermediate (69) (see also Ref. 80).

Compounds 72–74 are other systems with chiral metal centres that have been prepared photochemically (394, 186, 185, 379):

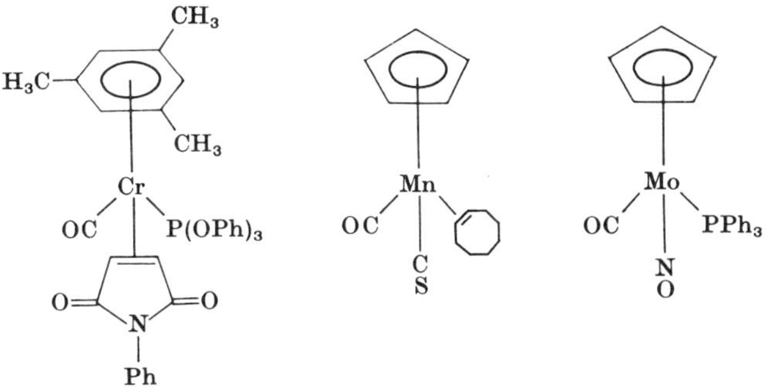

(72)   (73)   (74)

Although the photosubstitution of phosphines in phosphine–carbonylmetal complexes is an exceptional case (see, e.g., 69) the photochemistry of pentakis(trifluorophosphine)iron has found interesting synthetic applications. Substitution by conjugated dienes yields (π-diene)tris(trifluorophosphine)iron (0) complexes (349, 350):

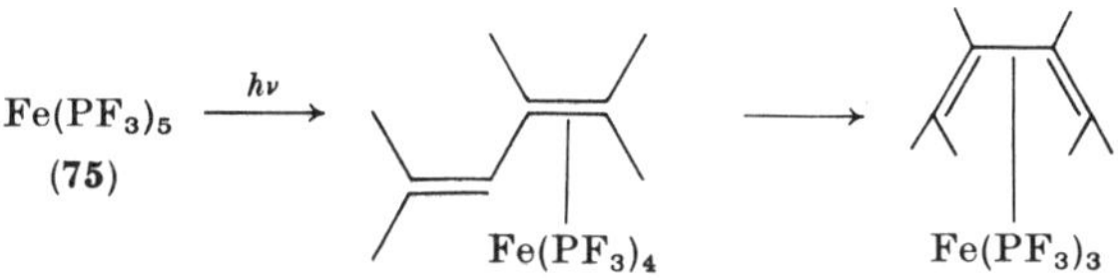

In some cases, the monosubstituted intermediates (π-diene)Fe(PF$_3$)$_4$ were identified by their mass and ^{1}H NMR spectra. Substitution of PF$_3$ in (75) has also been achieved with various monoolefins (347, 348) and with H$_2$ (352).

Ultraviolet irradiation of HCo(PF$_3$)$_4$ and a mixture of HCo(PF$_3$)$_4$ with HIr(PF$_3$)$_4$ resulted in elimination of PF$_3$ and formation of the bridged complexes (PF$_3$)$_3$Co(H)(PF$_2$)Co(PF$_3$)$_3$ and (PF$_3$)$_3$Co(H)(PF$_2$)-Ir(PF$_3$)$_3$ (353).

One major application of the photochemical formation of a free coordination site at a transition metal involves *isomerization reactions with hydrogen transfer via metal hydride intermediates* (see Section IV, A, 3, a). A typical example is the irradiation of Fe(CO)$_5$ in the presence of cyclohexa-1,4-dienetricarbonyliron. By labeling the methylene groups of the 1,4-diene with deuterium and analyzing the product (76), it was shown that the formation of the conjugated diene occurs via a π-allyl–hydridotricarbonyliron intermediate (9, *148*).

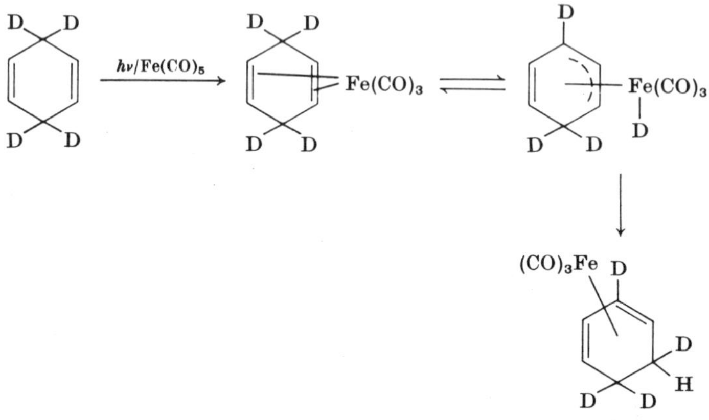

(76)

Several additional examples of *positional isomerization* should be mentioned (see also Refs. *57, 58, 387*,):

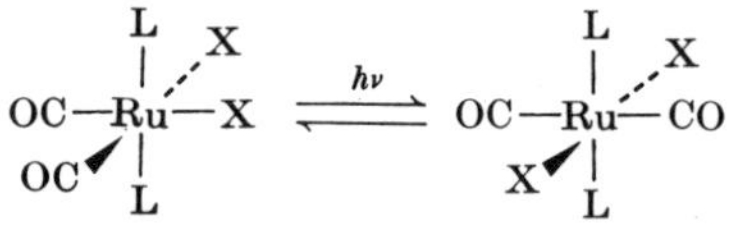

(L = PPh₃; X = I)

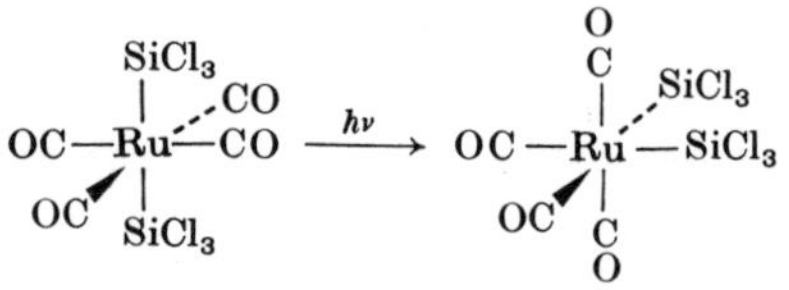

$$cis\text{-}M(CO)_4I_2 \xrightarrow{h\nu} trans\text{-}M(CO)_4I_2 \quad (422)$$

(M = Fe, Ru)

An interesting case of *photochemical cleavage of metal–metal bonds* (see also Refs. *609, 243, 251*) is given by the rhenium complex $[Re_2Cl_8]^{2-}$ (77) containing a metal–metal quadruple bond (*212*). Irradiation in CH_3CN solution at different wavelengths (366 nm ⩾ λ ⩾ 300 nm; λ = 254 nm) yielded $[ReCl_3(CH_3CN)_3]$ (79). A mechanism via complex

78 has been proposed for the formation of complex **79** both on long- and short-wavelength† UV photolysis:

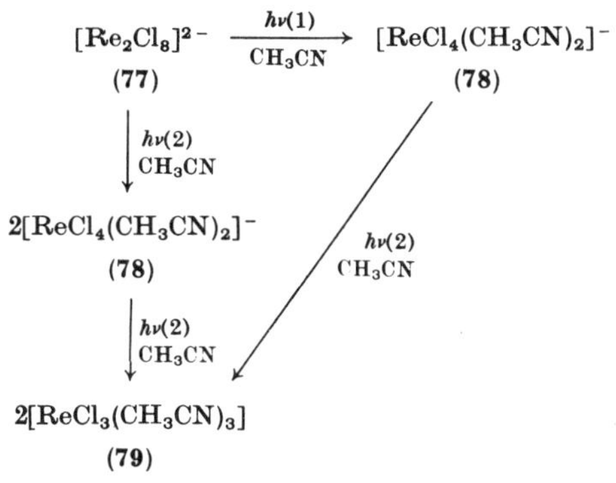
$[Re_2Cl_8]^{2-}$
(77)
$h\nu(1)$
CH_3CN
$[ReCl_4(CH_3CN)_2]^-$
(78)
$h\nu(2)$
CH_3CN
$2[ReCl_4(CH_3CN)_2]^-$
(78)
$h\nu(2)$
CH_3CN
$h\nu(2)$
CH_3CN
$2[ReCl_3(CH_3CN)_3]$
(79)

Like the cleavage of metal–metal bonds, the reverse reaction, i.e., the *formation of metal–metal bonds*, can be achieved photochemically.

Irradiation of cyclobutadiene–tricarbonyliron in THF at $-40°$ under nitrogen resulted in formation of the dinuclear complex **80**, for which a metal–metal triple bond was proposed (*201*):

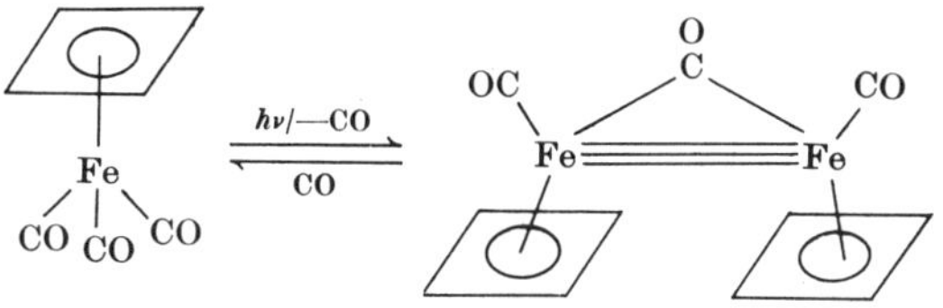
Fe
CO CO CO
$h\nu$/—CO
CO
O
C
OC
CO
Fe
Fe

(80)

Analogous complexes (**81**) were obtained from the irradiation of substituted cyclobutadiene–tricarbonyliron complexes:

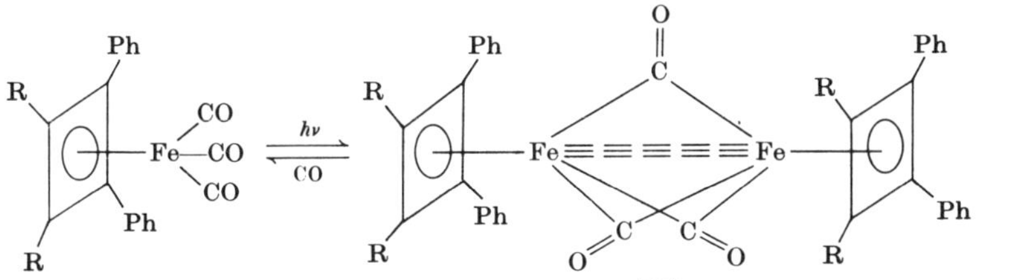
Ph
R
CO
Fe—CO
CO
Ph
R
$h\nu$
CO
Ph
R
O
C
Fe
Fe
Ph
R
C
C
O
O
R
Ph
Ph
R

(R = Ph, *t*-butyl)

(81)

† In the proposed mechanism, $h\nu(1)$ denotes $\lambda = 254$ *nm* and $h\nu(2)$ denotes 366 nm $\geq \lambda \geq$ 300 nm.

However, all CO groups were bridging (*391*). An X-ray analysis of (**81**) (R = *t*-butyl) showed an extremely short Fe—Fe distance (2.177 Å), which is consistent with an Fe—Fe triple bond.

Irradiation of $(CH_3)_5C_5Mo(CO)_3(CH)_3$ in THF gave complex **82** of stoichiometry $[(CH_3)_5C_5Mo(CO)_2]_2$ (*307*):

(**82**)

The unsubstituted cyclopentadienyl analog $(CO)_2(\pi\text{-cp})Mo$—Mo (π-cp)$(CO)_2$ was formed upon irradiation of $(CH_3)_2C$=CH—Ge—Mo(π-cp)$(CO)_2$ (*286*; see also Ref. *178*). A survey of reactions involving metal–metal bond formation is given in Table XI (see Section IV, C).

Some additional examples of *oxidative addition and reductive elimination* reactions should be mentioned to illustrate their scope and preparative importance.

Addition of R_3SiH to the coordinatively unsaturated species produced by photochemical expulsion of CO from a metal carbonyl compound provides a convenient route to the metal hydrides L_{n-1} $M(H)SiR_3$. Further reaction can lead to $L_{n-1}M(SiR_3)_2$. Reactions of this type have been carried out with R_3SiH and $Fe(CO)_5$, $C_6H_6Cr(CO)_3$, $C_5H_5Mn(CO)_3$, $C_5H_5Fe(CO)_2SiCl_3$, $C_5H_5Co(CO)_2$ (*283, 284*), $Fe_3(CO)_{12}$, $Ru_3(CO)_{12}$ (*322*), $Os_3(CO)_{12}$ (*321*), $C_5H_5Rh(CO)_2$ (*418*), and $Re_2(CO)_{10}$ (*267*). The compound $(CH_3)_3SnH$ reacts with $Os_3(CO)_{12}$ in an analogous manner (*321, 322*).

Binuclear complexes with R_2Si bridges result from the photoreaction of R_2SiH_2 with $W(CO)_6$ (*45*) and $Re_2(CO)_{10}$ (*266*), and of $(CH_3)_2HSi$—$SiH(CH_3)_2$ with $Fe(CO)_5$ (*355*).

Oxidative addition of C—H bonds to tungsten (*183, 220*) and of B—H bonds to iron (*536*) should be mentioned in this context (see also Ref. *524*).

The oxidative addition of 9,10-phenanthrenequinone to $Ir(PPh_3)_2$-COCl can be reversed photochemically (573):

$$(37)$$

Irradiation of $(\pi\text{-cp})Fe(CO)_2X$ in the presence of mercury led to insertion into the iron—X bond (404):

$$(\pi\text{-cp})Fe(CO)_2X + Hg \xrightarrow{h\nu} (\pi\text{-cp})Fe(CO)_2HgX$$
$$(X = Cl, Br, I)$$

Insertion of SnX_2 into metal–metal bonds has been observed with Group VI metal complexes (241):

$$[(\pi\text{-cp})M(CO)_3]_2 + SnX_2 \xrightarrow{h\nu} [(\pi\text{-cp})M(CO)_3]_2SnX_2$$
$$(M = Cr, Mo, W; X = F, Cl)$$

Photochemical elimination of N_2 from azide complexes or of triphenylphosphineoxide from nitrosylphosphine complexes leads to the isocyanide complexes **83–85** as a result of CO addition to nitrenes (379).

The photoreversal of the addition of olefins to nickel bis (1,2-ethenedithio-

lates) (*262*), the formation of diphenylacetylene from $\overset{Ph}{\underset{Ph}{\diagdown}}\!\!\!\triangleright\!\!=\!\!Mo(CO)_5$

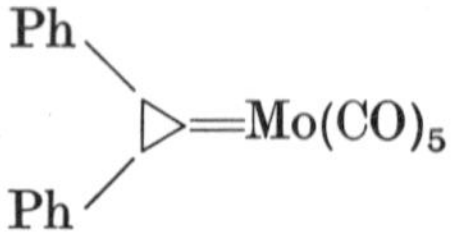

(*460*), the loss of PR_3 from $[R_3P(C_4H_4)Fe(CO)_2NO]^+PF_6^-$ (*176*), and the production of $[(C_6H_5)_3P]_2N[W(CO)_5Cl]$ from $[(C_6H_5)_3P]_2N[W_2Ni_3(CO)_{16}]$ (*480*) are further examples of *photoelimination* reactions.

The *photochemical decarbonylation* of metal–acyl compounds has generated preparative as well as mechanistic interest, owing to its close relation (principle of microscopic reversibility!) to the important CO insertion reaction (*607*), which has found wide industrial application. The stereochemistry of decarbonylation has been studied using the complex **86**. Separation of the two diastereomeric pairs of enantiomers of **86** led to an enrichment of 95/5 of one over the other (*18*). Photolysis under decarbonylation to **87** produced 88/12 mixtures of the corresponding diastereomeric pairs of **87**.

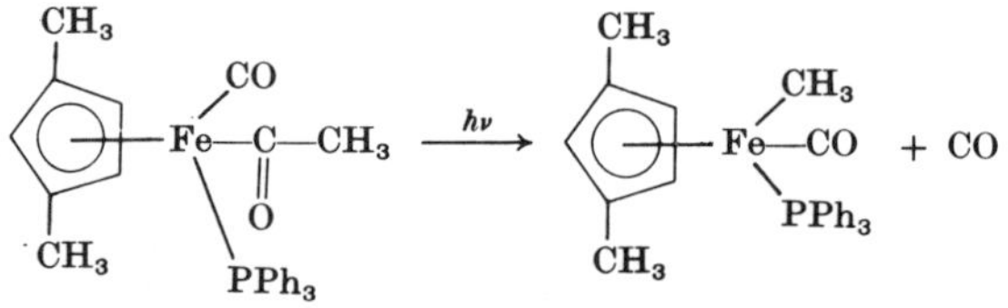

(86)   (87)

From this observation it becomes obvious that the loss of CO occurs with a high degree of stereoselectivity. Earlier work had already shown that in this process a terminal CO and not the acyl–CO group is eliminated, this is followed by migration of the alkyl group to the

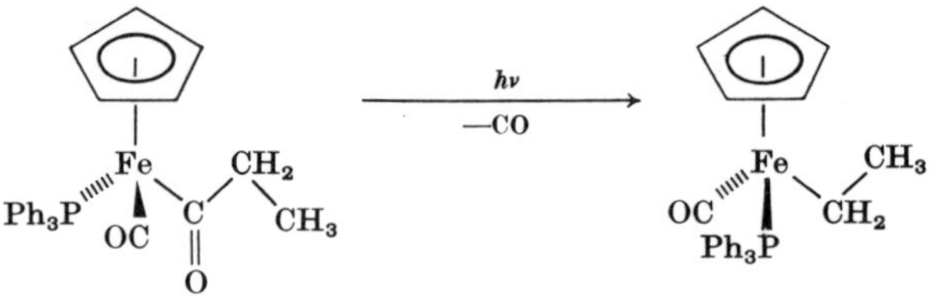

(88)   (89)

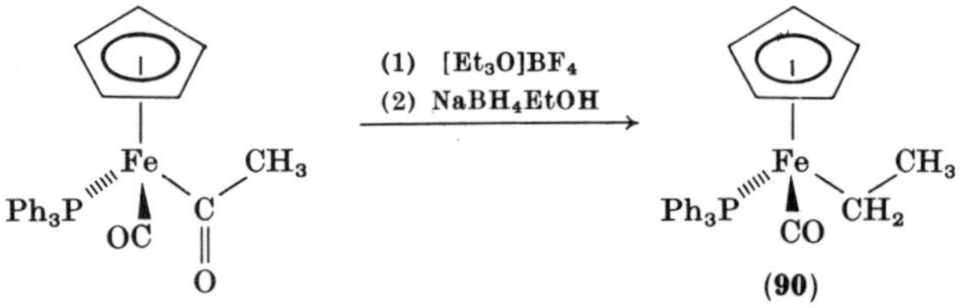

(90)

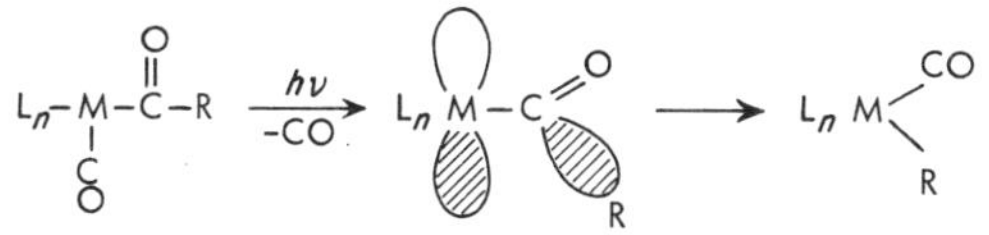

FIG. 13. Photochemical elimination of CO from metal–acyl compounds showing nucleophilic attack on CO—CH$_3$ σ-bond on metal p orbital.

free coordination site (4). This reaction occurs with inversion of configuration at the iron center, as demonstrated by Davison and Martinez, who used **88** of high optical purity (153). The configuration of photochemically formed **89** was determined by comparison with **90**.

The fact that photochemical decarbonylation involves inversion at the metal and retention of configuration at the alkyl group (54) is consistent with the following explanation: (i) The migrating alkyl group has a high preference for the vacant site produced by elimination of terminal CO; (ii) the cleavage of the acyl bond and the coordination of the alkyl group to the iron have to occur in a concerted fashion; and (iii) the reviewers suggest a nucleophilic attack of the CO–alkyl bond on the vacant site of the iron, as depicted in Fig. 13.

A 1,4-elimination of CO is observed on photolysis of tricarbonyl (7-norbornadienone) iron (**91**) (360, 361). Complex **91** is an example of the stabilization by a tricarbonyliron group of an organic moiety, which is highly unstable in the uncomplexed state. On irradiation, carbon monoxide is expelled and appreciable amounts of benzene are formed. At low temperatures quadricyclanone (**92**) was obtained in low yields:

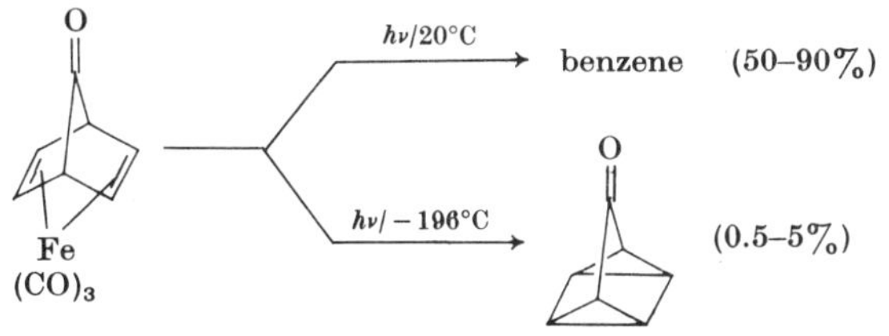

Other decarbonylation reactions of preparative interest include the formation of vinyliron compounds (406) and the preparation of h^5-cycloheptatrienyl tricarbonylmanganese (602) and of diazaallyl- and 2-azabutadienemanganese complexes (272, 308).

Carbonylation reactions are frequently connected with *skeletal rearrangements* and the interested reader is referred to work in this field especially to that by Aumann. Photolysis of Fe(CO)$_5$ in the presence of quadricyclane (**93**) results in skeletal rearrangement of the ligand as well as in CO insertion into a M—C bond (21). Although addition of the C-1—C-7 bond in **93** to $Fe(CO)_4$ followed by CO insertion is supposed to give **94**, the formation of **95** and **96** is explained by scission of the C-4—C-5 and C-5—C-6 bonds:

$$\text{(93)} \quad + Fe(CO)_5 \xrightarrow{h\nu} \text{(94)} \quad + \quad \text{(95)} \quad + \quad \text{(96)}$$

The same author (*19*) described the light-induced reaction of homo-semibullvalene (**97**) with $Fe(CO)_5$:

$$\text{(97)} \quad + Fe(CO)_5 \xrightarrow{h\nu}$$

Further examples for this type of reaction are the photochemical transformations of substituted vinylcyclopropanes, cyclopropylacetylenes, and methylenecyclopropanes in the presence of $Fe(CO)_5$ (*579, 576, 603*).

One of the most important functions of transition metals in organometallic catalysis is that of template for the *formation of C—C bonds* between coordinated olefins, dienes, etc. As pointed out in Section IV, A, photochemistry is valuable in investigating the mechanism of such reactions at low temperatures.

A well-known example of C—C bond formation in a thermal reaction is the production of the ferracyclopentane derivative (**98**). Recently, the photochemical reaction of $Fe(CO)_5$ and $Fe_3(CO)_{12}$ with C_2F_4 has been reported, providing a simple route to **98** (*188*):

$$\left.\begin{array}{l} Fe(CO)_5 \\ \\ Fe_3(CO)_{12} \end{array}\right\} \xrightarrow{h\nu/C_2F_4} \quad \text{(98)}$$

An intermediate $(C_2F_4)Fe(CO)_4$ complex that reacts readily with C_2F_4 to give **98** can be isolated (*189*). Olefins $CF_2{=}CFCl$, $CF_2{=}CFBr$, $CF_3CF{=}CF_2$, and $CF_3CH{=}CF_2$ and cyclic fluoroolefins failed to give ferracyclopentane derivatives.

As already mentioned in Section IV, A, 10, bis(butadiene)mono carbonyliron is a quite efficient catalyst for the mixed oligomerization of dienes and acrylate in the presence of PPh_3 leading to heptadienoic esters **(99)** (*92*):

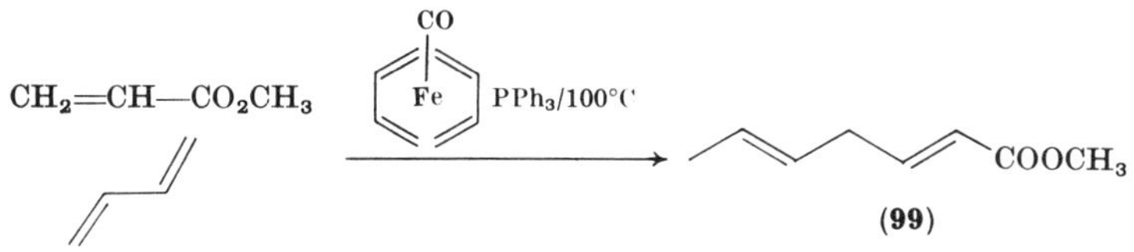

Some insight into the mechanism of this process has been provided by the photochemical reaction of butadienetricarbonyliron and acrylate, leading to **100** (*235*):

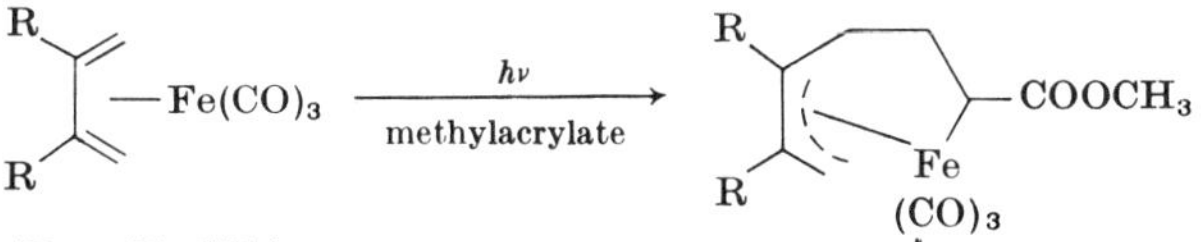

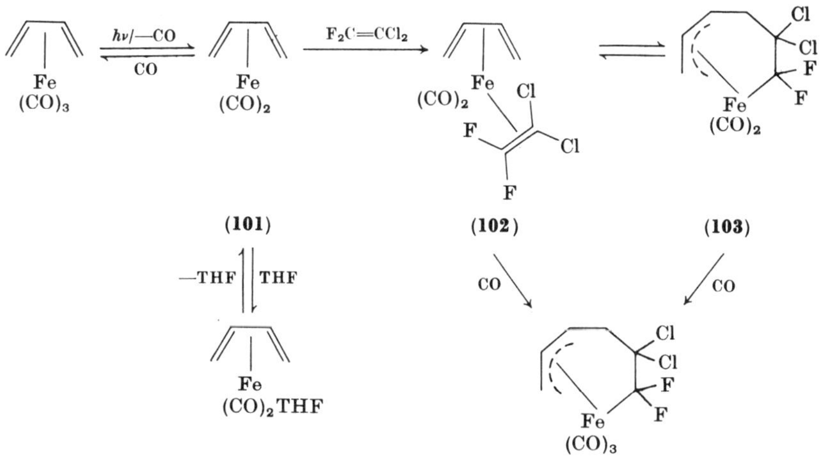

A similar reaction has been carried out using 1,1-dichloro-2,2-difluoro-ethylene (*302*). In the latter case it has been demonstrated by low-temperature irradiation of butadienetricarbonyliron in THF and subsequent addition of $CCl_2{=}CF_2$ in the dark that the carbon–carbon bond formation involves intermediate formation of **101** and **102**. It is still an open question whether C—C bond formation is triggered by the attack of an additional ligand (in this case of CO) following the principles of a S_N2 reaction or whether it occurs via the intermediate **103** (*325*).

Similar cases of C—C bond formation between fluoroolefins (or fluoroketones) with diene-, cyclobutadiene-, and related carbonyliron complexes have been reported by Green *et al.* (*63, 60, 62, 61, 233*; see also Ref. *98*).

The *photodimerization of norbornene* with Cu halides as catalysts is a well-known reaction (*561*). Recently, the dimerization has been studied in the presence of copper(I)triflate (triflate = trifluoromethanesulfonate) (*483, 484*). This catalyst shows several advantages over Cu halides: The yields of dimers (**106**) were improved significantly and the complex formed from Cu triflate and olefin is thermally stable and soluble in various organic solvents. Under these conditions the mechanism of photodimerization could be investigated in a wide range of olefin concentrations. The proposed mechanism for dimer formation is shown in Eqs. (38)–(40). Contrary to earlier proposals (*561*), the photochemical conversion of the bis(norbornene) complex (**104**) was found to be the key step on the way to **106**.

> The mechanistic details of the collapse of the photoexcited species are not provided by these studies. It is possible that the copper ion in (**104**) facilitates the required absorption of UV light by an otherwise weakly absorbing olefin, and may merely act as a template which promotes a concerted orbital-symmetry allowed photochemical $[2_{\pi s} + 2_{\pi s}]$ cycloaddition.
>
> The metal may alternatively participate in a stepwise process in which a σ-bonded intermediate is formed in a light-induced oxidative metallocycloaddition. In this process, the Cu(I)–bisolefin complex undergoes oxidation to a dialkylcopper intermediate (**105**), which collapses to product by reductive elimination (*484*).

A speculative suggestion for the intermediacy of (**105c**) dates back to 1969 (*332*). Copper triflate serves also as catalyst for the dimerization of cyclopentene, cyclohexene, cycloheptene (*485*), and mixed photodimerization of norbornene and cyclooctene (*484*).

Another important application of Cu_2Cl_2 has been in the improved photochemical synthesis of *trans*-cyclooctene from *cis*-cyclooctene (*164*).

Hardly any ligand in organometallic chemistry has attracted more interest then molecular nitrogen. Although there are numerous complexes containing N_2, successful attempts to reduce N_2 or to incorporate it into organic compounds have been rather scarce. This problem is far from solved; suggestions for a photochemical approach have been put forward (*203*).

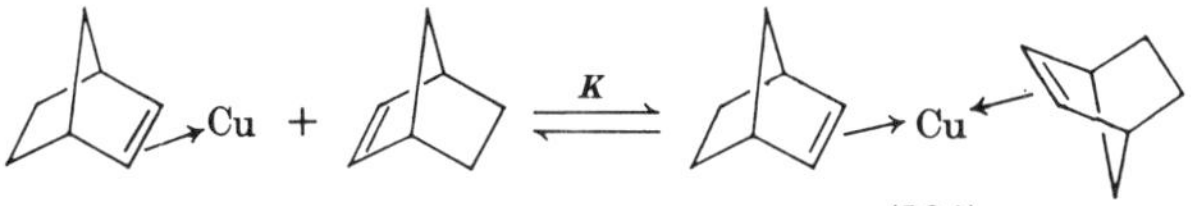

(38)

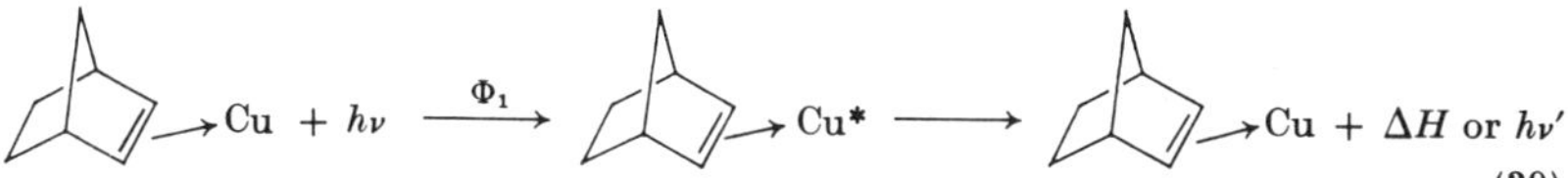

(104)

(39)

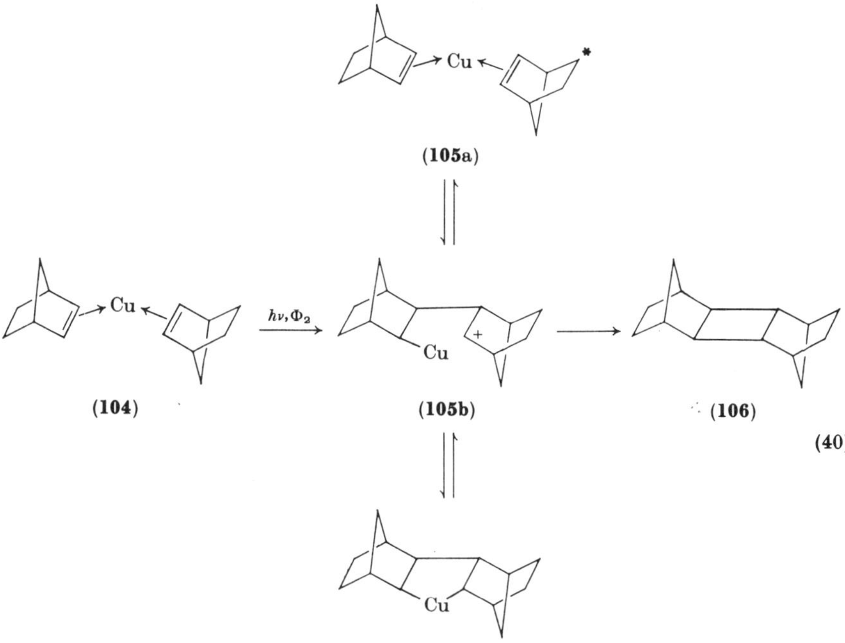

(105a)

(104)   (105b)   (106)

(40)

(105c)

Photochemical substitution of CO *by* N_2 (e.g., in arene–tricarbonyl-chromium) has been accomplished by Sellmann *et al.* (*529, 530*).

The rhenium N_2 complex (**109**) was not obtained by substitution of CO by N_2 but by introduction of hydrazine in **107** to give **108** and subsequent oxidation of **108** with hydrogen peroxide (*526*):

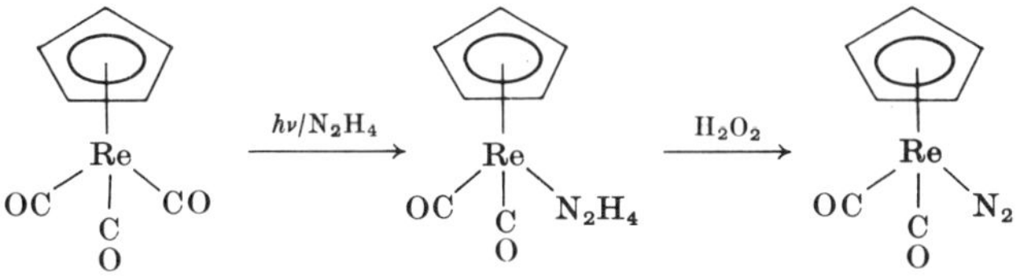

(107)   (108)   (109)

Photochemically produced hydrazine complexes have also been the starting materials for diimine complexes (*527, 528*).

The first reports on the photochemical reactions of complexes containing molecular nitrogen as a ligand, have been published. Irradiation of Re(I)–, Os(II)–, Fe(II)–, and Mo(0)–N_2 complexes in the presence of CO leads to N_2 expulsion and formation of metal carbonyl complexes (*146*).

The primary process in the photolysis of $FeH_2N_2(PEtPh_2)_3$ is the cleavage of the metal–dinitrogen bond (*329*). The resulting coordinatively unsaturated intermediate hydrogenates olefins and dienes. Photolysis in the presence of dimethylbutadiene yields, in addition to 2,3-dimethyl-2-butene, the diene–phosphineiron complex (**110**), in which the phosphine is coordinated by the 6π-electron system of a phenyl ring (and not by the phosphorus):

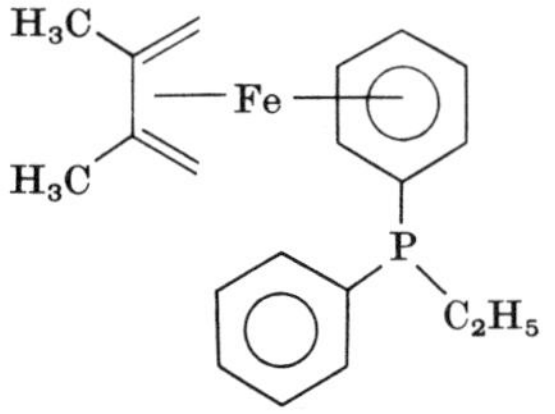

(**110**)

C. TABULAR SURVEY OF PHOTOCHEMICAL SUBSTITUTION REACTIONS OF METAL CARBONYL COMPOUNDS WITH n AND π DONORS AND OF PHOTOCHEMICAL FORMATION OF METAL–METAL BONDS

Reactions of preparative interest have been compiled in Tables IX–XI. Most of the information discussed in the pertinent text has also been included; however, the reader is advised to check also the corresponding sections for additional information.

TABLE IX

PHOTOCHEMICAL SUBSTITUTION REACTIONS OF METAL CARBONYL COMPOUNDS WITH n DONORS [a,b,c]

I. Coordination via N

Ammonia: $V(CO)_6{}^-$ (*462*)

Amines: $Mo(CO)_6$ (*565*); $Cr(CO)_5PPh_3$ (*523*); $Mo(CO)_5PPh_3$ (*523*); $W(CO)_5PPh_3$ (*523*)

Hydrazine: (π-cp)$Re(CO)_3$ (*526*)

Nitriles: cycloheptatriene $Cr(CO)_3$ (*12*); $Fe(CO)_5$ (*550*)

Table IX—*Continued*

Hydroxylamine
$Cr(CO)_6$
$Mo(CO)_6$ only degradation (89)
$W(CO)_6$

Thiourea derivatives: $[(\pi\text{-cp})Mo(CO)_3]_2$ (566); $Mo(CO)_6$ (566)

Azomethines: $C_5H_5V(CO)_4$ (88); $Cr(CO)_6$ (89, 589); $Mo(CO)_6$ (89, 589); $W(CO)_6$ (89)

Pyridine derivatives: $Cr(CO)_6$ (157, 459*); cycloheptatriene$Cr(CO)_3$ (12); $Mo(CO)_6$ (157, 565, 459*); $(\pi\text{-cp})Mo(CO)_3Cl$ (88); $Mo(CO)_5PPh_3$ (523); $W(CO)_6$ (157, 459*); $(\pi\text{-cp})W(CO)_3Cl$ (88); $W(CO)_5PPh_3$ (523); $C_6F_5Mn(CO)_5$ (419); Ru(carbonyl-octaethylporphyrin)pyridinate (265, 538)

Imidazole derivatives (L = 1,3-dimethyl-4-imidazoline-2-ylidene): $Cr(CO)_6$ (210, 40); $LCr(CO)_5$ (415); $Mo(CO)_6$ (40); $LMo(CO)_5$ (415); $W(CO)_6$ (210, 40); $LW(CO)_5$ (415)

Pyrazole derivatives: $Cr(CO)_6$ (210); $W(CO)_6$ (210); $Fe(CO)_5$ (317)

Tetrazole derivatives: $Cr(CO)_6$ (210); $W(CO)_6$ (210)

Thiazoles, isothiazoles, selenazoles: $Cr(CO)_6$ (210, 601, 40); $Mo(CO)_6$ (40, 601); $W(CO)_6$ (210, 601, 40)

Oxazoles: $Cr(CO)_6$ (210, 601, 40); $Mo(CO)_6$ (601, 40); $W(CO)_6$ (601, 40)

SCN^-: $Cr(CO)_6$ (478); $W(CO)_6$ (478); $Fe(CO)_5$ (478)

Other N- ligands

2,3-Diazobicyclo[2.2.1]heptene-2: $Cr(CO)_6$ (258, 256); $(\pi\text{-cp})Mn(CO)_3$ (256); trimethylbenzene–$Cr(CO)_3$ (258, 256)

$C_5H_5(CO)_2Ru\!-\!C_6H_4\!-\!N\!=\!N\!-\!C_6H_5$ (intramolecular: (86)

$(CO)_3CrC_6H_5\!-\!NHR$ (R = H, CH_3):$M(CO)_6$ (M = Cr, Mo, W): (371)

Benzocinnoline or phenanthridine
$Cr(CO)_6$
$Mo(CO)_6$ (323)
$W(CO)_6$

$Ph_2C\!=\!NLi$
$(\pi\text{-cp})Mo(CO)_3Cl$
$(\pi\text{-cp})W(CO)_3Cl$ (295)

(*continued*)

Table IX—*Continued*

II. Coordination via P

Aromatic Phosphines (monodentate)

$$\left.\begin{array}{l} V(CO)_6{}^- \\ Nb(CO)_6{}^- \\ Ta(CO)_6{}^- \end{array}\right\}(151)$$

Norbornadiene–Cr(CO)$_4$ (438); cycloheptatriene–Cr(CO)$_3$ (12); (C$_6$H$_5$CO$_2$CH$_3$) Cr(CO)$_2$L (L = CO, CS) (279, 277, 278); [C$_6$H$_4$(CO$_2$CH$_3$)$_2$]Cr(CO)$_3$ (551); (C$_6$H$_5$CO$_2$–menthyl)Cr(CO)$_3$ (278); (C$_6$H$_5$CO$_2$H)Cr(CO)$_3$ (278); (t-Bu— C$_6$H$_4$CO$_2$CH$_3$)Cr(CO)$_3$ (278); (t-Bu—C$_6$H$_4$CO$_2$–menthyl)Cr(CO)$_2$ (278); (CH$_3$C$_6$H$_5$)Cr(CO)$_3$ (370); (Et$_2$N—C$_6$H$_5$)Cr(CO)$_3$ (370); (CO)$_5$Cr[CR′R″] (R′ = OCH$_3$, OC$_2$H$_5$; R″ = CH$_3$, C$_2$H$_5$) (195); Cr(CO)$_5$PPh$_3$ (523); Mo(CO)$_6$ (38*); CF$_3$Mo(CO)$_3$(π-cp) (310); Mo(CO)$_5$PPh$_3$ (523); Mo(CO)$_5$NHC$_5$H$_{10}$ (523); W(CO)$_6$ (83); W(CO)$_5$PPh$_3$ (523); W(CO)$_5$NC$_5$H$_5$ (523); [Ph$_3$PNPPh$_3$] [CF$_3$CO$_2$W(CO)$_5$] (493); (CO)$_5$W[CR′R″] (195); (π-RC$_5$H$_4$)Mn(CO)$_3$ (R = C$_2$H$_5$, Ph—CH$_2$, Cl, Br, I, COOCH$_3$, COOH, COCH$_3$) (356, 304); indenyl– and fluorenyl–Mn(CO)$_3$ (305); pyrrolyl–Mn(CO)$_3$ (306); (π-cp)Mn(CO)$_2$R (R = CO, CS) (87*, 129*); Mn$_2$(CO)$_{10}$ (383, 357*); Mn (CO)$_5$Br (385*); Re$_2$(CO)$_{10}$ (384*, 535*); Fe(CO)$_5$ (497, 550, 121); R—Fe(π-cp)(CO)$_2$ (R = CF$_3$, CF$_3$CO, C$_2$F$_5$, (CF$_3$)$_2$CF) (310); [Fe(CO)$_3$SC$_6$H$_5$]$_2$ (156); (π-cp)Fe(CO)$_2$CH$_2$OCH$_3$ (152); (π-cp)Fe(CO)$_2$SnR$_3$ (139, 142); (π-cp)Fe(CO)$_2$SiR$_3$ (123); (π-cp)Fe(CO)$_2$—σ— C$_5$H$_4$—Fe(π-cp) (403); (π-cp)Fe(CO)$_2$(C$_6$H$_4$F) (397, 408); Ru$_3$(CO)$_{12}$ (286)

Aliphatic phosphines (monodentate): V(CO)$_6{}^-$ (151); (π-cp)V(CO)$_4$ (199); Cr(CO)$_6$ (377*, 376*, 551, 511, 515, 517); [C$_6$H$_4$(CO$_2$CH$_3$)$_2$]Cr(CO)$_3$ (551); mesitylene– Cr(CO)$_3$ (551); hexamethylbenzene–Cr(CO)$_3$ (551); benzene Cr(CO)$_3$ (551); Mo(CO)$_6$ (377*, 376*, 551, 511, 515, 517); W(CO)$_6$ (377*, 376*, 551, 511, 515, 517); (π-R$_5$C$_5$)Mn(CO)$_3$ (R = CH$_3$, H) (516, 517); Fe(CO)$_5$ (497, 550*, 517, 518, 121); [2-5-η-(5-phenylpenta-2,4-dienal)]–Fe(CO)$_3$ (75)

Aromatic phosphines (bidentate)

$$\left.\begin{array}{l} V(CO)_6{}^- \\ Nb(CO)_6{}^- \\ Ta(CO)_6{}^- \end{array}\right\}(151)$$

Cr(CO)$_6$ (487); Mo(CO)$_6$ (487); CF$_3$Mo(CO)$_3$(π-cp) (310); R—Fe(π-cp)(CO)$_2$ [R = C$_2$F$_5$, (CF$_3$)$_2$CF] (310); (π-cp)$_2$Fe$_2$(CO)$_4$ (108, 109); [(π-cp)Fe(CO)$_2$(OCMe$_2$)]$^+$ (82); W(CO)$_6$ (487); Mn$_2$(CO)$_{10}$ (464, 135); Mn(CO)$_5$Br (464); (π-cp)Mn(CO)$_3$ (228); Re$_2$(CO)$_{10}$ (135); Fe(CO)$_5$ (138, 139, 121); Fe$_3$(CO)$_{12}$ (138); [Fe(CO)$_3$- SC$_6$H$_5$]$_2$ (156); (π-cp)Fe(CO)$_2$SnR$_3$ (139); Ru$_3$(CO)$_{12}$ (137)

Aliphatic phosphines (bidentate): Cr(CO)$_6$ (73, 46); Mo(CO)$_6$ (73); W(CO)$_6$ (73); Fe(CO)$_5$ (7)

Polydentate phosphines: Cr(CO)$_6$ (130); (π-cp)Mo(CO)$_3$Cl (309); W(CO)$_6$ (130); Mn$_2$(CO)$_{10}$ (182); (π-cp)Mn(CO)$_3$ (311, 313); (π-cp)Fe(CO)$_2$Br (309); Co$_2$(CO)$_8$ (431*)

Table IX—*Continued*

Phosphorus ylides: $W(CO)_6$ (*292, 539*); $Mo(CO)_6$ (*539*); $Cr(CO)_6$ (*539*)

Phosphites: $Cr(CO)_6$ (*271, 377, 376**); $(C_6H_6)Cr(CO)_3$ (*394*); mesitylene–$Cr(CO)_3$ (*551*); hexamethylbenzene–$Cr(CO)_3$ (*551*); $(C_6H_5—CO_2R)Cr(CO)_2L$ (R = H, CH_3; L = CO, CS) (*278, 276, 279*); $(CH_3—C_6H_4CO_2R)Cr(CO)_2L$ (R = CH_3, H; L = CO, CS) (*278, 276, 279*); cycloheptatriene–$Cr(CO)_3$ (*12*); $Mo(CO)_6$ (*271, 377*); $W(CO)_6$ (*271, 377, 376*, 83, 551*); (π-cp)$Mn(CO)_2R$ (R = CO, CS) (*129**); (π-$C_5H_3R^1R^2$)$Mn(CO)_3$ (R^1 = CH_3, R^2 = CO_2CH_3) (*366**); $CH_3CO—Mn(CO)_3$-$[P(OCH_3)_3]_2$ (*562*); $Fe(CO)_5$ (*497, 550*, 121*); $[Fe(CO)_3SC_6H_5]_2$ (*156**); $(C_4H_4)Fe(CO)_3$ (*201**)

PH_3: $Cr(CO)_6$ (*198**); $Mn_2(CO)_{10}$ (*197*); $ONCo(CO)_3$ (*482*)

PF_3 (and other PX_3; X = halogen): $V(CO)_6^-$ (*346**); $Cr(CO)_6$ (*390*, 551*); $(PF_3)_3Cr(CO)_3$ (*365**); $(PF_3)_3Mo(CO)_3$ (*365**); $(PF_3)_3W(CO)_3$ (*365**); (π-cp)Mn-$(CO)_3$ (*390*, 351, 497**); $Fe(CO)_5$ (*571*, 497**); butadiene–$Fe(CO)_3$ (*596**); cyclohexadiene–$Fe(CO)_3$ (*595**); trimethylenemethane–$Fe(CO)_3$ (*116**); $RCo(CO)_4$ (R = CF_3, C_2F_5, C_3F_7) (*570**)

$PX_{3-n}R_n$ (X = halogen): $Cr(CO)_6$ (*180, 540*, 377**); $Mo(CO)_6$ (*180, 540*, 377**); $W(CO)_6$ (*180, 540*, 377*, 213*); (π-cp)$Mn(CO)_3$ (*314**)

$P(NR_3)_3$: $Cr(CO)_6$ (*312**); $W(CO)_6$ (*312**)

$PR_{3-n}L_n$ (L = organometallic derivatives; n = 1–3): $Cr(CO)_6$ (*511*, 515, 517*); $Mo(CO)_6$ (*511*, 515, 517*); $W(CO)_6$ (*511*, 515, 517*); (π-C_5H_4R)$Mn(CO)_3$ (R = H, CH_3) (*516, 517*); $Fe(CO)_5$ (*518, 517*)

Triferrocenylphosphine: π-cp $Mn(CO)_3$ (*401*); $C_6H_6Cr(CO)_3$ (*401*)

III. Coordination via As, Sb, Bi

Arsines, stibines, bismuthines (monodentate): $V(CO)_6^-$ (*151*); $Cr(CO)_6$ (*510, 192, 551, 508, 514, 84*); $[C_6H_4(CO_2CH_3)_2]Cr(CO)_3$ (*551*); $Mo(CO)_6$ (*38*, 508, 514, 192, 84, 510*); $W(CO)_6$ (*510, 84, 192, 551, 508, 514, 83*); norbornadiene–$W(CO)_4$ (*312*); fluorenyl– and indenyl–$Mn(CO)_3$ (*305*); pyrrolyl–$Mn(CO)_3$ (*306*); $Mn_2(CO)_{10}$ (*357**); (π-cp)$Mn(CO)_2CS$ (*129*); $Re_2(CO)_{10}$ (*535**); $Fe(CO)_5$ (*550, 192, 139, 121*); (π-cp)$Fe(CO)_2SnR_3$ (*139, 142*); $Fe_3(CO)_{12}$ (*112*)

Arsines (bidentate): $Cr(CO)_6$ (*487**); $Mo(CO)_6$ (*487**); $W(CO)_6$ (*487**); (π-C_5H_4R)-$Mn(CO)_3$ (R = H, CH_3) (*486*); $Mn_2(CO)_{10}$ (*135*); $Re_2(CO)_{10}$ (*135*)

$As(NR_3)_3$: $Cr(CO)_6$ (*312*); $W(CO)_6$ (*312*); norbornadiene–$W(CO)_4$ (*312*); $Fe(CO)_5$ (*139, 141*); $Fe_3(CO)_{12}$ (*112, 140*); $Ru_3(CO)_{12}$ (*137*)

Arsines and stibines with organometallic ligands: $Cr(CO)_6$ (*509, 510, 514*); $Mo(CO)_6$ (*509, 510, 514*); $W(CO)_6$ (*509, 510, 514*); (π-cp)$Mn(CO)_3$ (*514*); $Fe(CO)_5$ (*514*)

(*continued*)

Table IX—*Continued*

$AsR_{3-n}X_n$ (X = halogen): $Cr(CO)_6$ (*180, 551*); $Mo(CO)_6$ (*180*); $W(CO)_6$ (*180*)

Mixed phosphine–arsine ligands (coordination via P and As): $Fe(CO)_5$ (*112*); $Fe_2(CO)_9$ (*112*)

IV. Coordination via O

Alcohols; ketones: $W(CO)_6$ (*83, 620*)

Ethers: $V(CO)_6^-$ (*71*); $W(CO)_6$ (*83*)

Other oxygen compounds: ethyl acetate + $Mo(CO)_6$ (*33*); benzisoxazole + $W(CO)_6$ (*601*)

V. Coordination via S, Se, Te

Thioethers, Thiophenols: $Cr(CO)_6$ (*519*); $Mo(CO)_6$ (*519*); $W(CO)_6$ (*519, 83*)

$R_3P{=}S$ and $R_3As{=}S$: $Cr(CO)_6$ (*2, 3, 369*); $Mo(CO)_6$ (*369*); $W(CO)_6$ (*2, 3*)

R—S—R′ (R = metalorganic ligand; R′ = R, alkyl, aryl): $(\pi\text{-cp})V(CO)_4$ (*179*); $Cr(CO)_6$ (*181, 519*); $Mo(CO)_6$ (*179, 519, 324*); $W(CO)_6$ (*179, 519*)

Other sulfur ligands: thiazolidine-2-thione + $M(CO)_6$ (M = Cr, Mo, W) (*162*); 2-methyl-4,5-benzothiazoline + $Cr(CO)_6$ (*601*); thioformaldehyde trimer: $Cr(CO)_6$ (*495*); $Mo(CO)_6$ (*495*); thiomorpholin-3-thione: $Cr(CO)_6$ (*163*); $Mo(CO)_6$ (*163*); $W(CO)_6$ (*163*); thiomorpholin-3-one: $Cr(CO)_6$ (*432**); $Mo(CO)_6$ (*432**); $W(CO)_6$ (*432**); thiocyclobutene: $Fe(CO)_5$ (*555, 556*); $[M(CO)_5NCS]^-$ (M = Cr,W): $Mo(CO)_6$ (*43*); $W(CO)_6$ (*43*); $(\pi\text{-cp})Ru(CO)_2SR$: $(\pi\text{-cp})Ru(CO)_2SR$ (*319*)

Selenides and tellurides
$Cr(CO)_6$
$Mo(CO)_6$ (*512, 513, 520*)
$W(CO)_6$

VI. Coordination via C

^{13}CO and $C^{18}O$: $(\pi\text{-cp})V(CO)_4$ (*99*); $Mo(CO)_5NHC_5H_{10}$ (*145*); $Mo(CO)_5PPh_3$ (*523*); $W(CO)_5NHC_5H_{10}$ (*523*); $W(CO)_5PPh_3$ (*523*); $Mn_2(CO)_{10}$ (*296, 65*); $Mn(CO)_5Br$ (*47*); $(\pi\text{-cp})Mn(CO)_3$ (*99*); $Fe(CO)_5$ (*296*, 411*, 66*); $Fe_3(CO)_{12}$ (*296**); butadiene–$Fe(CO)_3$ (*596**); $Co_2(CO)_8$ (*296**); $Co_4(CO)_{12}$ (*296**); $(\pi\text{-cp})Co(CO)_2$ (*99**)

CN^-: $V(CO)_6^-$ (*462*); $Cr(CO)_6$ (*478*); $W(CO)_6$ (*478*); $(\pi\text{-}(CH_3)_5C_5)Mn(CO)_3$ (*166*); $Fe(CO)_5$ (*478*)

Carbenes: $Fe(CO)_5$ (*193, 194*)

Table IX—*Continued*

VII. Coordination of Special Ligands

Ligand	Starting material	Ref.
NO	$(\pi\text{-cp})V(CO)_4$	*259*
	$Cr(CO)_6$	*259, 553*
	$Fe(CO)_5$	*259*
	$Mn_2(CO)_{10}$	*259, 260*
H_2	$Fe(CO)_3(PPh_3)_2$	*572*
	$Fe_3(CO)_{10}NSi(CH_3)_3$	*330*
	$Ru(CO)_3(PPh_3)_2$	*572*
	$Os(CO)_3(PPh_3)_2$	*572*
SiI_4	$Fe(CO)_5$	*494*
Carboranes,	$Cr(CO)_6$	*598, 42, 532*
boranes,	$Mo(CO)_6$	*598, 42, 532, 175*
phosphacarboranes,	$W(CO)_6$	*598, 42, 532, 175*
azacarboranes	$Re(CO)_5Br$	*209*
	$Fe(CO)_5$	*154, 207, 41*
	$(\pi\text{-cp})Fe(CO)_2I$	*209*
Thio- and seleno- ketocarbenes	$Fe(CO)_5$	*501*

[a] In Parts I–VI of this table, references are given in parentheses with compounds.

[b] $(\pi\text{-cp}) = [1\text{-}5\text{-}\eta\text{-cyclopentadienyl}]$.

[c] * indicates substitution by more than one ligand.

TABLE X

Photochemical Substitution Reactions of Metal Carbonyl Compounds
with π Donors[a,b]

Ligand	$M(CO)_n$	Ref.
I. Monoolefins: $M(CO)_{n-1}L$, $M(CO)_{n-2}L_2$ (L = olefin)		
Ethylene	$(\pi\text{-cp})_2MoCO$	*558*
	$Ph_3SnFe(CO)_2(\pi\text{-cp})$	*342*
	$Ru_3(CO)_{12}$	*289*
Tetrafluoroethylene (and other	$Fe(CO)_5, Fe_3(CO)_{12}$	*188, 190*
fluorinated olefins)	$Os(CO)_3[P(OCH_3)_3]_2$	*122*
Tetramethoxyethylene	$(\pi\text{-cp})Mn(CO)_3$	*254*
Cyclopentene	$(H_3C)_6C_6Cr(CO)_3$	*13*
Cyclooctene	$(\pi\text{-}C_6H_5CO_2CH_3)Cr(CO)_3$	*276, 278*
	Cycloheptatriene $Cr(CO)_3$	*12*
	$(\pi\text{-cp})Mn(CO)_2CS$	*186, 185*
	$(\pi\text{-cp})MnCO(CS)_2$	*186, 185*
2-Vinylaziridine	$Fe(CO)_5$	*21*

(*continued*)

Table X—*Continued*

Ligand	$M(CO)_n$	Ref.
2-Vinyloxirane	$Fe(CO)_5$	*21*
Vinylcyclopropane	$Fe(CO)_5$	*20*
Homosemibullvalene	$Fe(CO)_5$	*19*
Allylchloride	$Na[V(CO)_6]$	*496*
Allyltrifluoroacetate	$Mo(CO)_6$	*155*
	$W(CO)_6$	*155*
Maleic acid	$[(CH_3)_6C_6]Cr(CO)_3$	*13*
Fumaric acid	$[(CH_3)_6C_6]Cr(CO)_3$	*13*
Dimethylmaleate	$(C_4H_4)Fe(CO)_3$	*461*
Dimethylfumarate	$(C_4H_4)Fe(CO)_3$	*594*
Fumaric acid dinitrile	$Cr(CO)_6$	*192*
	$Mo(CO)_6$	*192*
	$W(CO)_6$	*192*
Maleimide	$(C_6H_6)Cr(CO)_3$	*394*
Maleic anhydride	$[(CH_3)_6C_6]Cr(CO)_3$	*13*
	$(\pi\text{-cp})Mn(CO)_3$	*257*
	Substituted $ArCr(CO)_3$	*255, 257*
Acrylonitrile	$(\pi\text{-cp})_2MoCO$	*558*
Acrylic acid	$Fe(CO)_5$	*121*
Methylacrylate	$Fe(CO)_5$	*237*
Endic anhydride	$[(CH_3)_6C_6]Cr(CO)_3$	*13*
Citraconic anhydride	$[(CH_3)_6C_6]Cr(CO)_3$	*13*
trans-Benzalacetophenone	$Fe(CO)_5$	*333*
(3-Butenyl)diphenylphosphine	$Mo(CO)_6$	*211*
	$Mo(CO)_4$norbornadiene	*211*
	$Mo(CO)_4Cl_2$	*211*
	$Mo(CO)_3$cycloheptatriene	*211*
$\begin{array}{c} (CF_2)_n \\ \mid \;\; \mid \\ Ph_2P\text{—}C{=}C\text{—}PPh_2 \end{array}$ $(n = 2\text{–}4)$	$Ru_3(CO)_{12}$	*137*
Thiacyclobutene	$Fe(CO)_5$	*555, 556*
1-Methyl-2R-1,2-dihydropyridazine-3,6-dione	$Fe(CO)_5$	*407*
$(\pi\text{-cp})Fe(CO)_2\text{—}CH_2CR_2\text{—}$ $CH{=}CH_2$	Intramolecular	*234*

II. Dienes: $LM(CO)_{n-2}$, etc. (L = diene)

1,3-Butadiene	$Fe(CO)_5^*$	*326, 327, 340*
	$R_3MFe(CO)_2(\pi\text{-cp})$ (M = Sn,	
	Pb, Ge; R = CH_3, Ph)	*399, 398*
	$(\pi\text{-cp})Mn(CO)_3$	*196*
Isoprene	$Fe(CO)_5^*$	*327, 326*
1,3-Pentadiene	$[C_7H_9Fe(CO)_3]BF_4$	*16*
***trans,trans*-2,4-Hexadiene**	$Cr(CO)_6$	*328*
	$Fe(CO)_5^*$	*92*
Dimethylbutadiene	$Fe(CO)_5^*$	*326*

Table X—*Continued*

Ligand	$M(CO)_n$	Ref.
	$FeH_2N_2(PEt Ph_2)_3$	*329*
Cyclopentadiene	$(\pi\text{-cp})Mn(CO)_3$	*39*
1,3-Cyclohexadiene	$Fe(CO)_5^*$	*327, 326*
	$[C_6H_7Fe(CO)_3]BF_4$	*16*
	$[C_7H_9Fe(CO)_3]BF_4$	*16, 288*
	$[C_8H_{11}Fe(CO)_3]BF_4$	*10*
	$(\pi\text{-cp})Mn(CO)_3$	*196*
1,3-Cyclohexadiene	$[CH_3OC_6H_6Fe(CO)_3]BF_4$	*16*
1,3-Cycloheptadiene	$[C_6H_7Fe(CO)_3]BF_4$	*16*
	$[C_7H_9Fe(CO)_3]BF_4$	*16*
1,5-Cyclooctadiene	$Cr(CO)_6$	*315*
	$Fe(CO)_5$	*336, 161*
	$Os_3(CO)_{12}$	*127*
1,3-Cyclooctadiene	$Fe(CO)_5$	*336, 161*
	$Cr(CO)_6$	*315*
	$Ru_3(CO)_{12}$	*127*
	$Os_3(CO)_{12}$	*127*
Bicyclo[5.1.0]-2,4-octadiene	$Fe(CO)_5$	*77*
syn-7-Acetoxybenzonorborn-adiene	$[(H_3C)_5C_5]Mn(CO)_3$	*600*
Sorbic acid (methyl ester)	$Fe(CO)_5^*$	*326, 92*
Muconic esters	$Fe(CO)_5^*$	*92*
Dihydroacetophenone	$Fe(CO)_5$	*274*
Dihydrobenzaldehyde	$Fe(CO)_5$	*274*
α-Pyrone	$(\pi\text{-cp})Co(CO)_2$	*473*
Thiophene-1,1-dioxide	$Fe(CO)_5$	*114*
Alkoxydivinylborane–$Fe(CO)_4$	Intramolecular	*261*

III. Cyclic trienes and tetraenes

Ligand	Starting material	Ref.
Cycloheptatriene	$C_4H_4Fe(CO)_3$	*594*
	$(\pi\text{-}C_5H_4R)\text{—}Mn(CO)_3$ (R = H, CH_3)	*423*
1,3,5-Cyclooctatriene	$(\pi\text{-cp})Mn(CO)_3$	*423*
cis-Bicyclo[6.1.0]-nonatriene	$Fe(CO)_5$	*458, 498*
Bicyclo[6.2.0]-1,3,5,-decatriene	$Fe(CO)_5$	*498*
Cyclooctatetraene	$(\pi\text{-cp})Mn(CO)_3$	*423*
	$Os_3(CO)_{12}$	*85*
N-Carbethoxyazepine	$C_4H_4Fe(CO)_3$	*594*

IV. Acetylenes

Ligand		Ref.
Cyclooctyne	$Mo(CO)_6$	*343*
	$Fe(CO)_5$	*344*

(*continued*)

Table X—*Continued*

Ligand	Starting material	Ref.
R—C≡C—R (R = H, CH_3, Ph)	$(\pi\text{-cp})_2MoCO$	*558*
(R = Ph)	$(\pi\text{-cp})W(CO)_3Ph$	*56*
	$[(\pi\text{-cp})Fe(CO)_2]_2$	*400*
	Pyrrolyl–$Mn(CO)_3$	*306*
Dimethylacetylenedicarboxylate	$C_4H_4Fe(CO)_3$	*594*

V. Aromatic compounds

Substituted styrenes	$Fe(CO)_5$	*581, 577, 578, 580*
Benzocyclobutadiene–$Fe(CO)_3$	$Fe(CO)_5$	*575*
Hexaalkyborazole	$Cr(CO)_6$	*160*
Benzene	$(\pi\text{-cp})Mn(CO)_3$	*423, 196*

VI. Heterodienes and aromatic heterocycles

Pinocarvone	$Fe(CO)_5$	*333*
Pulegone	$Fe(CO)_5$	*333*
Isobutylidene–Meldrums acid–$Fe(CO)_4$	Intramolecular	*338*
β-Substituted vinylketones	$Fe(CO)_5$	*481, 474, 405*
R—CH=CHCOFe(CO)$_2$(π-cp)	Intramolecular	*406*
Ph_3P=CH—CR=CH_2	$Mo(CO)_6$	*539*
2,4,6-Triphenylphosphorine	$Cr(CO)_6$ } $Mo(CO)_6$ } $W(CO)_6$ }	*158, 159*

[a] $(\pi\text{-cp}) = [1\text{-}5\text{-}\eta\text{-cyclopentadienyl}]$.
[b] * indicates substitution by more than one ligand.

TABLE XI
PHOTOCHEMICAL FORMATION OF METAL–METAL BONDS[a]

Starting material	Product	Ref.
I. M—M bonds		
$Cr(CO)_5PPh_2H$	$Cr_2(CO)_8(PPh_2)_2$	*564*
$(CH_3)_5C_5Mo(CO)_3CH_3$	$[(CH_3)_5C_5Mo(CO)_2]_2$	*307*
$Mo(CO)_5PPh_2H$	$Mo_2(CO)_8(PPh_2)_2$	*564*
$(\pi\text{-cp})Mo(CO)_3Cl$	$[(\pi\text{-cp})Mo(CO)_2Cl]_2$	*5*
$(\pi\text{-cp})Mo(CO)_2PR_3Cl$	$[(\pi\text{-cp})Mo(CO)_2Cl]_2$	*6*
$(CH_3)_2(CH_2$=CH)GeMo(π-cp)(CO)$_2$	$[(\pi\text{-cp})Mo(CO)_2]_2$	*286*
$W(CO)_5PPh_2H$	$W_2(CO)_8(PPh_2)_2$	*564*
$(\pi\text{-cp})W(CO)_3Cl$	$[(\pi\text{-cp})W(CO)_2Cl]_2$	*5*
$(\pi\text{-cp})W(CO)_2PR_3Cl$	$[(\pi\text{-cp})W(CO)_2Cl]_2$	*6*
$(CH_3)_2ClGeMn(CO)_5$	$[(CH_3)_2GeMn(CO)_4]_2$	*144, 286*
$HGePh_2Mn(CO)_5$	$[Ph_2GeMn(CO)_4]_2$	*120*

Table XI—*Continued*

Starting material	Product	Ref.
$(\pi\text{-cp})Re(CO)_3$	$(\pi\text{-cp})_2Re_2(CO)_5$	*206*
$[Fe(CO)_4PPh_2]Li$	$Fe_2(CO)_6(PPh_2)_2$	*564*
$[(\pi\text{-cp})Fe(CO)_2Cl]$	$[(\pi\text{-cp})Fe(CO)_2]_2$	*5*
$[(\pi\text{-cp})FeCOP(OPh)_3]Br$	$[(\pi\text{-cp})Fe(CO)_2]_2$	*6*
$(CH_3)_2(CH_2{=}CH)GeFe(\pi\text{-cp})(CO)_2$	$[(\pi\text{-cp})FeCO]_2COGe(CH_3)_2$	*287, 285*
$(CH_3)_2ClGeFe(\pi\text{-cp})(CO)_2$	$[(CH_3)_2GeFe(\pi\text{-cp})CO]_2CO$	*144*
$Fe(CO)_4PR_2H$	$Fe_2(CO)_6(PR_2)_2$	*563*
$(Cot)Fe(CO)_3$	$(Cot)Fe_2(CO)_5$	*521*
$C_4H_4Fe(CO)_3$	$(C_4H_4)_2Fe_2(CO)_3$	*201*
$(Ph_4C_4)Fe(CO)_3$	$(Ph_4C_4)_2Fe_2(CO)_3$	*391*
$[Ph_2(t\text{-but})_2C_4]Fe(CO)_3$	$[Ph_2(t\text{-but})_2C_4]_2Fe_2(CO)_3$	*391*
$Fe(CO)_5$ (prot. solvent)	$[HFe_3(CO)_{11}]^-$	*290*
$(\pi\text{-cp})Fe(CO)_2P(CF_3)_2$	$[(\pi\text{-cp})Fe(CO)P(CF_3)_2]_2$	*168, 167*
$(\pi\text{-cp})Fe(CO)_2P(O)(CF_3)_2$	$(\pi\text{-cp})_2Fe_3(CO)_2[OP(CF_3)_2]_4$	*168, 167*
$(CO)_4Fe{-}P(CH_3)_2{-}Fe(CO)_2(\pi\text{-cp})$	$(\pi\text{-cp})Fe_2(CO)_5P(CH_3)_2$	*176a*
$(CO)_4Fe{-}PPh_2{-}Fe(CO)_2(\pi\text{-cp})$	$(\pi\text{-cp})Fe_2(CO)_5PPh_2$	*244, 242, 245*
$(CH_3)_2ClCo(CO)_4$	$[(CH_3)_2GeCo(CO)_3]_2$	*144*
$(\pi\text{-cp})Co(CO)_2$	$(\pi\text{-cp})_2Co_2(CO)_3$	*587*
$[(CO)_4Co]_2GePh_2$	$[(CO)_7Co_2]GePh_2$	*118*
$Ru(CO)_5$	$Ru_3(CO)_{12}$	*101*
$(\pi\text{-cp})Ru(CO)_2SR$	$[(\pi\text{-cp})Ru(CO)SR]_2$	*319*
$[FeRu(\pi\text{-cp})_2(CO)_4PPh_2]B(C_6H_5)_4$	$[Ru(\pi\text{-cp})(CO)_2]_2$	*242*
$Os(CO)_5$	$Os_3(CO)_{12}$	*101*
	$Os_2(CO)_9$	*389*

II. Formation of M—M' bonds

M	M'	Ref.
Ge	$Cr(CO)_6$	*605*
	$Mo(CO)_6$	*605*
	$W(CO)_6$	*605*
	$(\pi\text{-cp})Mn(CO)_3$	*477*
	$Ru_3(CO)_{12}$	*320*
	$Os_3(CO)_{12}$	*320*
Sn	$(\pi\text{-cp})Mn(CO)_3$	*477*
Si	$Ru_3(CO)_{12}$	*76*
	$Os_3(CO)_{12}$	*76*
Mo	$(CO)_4Fe{-}As(CH_3)_2{-}ML_n$	*177*
W	—	*177*
Mn	—	*177*
Co	—	*177*

^{a} $(\pi\text{-cp})$ = [1-5-η-cyclopentadienyl].

^{b} Cot = Cyclooctatetraene.

V. Biological Applications of Organometallic Photochemistry

The light reaction of photosynthesis in green plants and bacteria appears to be the most sophisticated and efficient application of some of the basic photochemical principles discussed in the previous sections. The turnover of material in photosynthesis has been estimated to exceed the sum of all industrial processes by $\sim10^2$ (490). All our fossil fuels result from that source. An organometallic photochemist, interested in the conversion of "dependable" sun energy (250, 455) and its storage in the form of chemical compounds, should, therefore, try to learn how nature makes photochemical use of metal-containing systems.

Here we focus attention on those parts of the photosynthetic light reaction involving transition metals and on a number of applications of photochemistry to other branches of biochemistry. The photochemical study of hemoglobin and vitamin B_{12} and its model compounds is important in understanding the function of these systems, although they do not act photochemically in nature. By discussing these examples, we hope to demonstrate the importance of the mechanistic principles outlined earlier, and to show, from the viewpoint of an organometallic chemist, the role they play in biological systems.

A. DEACTIVATION OF EXCITED STATES

In Section III we briefly discussed the effect of transition metals on the lifetime and intersystem crossing rates of systems. The significance of deactivation processes (leading to stability toward light) induced by incorporation of transition metals into bio-organic molecules is demonstrated by the following two examples.

The importance of the fact that chlorophyll, but not hemoglobin, undergoes certain photochemical reactions becomes obvious when the resistance of our blood toward light is considered. Incorporation of iron instead of magnesium into the porphyrin framework must result in significantly accelerated energy dissipation of electronically excited states.

This can be also deduced from a comparison of excited-state emission and lifetime data for some porphyrin complexes (40a). Magnesium etioporphyrin shows strong fluorescence as well as phosphorescence, the triplet lifetime being 160 msec. By contrast, iron(II) mesoporphyrin dimethyl ester does not fluoresce; it shows very weak phosphorescence and the triplet lifetime is < 0.5 msec. The metal-free mesoporphyrin dimethyl ester shows weak phosphorescence and its triplet state lifetime is 14 msec (40a). An explanation of these phenomena can be sought in low-lying d-d states of the iron system that accept the energy of porphyrin-localized excited states and dissipate it efficiently. (See Section IV, D for further discussion of such internal conversion phenomena.)

The relevance of iron for the behavior of our blood toward light becomes clear if it is no longer incorporated into the porphyrin in sufficient amounts. The photoexcited porphyrin is then a very efficient sensitizer for the production by energy transfer of oxygen in its reactive singlet state. Singlet oxygen attacks cholesterol in membranes of red blood cells and the resulting hydroperoxide (*491*) causes their hemolysis. The result is a disease known as erythropoietic protoporphyria (*170*). The most efficient cure is a large dose of β-carotene, which functions as a quencher.

Another case may illustrate how efficiently nature makes use of radiationless deactivation. The very light-resistant violet color of the primaries of the African Turaco bird is due to the incorporation of copper into a porphyrin system, which probably results in very efficient radiationless deactivation (*472*). Copper is widely used in dye chemistry to increase the light resistance of dyes (*632, 191*).

B. PHOTOSYNTHESIS

Energy transfer and electron transfer, which we have discussed in Section III, is of the utmost importance in photosynthesis (*455*). The photosynthetic apparatus of green plants consists of two different photosystems (PS I and PS II).† The PS are basically made up of several hundred chlorophyll (Chl) molecules, which absorb the light (antenna chlorophyll) and guide the electronic excitation to one reactive center (RC). Efficient energy transfer (by excitation interaction) in the antenna chlorophyll is thought to be arranged by lining up the chlorophyll molecules via coordination of the 9-keto group of ring V at one Chl (**111**) to the Mg of the next (*293*). The RC's, which are usually denoted by P (from pigment) with the appropriate indication of their electronic absorption maxima, contain special Chl complexes (*293*). In the case of PS I, the RC probably contains two Chl molecules. The absorption of the RC is shifted to the red, compared with that of the antenna Chl, and is denoted by P_{700}. Photosystem I is involved in the reduction of CO_2 by electron transfer via several "go-betweens," e.g., nicotinamide adenine dinucleotide phosphate (NADP); PS II oxidizes water to molecular oxygen. Photosystems I and II act

† Photosynthetic bacteria operate with a third and different system (*489*).

(R = CH$_3$: chlorophyll a or CHO: chlorophyll b)

(111)

in "tandem" as shown in Fig. 14 (*100*). The primary photoreaction of PS I and II can be described by the general equation

$$D\cdots\overset{..}{P}\cdots A \underset{k_{-1}}{\overset{h\nu/k_1}{\rightleftharpoons}} D\cdots\overset{.}{P}{}^+\cdots A^{(-)} \xrightarrow{k_2} D^+\cdots\overset{..}{P}\cdots A^- \tag{41}$$

where D is an electron donor and A is an acceptor. The most probable candidate for A in PS I is a bound form of ferredoxin (X in Fig. 14), since the appearance of the Chl-radical cation in P$_{700}$ (*293, 488*) is paralleled by the formation of a reduced form of ferredoxin (*184, 574, 294*). Cytochrome f and plastocyanin† have been discussed as candidates for D in PS I (*455, 574*). The situation appears less clear for PS II, but A may be a compound related to a carotenoprotein (Q = C$_{550}$ in Fig. 14). Cytochrome b seems to be involved as an electron donor in PS II, but not necessarily as the primary one. Evidence for the back-reaction [k_{-1} in Eq. (41)] has been obtained for PS II (*100*). So far as the complex question of the multiplicity of the excited state of Chl undergoing electron transfer is concerned, the reader is referred to Refs. *574, 294, 456*, and *380*.

The key role of iron-containing systems in electron transfer processes in photosynthesis can be summarized by a very crude simplification:

$$\text{Chl} + \text{Fe}_A{}^n \xrightarrow{h\nu} \text{Chl}^{.+} + \text{Fe}_A{}^{n-1}$$
$$\text{Chl}^{.+} + \text{Fe}_D{}^{n-1} \longrightarrow \text{Chl} + \text{Fe}_D{}^n$$
$$(n = \text{oxidation state})$$

† Plastocyanin is a copper-containing protein.

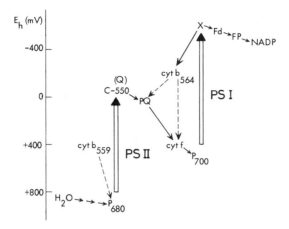

FIG. 14. Photosynthetic electron transport scheme for green plants. Dashed arrows indicate uncertain pathways. Fd = ferredoxin, FP = flavoprotein. [From Butler (*100*).]

In this context the reader should be reminded of the quenching process, involving reversible electron transfer with iron (*53*) (discussed in Section III, E, 4).

$$[Ru(dipy)_3^{2+}]^* + Fe^{3+} \longrightarrow [Ru(dipy)_3^{3+}] + Fe^{2+}$$
$$[Ru(dipy)_3^{3+}] + Fe^{2+} \longrightarrow [Ru(dipy)_3^{2+}] + Fe^{3+}$$

In general the large body of knowledge of quenching processes, involving inorganic (*28*) and organometallic compounds (*52, 226, 609*) should become valuable in understanding the intimate details of such processes in photosynthesis.

The iron in cytochromes is well "protected" by the porphyrin and protein (*171*) so that one wonders how electron transfer involving the iron occurs. It has been suggested (*171*) (e.g., for cytochrome c) that easily accessible portions of the porphyrin system function as the areas where electron transfer occurs and that by subsequent rapid internal transfer the iron reaches its new oxidation state:

$$Fe(II)\ cyt\ c \xrightarrow{\ -e^-\ } [Fe(II)\ cyt\ c]^{\cdot +} \xrightarrow[\text{transfer}]{\text{internal}} [Fe(III)\ cyt\ c]$$

A recent paper by Whitten *et al.* (*79*) on Ru-porphyrin complexes has demonstrated the influence of axial ligands in determining whether primary electron transfer involves the ring or the metal. We will return to this question in discussion of vitamin B_{12} photochemistry.

C. Hemoglobin and Related Systems

Direct absorption of light by porphyrin-containing transition metals does not play an important role in nature. However, the early observation by Haldane and Lorrain-Smith (246) of the light sensitivity of the hemoglobin–CO complex has led to detailed investigations of the photochemical behavior of a variety of hemoglobin complexes and related systems containing small molecules. These studies have led to a better understanding of details of the biological action of these systems.

The structure of hemoglobin (Hb) consists of two different pairs of subunits, each one containing a polypeptide chain and a heme (112) group. The chains are held together by weak interactions (hydrogen bonds, salt bridges, etc). Each heme group contains an iron atom åt the center of a fully conjugated porphyrin and is bound to an imidazole N of a histidine of the polypeptide chain (112a). The sixth coordination site of the iron which is unoccupied in deoxyhemoglobin, can be occupied by O_2, CO, or other ligands. The formal oxidation state of the iron is II, both in oxy- and in deoxy-Hb (15).

(112)   (112a)

$(X = O_2, CO, NO, CN^-, etc.)$

Myoglobin (Mb) is a monomeric protein, consisting of a single subunit, similar in properties and structure to an Hb subunit. Its behavior is correspondingly simpler than that of Hb.

An essential property of Hb is that it is saturated with oxygen in the lungs but releases much of its oxygen in the tissues. This shift of equilibrium must occur despite relatively small differences in the concentration of O_2 between lungs and tissues. It is achieved by the interaction between the heme groups on binding oxygen: The more oxygen already bound, the more likely other oxygen molecules are to bind, until the Hb is saturated.† The mechanism of this cooperative effect has been one of the chief problems of Hb research, and an important contribution to its understanding has been made by photochemistry.

The "bioinorganochemically" minded reader, who worked himself through the section dealing with metal–CO cleavage, will now be rewarded for his endurance. The readily reversible photochemistry of HbCO allows the study of the equilibrium between liganded and unliganded forms of Hb. Gibson showed that flash photolysis of Hb could be used to observe transient species in the Hb/HbCO equilibrium (222). However, the interpretation of these results is complicated by the splitting of Hb into dimers, etc. (224, 14). Nevertheless, the method does allow the study of conformational changes if a sufficiently short time scale is used. Alpert et al. (10) have followed spectroscopic changes after laser photodissociation of CO from Hb and Mb (rate constant, ca 10^7 sec^{-1}; activation energy, ca. 1.1 kcal mole^{-1}). The spectra are assigned to conformational changes in the tertiary and quaternary‡ (Hb only) structures of the molecules. The pH dependence of the rate constants has been correlated with cooperativity of the proteins.

The quantum yields for ligand dissociation from heme proteins were first measured in the pioneering work of Warburg (592, 593). From these and later measurements (94, 412), it was established that the quantum yield for CO dissociation from MbCO was 0.85–1.0 and for HbCO varied from 0.25 to 0.70 according to the conditions. The greater sensitivity of HbCO than MbCO to conditions is not surprising, since Mb exhibits neither the cooperativity nor the pH sensitivity (Bohr effect) of Hb. Other Hb complexes could also be photodissociated, but the much lower quantum yields for these reactions have not been explained (e.g., Hb—O_2:Φ = 0.008; Mb—O_2:Φ = 0.03) (223).

Quantum yields have been measured not only by conventional relaxation methods (593) but also by steady-state (90, 59) and pulse methods (91). Steady-state illumination reduces the affinity of Hb for CO, while leaving the cooperativity and Bohr effect unaffected (90). The observed kinetics have been interpreted by assuming that light influences ligand binding but not the protein-modulated interactions (554). The pulse method (91) has been used to establish that the quantum yield is independent of the fractional saturation with CO and of the molecular conformation.

The structural basis of the photochemical observations have been

† This behavior is reflected in a sigmoid O_2 binding curve (15).

‡ The folding of the individual subunits is called the tertiary structure; the relative arrangement of the subunits is the quaternary structure.

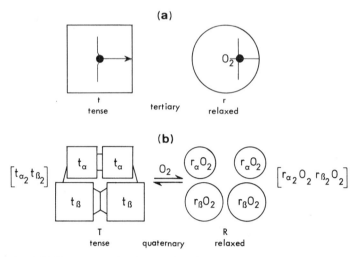

FIG. 15. (a) Diagrammatic sketch showing the relaxation of tension at the heme on going from tertiary deoxy to the oxy structure. (b) Concerted change of tertiary and quaternary structure on binding oxygen or CO to the ferrous sub-units. The clamp between the β subunits in the T structure represents 2,3-diphosphoglycerate; the other links represent salt bridges. [From Perutz (*426*).]

worked out by Perutz *et al.* (*427*, *426*). They have shown by X-ray diffraction that the structures of deoxy- and oxy-Hb (or HbCO) are significantly different. In deoxy-Hb, the iron atom lies 0.65 Å out of the plane of the porphyrin ring in the direction of the histidine, whereas in HbO_2 the iron atom is close to an in-plane arrangement. The Fe—N (histidine) distance probably is longer in Hb than in HbO_2. There are also important differences in the tertiary and quaternary structure. Heme–heme interaction has been shown to be linked to a change in tension at the heme–iron atom, brought about by a transition between the two alternative quaternary structures (*263a*, *426*, *425*) (Fig. 15).

D. VITAMIN B_{12} PHOTOCHEMISTRY

After this extensive discussion of the biological applications of the principles of M—CO bond cleavage, we would like to turn to the photochemistry of vitamin B_{12} and of some model compounds as an example that demonstrates the biological relevance of homolytic M–alkyl bond cleavage (as discussed in Section IV, A, 9).

The structure of vitamin B_{12} was solved by Hodgkin *et al.* (*136*) and is shown in Fig. 16. The central cobalt atom is surrounded by the

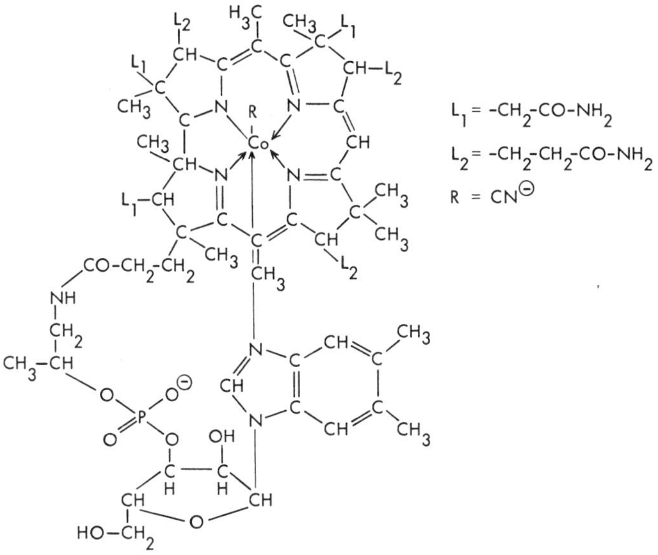

L1 = -CH2-CO-NH2
L2 = -CH2-CH2-CO-NH2
R = CN⊖

FIG. 16. Structure of vitamin B_{12}.

corrin ring, which is not fully conjugated. When the "lower" axial ligand is α-5,6-dimethylbenzimidazole nucleotide, the system is called *cobalamin*. The "upper" axial ligand R is variable and can be CH_3 (methylcobalamin), H_2O, OH^-, CN^- (vitamin B_{12}), 5′-deoxyadenosine, etc. The last-named compound is also called coenzyme B_{12}. The cobalt atom is normally in oxidation state III (B_{12a}) but can be reduced to oxidation state II (B_{12r}) or I (B_{12s}) (*78*):

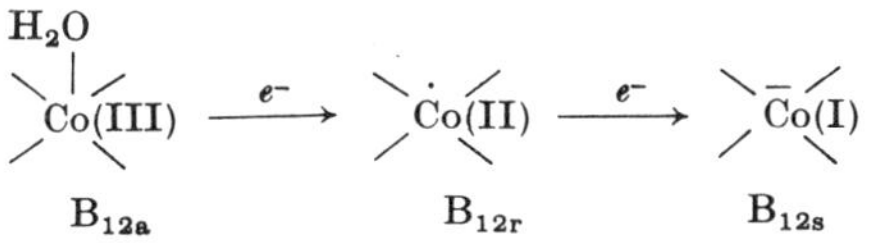

H2O
Co(III)
e⁻
Co(II)
e⁻
Co(I)
B12a
B12r
B12s

Whereas coenzyme B_{12} is a cofactor (i.e., acts in conjunction with an apoenzyme) for hydrogen-transfering enzymes, methylcobalamin is a cofactor for enzymes involved in the metabolism of one-carbon fragments. Detailed information is found in excellent reviews (*78*, *453*, *608*).

The benzimidazole side chain can be cleaved at the phosphate, leading to the cobinamide, in which H_2O normally functions as the lower axial ligand.

Methylcobalamin and related species show surprisingly high (*169*) thermal stability, but they can be readily cleaved photochemically to B_{12r} and alkyl radicals. The products of photolysis vary according to the presence or absence of O_2 (78). Aerobic photolysis of methylcobalamin results in the formation of formaldehyde and in photoaquation; in the absence of O_2, methane and ethane are generated:

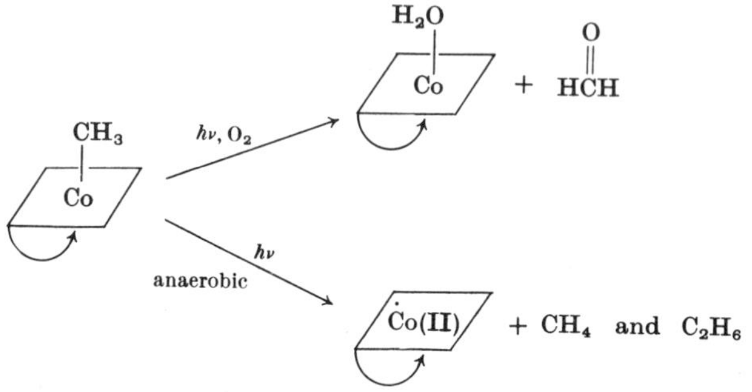

Photolysis of CD_3–cobalamin gave 89% C_2D_6 and 11% CH_3—CD_3, indicating that the methyl groups of methylcobalamin are the main source of the ethane formed (504). The ratio $1:2.2$ of methane to ethane in methylcobalamin photolysis can be affected by exchanging the lower axial ligand. Increase of electron density at cobalt by using CN^- as lower axial ligand results in the sole formation of methane. A mechanism involving CH_3^- formation has been proposed for this observation (504):

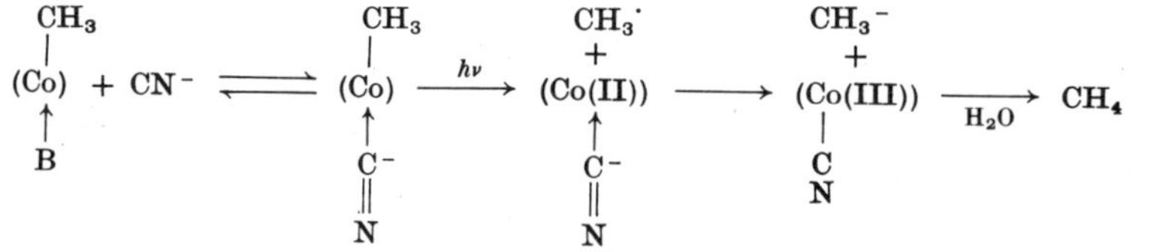

The quantum yields of methylcobalamin photolysis under oxygen are between 0.2 and 0.5, depending slightly on pH and the wavelength of incident light (452, 559). Quantum yields under anaerobic conditions are much smaller; this observation has been attributed to the fast recombination of B_{12r} and the alkyl radicals formed initially. Attempts to correlate the nature of the alkyl ligand with the quantum yield of the Co—R cleavage under anaerobic conditions have to be treated with some caution, owing to the sensitivity of quantum yields toward slight traces of oxygen (78).

A system with an exceptional dependence on oxygen is coenzyme B_{12}, from which 8,5'-cyclic adenosine is formed at low O_2 concentrations

(*364*), and adenosine-5'-aldehyde at higher O_2 concentrations (*264*). Further information on photochemistry of B_{12}-related systems is found in Refs. *364, 208, 588, 451, 452, 559, 64, 421, 78, 450*, and *70*.

It is generally agreed that the homolytic Co—R bond cleavage results from excitation to a R → Co CT state (note that ligand R is formally in the R^- state in the ground state):

$$Co(III)—R \xrightarrow{h\nu} Co^{\cdot}(II) + R^{\cdot}$$

The corresponding transition has been localized in cobaloximes (see below) in the 400–450 nm area with extinction coefficients of $1-2 \times 10^3$ (*502*). It has been argued that in cobalamins this CT transition would be hidden underneath the strong intraligand π-π^* transition of the corrin π-system. Such π-π^* transitions are observed around 550 nm ($\varepsilon \sim 10^4$) and around 360 nm ($\varepsilon \sim 3 \times 10^4$) (*450*).

It appears to the reviewers as an open question whether the direct excitation of the R—Co CT band in cobalamins [suggested by Schrauzer (*502*)] would allow the relatively high quantum yields of aerobic R—Co bond cleavage to be explained. Such an assumption would demand at least unity efficiency for this process as can be seen from the ratio of extinction coefficients. In contrast to views expressed by Schrauzer (*502*), absorption of light in π-π^* transitions and internal conversion to the CT excited state, as suggested by Pratt (*450*) and Vogler (*583*), may play an important role:

$$B_{12a} \xrightarrow{h\nu} B_{12a}(\pi\text{-}\pi^*) \xrightarrow[\text{conversion}]{\text{internal}} B_{12a}(R{\rightarrow}Co\ CT) \longrightarrow B_{12r} + R^{\cdot} \quad (43)$$

In this way the quantum yields may be simply explained. However, as will be seen from the discussion of the photoreaction of cobaloximes with oxygen (see Section V, E) the mechanistic details of the photoreaction of cobalamins with oxygen are still somewhat unclear. Arguments for internal conversion according to Eq. (43), which are based on the high values of quantum yields under aerobic conditions (*450*), are therefore still somewhat uncertain.

However, on the basis of a systematic study of the effect of different transition metals on the luminescence properties of synthetic metallocorrins, strong evidence has been produced recently that d-d states in dicyanocobalt(III)corrin are below states localized in the corrin ring system (*209a*). Only exceedingly weak fluorescence from the corrin ring is observed in this case. Strong fluorescence occurs with those metallocorrins that have d-d states, whose energy is above that of the corrin-localized states. The intramolecular quenching of emission, e.g., in the Co complex appears to be a reliable criterion for internal conversion to

d-d states, from which chemical reactions at the metal center may originate (209a).

The extent to which internal conversion occurs from π-π^* states in cobalamins to d-d states in competition with the R—Co CT state remains unclear. Final clarification of this point appears all the more desirable since the major biological application of B_{12} photochemistry has been to examine the involvement of radicals in enzymatic reactions. Existing or missing parallels between enzymic reactions and the photochemical behavior of cobalamins and cobaloximes toward model substrates have been used as guidelines to argue for or against radicals as enzymic intermediates (78, 506). A typical example is the controversial discussion about the mechanism of action of coenzyme B_{12} in dioldehydrase and related systems (78, 503, 531, 81).

It should be pointed out that another very significant method for investigating the oxidation state of cobalt in enzymic reactions is that of "spin labeling" by introduction of nitroxyl radicals (78). As a typical example the formation of labeled cobinamide is shown:

(44)

Aerobic photolysis of the methyl(nitroxide)cobinamide produces the aquo(nitroxide)cobinamide according to ESR evidence. However, anaerobic photolysis generates the cobalt(II) compound [Eq. (44)]. Similar results have been obtained with spin-labeled cobinamide coenzyme (Scheme 3). The cobalt(II) species does not give any ESR signal, probably owing to spin pairing of the two odd electrons (363). Admission of oxygen regenerates the signal of Co(III) (as aquo complex) completely.

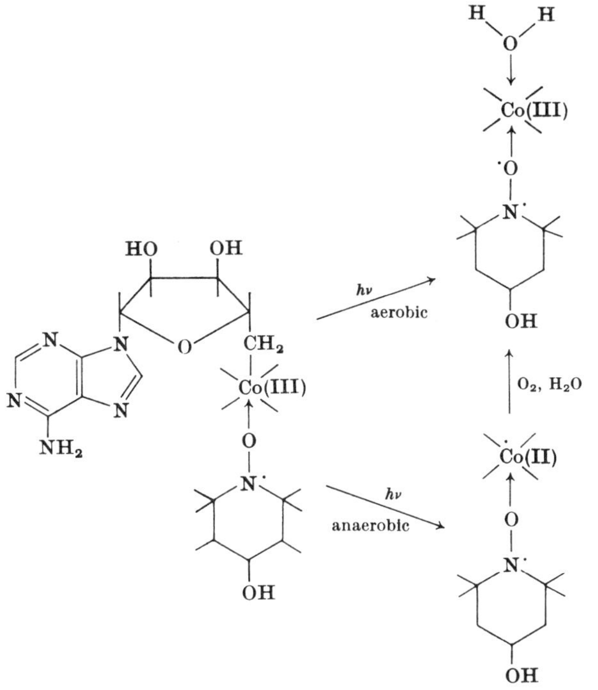

SCHEME 3

E. PHOTOCHEMISTRY OF COBALOXIMES

One of the most exciting aspects of B_{12} photochemistry (and chemistry in general) is that much of it can be studied using relatively simple molecules as models. Some of these model complexes are listed in Fig. 17.

The "equatorial" corrin system of B_{12} is replaced in most of these model complexes by the bis(dimethylglyoximato) (499) or, less frequently, by bis(salicylideneiminato) system (48) and derivatives thereof (169). It was convincingly demonstrated by Schrauzer (499, 500, 502) on the basis of its analogous chemical reactions, spectra, and electronic structure (derived from EHMO calculations) that the properties of cobalt(III)bis(dimethylglyoximato) complexes of the general structure (113a) very closely resemble those of cobalamins. They have therefore been named cobaloximes.

The photochemistry of cobaloximes (169) also parallels that of cobalamins (78, 502, 504, 505). However, the quantum yields for Co—R cleavage under aerobic conditions are smaller by a factor of 10

Fɪɢ. 17. Inorganic model systems for B_{12}: (113a) bis(dimethylglyoxime) cobalt (or cobaloxime) with axial alkyl (R) and base (B) ligands; (113b) BF_2^- cobaloxime; (113c) diacetylmonoximeiminodiacetylmonoximatoiminopropane 1,3-cobalt; (113d) N, N'-ethylenebis(salicylideneiminato)cobalt. [From Brown (78).]

for the cobaloximes (502). The wavelength of irradiation was in the 400–450 nm region; the electronic spectra of cobaloximes show an absorption maximum in this region, which has been attributed to an alkyl-to-cobalt charge transfer (502). It has been argued that the difference in quantum yields is due to the lower strength of the Co—C bond in cobalamins as compared with the cobaloximes.† Photodissociation of the bond between cobalt and the lower axial ligand B appears to be an important process in cobaloxime photochemistry, according to recent observations by Giannotti et al. (216) (under aerobic conditions):

$$\text{(46)}$$

† "A slightly greater partial positive charge on cobalt" in the cobaloximes in comparison to the cobalamins, as derived from EHMO calculations, has been correlated with "slightly more stable" axial bonds in the cobaloximes (502).

Another explanation could be a different ratio of photodealkylation and cleavage of the Co—B bond in cobaloximes and cobalamins, due to different rates of internal conversion processes.

How complicated it may be to discover the detailed mechanism of B_{12} photodealkylation becomes obvious from a recent ESR study of anaerobic alkylcobaloxime photochemistry (214). Of the cobaloximes investigated which all contained pyridine as lower axial ligand B, those with R = methyl and benzyl abstracted an electron from the solvent ($CHCl_3$ with ~0.75% C_2H_5OH at −20°C) in the primary photochemical process, leading to a Co(II) species still containing both axial ligands. On the other hand, the cobaloximes with R = isopropyl, isobutyl, n-pentyl, and cyclohexyl at −160°C gave the corresponding radicals R and the dealkylated Co(II) species. On warming up the samples to −100°C further changes occurred, which were ascribed to "loss of the pryidine ligand or substitution of a solvent molecule in the alkyl position." This work provides the first direct evidence for the production of cobalt(II) species as the primary products of anaerobic cobaloxime photochemistry. However, it should be noted, that the formation of (114) and (115) results from anaerobic irradiation of cyclohexylpyridinatocobaloxime in $CDCl_3$ (173).

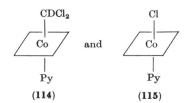

(114)   (115)

The effect of R on the course of photolysis resembles observations by Schrauzer (502): The photolysis rates of alkylcobaloximes increased in the order $CH_3 < C_2H_5 < CH_2CH(CH_3)_2 < C_6H_{11}$.

The aerobic photochemistry of cobaloximes confronts us with another set of problems. In a series of papers it was shown by Giannotti and his collaborators (219, 174, 205, 221, 381, 216) that alkylperoxycobaloximes (116) can be isolated from the photoreaction (in the range of −20° to 120°C) of alkylcobaloximes with O_2:

(116)

For R $= -CH(CH_3)$ $(C_6H_4CH_3)$, the structure was confirmed by X-ray diffraction (113). Benzyl- and allylcobaloximes also produced **116** in a thermal reaction at ambient or slightly elevated temperatures (174). Anaerobic decomposition of compound **116** by further irradiation or heating gave alcohols, aldehydes, and/or ketones depending on the solvent and R (215).

Since similar peroxy compounds could play a role in the aerobic cobalamin photolysis (see section V, D), the mechanism of alkyl-peroxycobaloxime formation deserves special interest. The following scheme has been proposed recently for methylcobaloxime (216):

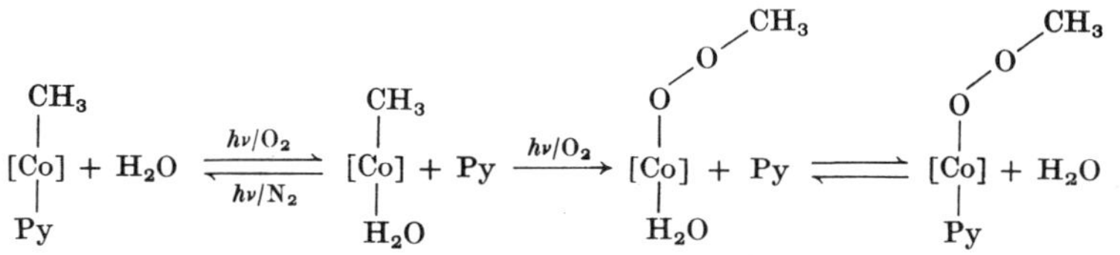

$$
\underset{\underset{Py}{\overset{|}{|}}}{\overset{\overset{CH_3}{\overset{|}{|}}}{[Co]}} + H_2O \underset{h\nu/N_2}{\overset{h\nu/O_2}{\rightleftharpoons}} \underset{\underset{H_2O}{\overset{|}{|}}}{\overset{\overset{CH_3}{\overset{|}{|}}}{[Co]}} + Py \overset{h\nu/O_2}{\longrightarrow} \underset{\underset{H_2O}{\overset{|}{|}}}{\overset{\overset{O^{\diagup O\diagdown CH_3}}{\overset{|}{|}}}{[Co]}} + Py \rightleftharpoons \underset{\underset{Py}{\overset{|}{|}}}{\overset{\overset{O^{\diagup O\diagdown CH_3}}{\overset{|}{|}}}{[Co]}} + H_2O \qquad (47)
$$

The first detectable photochemical reaction of alkylcobaloximes (in alcoholic solution) in the presence of water generally appears to be the expulsion of the lower axial ligand B and the rapid formation of the alkyl(aquo)cobaloxime (216), as demonstrated for a variety of B species, including spin-labeled amines (221, 216)† The insertion of O_2 into the Co—R bond of the aquo complex needs another light quantum. The suitability of the aquocomplex for this second step parallels the general enhancement of relative aerobic photolysis rates of alkylcobaloximes on decreasing the basicity of B, rendering the aquo complexes the most photoreactive systems (502). This phenomenon may be owing to a favorable ratio of internal conversion leading to Co—R cleavage, as compared with that leading to ligand (e.g., B) expulsion and other processes.

The effect of water as axial ligand can be rationalized using a very simplified MO picture (Fig. 18). Let us consider for the axial bonding of B and R in cobaloximes only the cobalt $3d_{z^2}$ and $4p_z$ and the R and B σ-orbitals. In the first step we produce a symmetric and an antisymmetric combination by letting the two σ-orbitals interact. The energy and composition of the two linear combinations will be quite different, according to whether the interaction is weak (Fig. 18A) or strong (Fig. 18B). Whereas in the former case the transition $\sigma_2 \rightarrow \sigma_3*$ has substantial $CH_3 \rightarrow Co$ charge-transfer character, this action is

† Apparently, it has only been demonstrated that photoaquation *precedes* O_2 insertion under aerobic conditions in the case of methylcobaloxime.

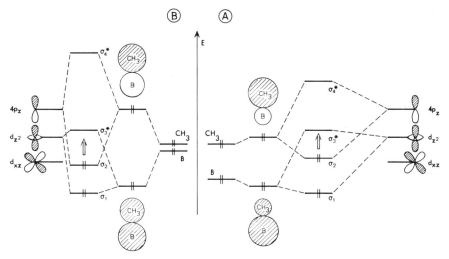

FIG. 18. Simplified MO scheme demonstrating the interaction among CH_3, base B, and cobalt orbitals for (A) weak interaction of CH_3 and B and for (B) strong interaction of CH_3 and B in cobaloximes.

somewhat reduced in the latter case. Thus, with weak interaction and B = H_2O (I.P., 12.6 eV), formation of CH_3· would be more probable, but with strong interaction and B = pyridine (I.P., 9.3 eV), less reaction resulting from homolytic Co—CH_3 cleavage and more dissociation of the Co–pyridine bond (as well as of Co—$CH_3 \to Co^+ + CH_3^-$) would be expected. In addition, it is not obvious whether d-d transitions would become more important if the CT-transition energy is raised somewhat (Fig. 18B); e.g., $d_{xz} \to d_{z^2}$ may come into play in this case, especially if pyridine, for instance, shifted MO's with strong d_{xz} contribution by π-interaction. Excitation of $d_{xz} \to d_{z^2}$ transition would increase the repulsion between Co and R and B, thus accelerating, for instance, photoaquation.

In the case of a CT process preceding O_2 insertion, the overall reaction would be as follows:

$$\begin{array}{c} R \\ | \\ \boxed{Co(III)} \end{array} \xrightarrow{h\nu} \begin{array}{c} \\ \boxed{\dot{C}o(II)} \end{array} + R^{\cdot} \qquad (48)$$
$$\begin{array}{c} | \\ H_2O \end{array} \qquad\qquad \begin{array}{c} | \\ H_2O \end{array}$$

$$R^{\cdot} + O_2 \longrightarrow R{-}O{-}O^{\cdot} \qquad (49)$$

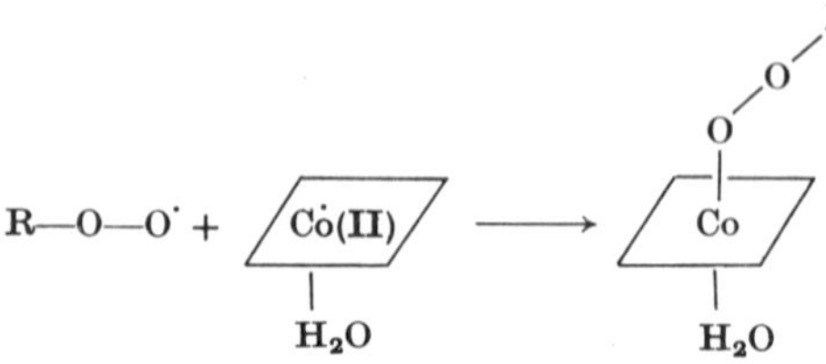

$$R\text{—}O\text{—}O^{\cdot} + \boxed{\dot{Co}(II)} \longrightarrow \boxed{Co} \qquad (50)$$

Since substantial quantum yields for Co—R cleavage in cobalamins (*450*) as well as cobaloximes (*502*) can only be observed under aerobic conditions, we have no direct information whether all or only a fraction of Co—OO—R is formed according to the mechanism shown in Eqs. (48)–(50). In order to challenge investigators in this field to produce more quantitative data, the reviewers would like to suggest two further

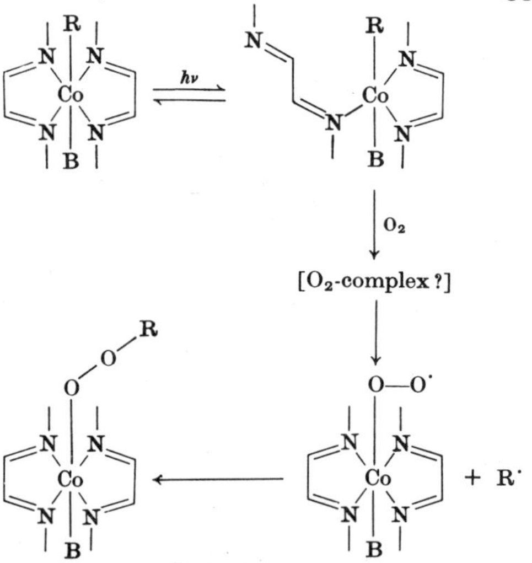

SCHEME 4

mechanisms for the formation of Co—OO—R (Scheme 4). Photochemically triggered partial decoordination of one "equatorial" nitrogen (e.g., as a result of a d_{xz} or $d_{yz} \rightarrow d_{x^2-y^2}$ transition)† could allow an attack of O_2 on Co, resulting in the formation of a radical pair and its subsequent combination as follows:

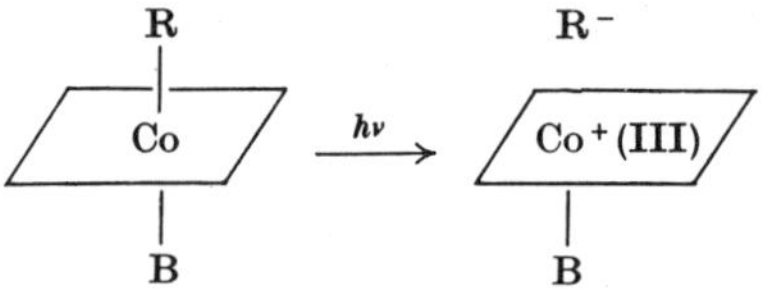

† The fast back-coordination of the nitrogen in the absence of oxygen could be a very efficient pathway of radiationless deactivation.

$$R^- + O_2 \longrightarrow R\text{---}OO^-$$

Dissociation of a Co—R bond to Co^+ and R^- (as a result of a d-d excitation or electron transfer, see page 70) with concomitant attack of O_2 on R^- and combination of the two ionic species could also lead to the peroxy compound. Note that the high basicity of B leads to exclusive formation of CH_4 from Co—CH_3 (via $CH_3^- + H_2O$) (504), and also supports photodissociation into $Co^+ + R^-$, according to Fig. 18B.

This mechanism would not be important for aquo complexes. In protic media, alkane formation would compete with oxygen insertion.

As a third alternative mechanism, one suggested by Dodd and Johnson (169), should be considered:

$$RCo(dmgH)_2py \xrightarrow{h\nu} RCo(dmgH)_2 + py$$
$$RCo(dmgH)_2 + O_2 \rightleftharpoons RCo(dmgH)_2OO\cdot \text{ (cis ?)}$$
$$RCo(dmgH)_2OO\cdot + py \longrightarrow ROO\ Co(dmgH)_2py$$

Although, in general, preference is given to a radical mechanism for the formation of alkylperoxycobaloximes (381, 216), attempts to find chemical evidence for the ionic or radical character of R in the course of reaction failed to give an unequivocal answer (381). The stereochemistry of R in the reaction with O_2 is also a controversial issue. Although in some cases compounds with optically active carbon attached to cobalt did not racemize in the insertion reaction (205, 218), complete racemization was observed with other systems (282). Whether these discrepancies are due to "ghost" rotations of the highly colored solutions that are opaque to light (282) or due to a difference in reaction mechanism remains obscure.

Other interesting photoreactions of alkylcobaloximes which we would like to mention, are the photochemical insertion of sulfur (217) and the β-elimination of olefins (173) observed in the anaerobic photolysis of cobaloximes:

$$\underset{\underset{Py}{|}}{\overset{\overset{R}{|}}{Co}} \quad \xrightarrow{h\nu/S_8} \quad \underset{\underset{Py}{|}}{\overset{\overset{\overset{R}{|}}{S_4}}{Co}} \tag{51}$$

$$\underset{\underset{Py}{|}}{\overset{\overset{\overset{\overset{R'CH_2}{|}}{RCH}}{|}}{Co}} \quad \xrightarrow{h\nu} \quad R-CH=CH-R' \; + \; \left[\underset{\underset{Py}{|}}{\overset{\overset{H}{|}}{Co}}\right] \quad \xrightarrow{Ph-C\equiv C-R} \quad \underset{\underset{Py}{|}}{\overset{\overset{Ph}{\underset{C=C}{\diagdown}}\overset{R}{\diagup}}{Co}}^{H} \tag{52}$$

Summaries of the literature concerning the photochemistry of other "model" complexes are found in Refs. *124*, *125*, and *117*.

In summing up, it can be stated that the photochemistry of vitamin B_{12} model complexes is an extremely exciting and active field. Two factors have been of special importance: (a) The close relationship in chemistry and spectroscopy between "model" systems and the "real" B_{12} is an ideal situation for fruitful cooperation among inorganic, organometallic, photo-, and biochemistry; and (b) The environmental "mercury" problem, which is caused by the alkylation of mercury by B_{12} (*78*), has led to much (hopefully long-lasting) attention for this field of biochemistry.

However, we should not close this section without a quotation from D. G. Brown (*78*): "The inorganic chemist has profited more from B_{12} than B_{12} has profited from him!" Much remains to be done!

VI. Conclusions

Photochemistry is now well-established as one of the important techniques in the arsenal of the organometallic chemist. This is especially true of the preparative field, as is demonstrated by the numerous photochemical syntheses and transformations of organometallic compounds.

The possibility of working at low temperatures opens up a route to thermally unstable compounds. The preparation of catalytically active compounds and the clarification of reaction mechanisms by synthetic and/or spectroscopic identification of intermediates are two important applications of the low-temperature techniques.

The knowledge of primary processes in organometallic photochemistry has greatly increased in recent years. For instance, it provides useful information about the structure and reactivity of coordinatively unsaturated species, about energy transfer, and about the connection between electronic excitation and chemical reactivity. The knowledge gained from model compounds will surely be used more extensively in understanding the more complicated systems, in particular for bioinorganic species.

There are many contacts between organometallic photochemistry and areas of "conventional" chemistry, biology, medicine, and technology. A few examples should give an idea of the varied possibilities.

The photochemical reactivity of organometallic systems is subtly dependent on small changes in the conditions. For instance, the dramatic effect of a change of substitutents on the ratio of internal conversion to chemical reaction provides a challenging exercise for the theoretician. The theoretical understanding of radiationless processes in transition metal complexes (472a) will have particular importance in the future.

Application of laser radiation in organometallic photochemistry is as yet little explored. There are already some examples of the preparative use of intensity-dependent reactions (613). In addition the application of lasers will give new stimulus to the partial photoresolution of racemates of transition metal compounds (487a, 626a).

At a time when the use of solar energy is receiving so much attention (250), it is essential to have a better understanding of the role that bioinorganic compounds play in the collection, transmission, and chemical conversion of solar energy.

The application of photochemistry to organic synthesis has been furthered by the use of transition metals. The introduction of transition metals allows particular spatial configurations to be fixed by coordination of transition metals to selected centers. In this way photochemical bond formation is rendered possible between centers whose approach is normally highly improbable. An outstanding example is the final ring closure in Eschenmoser's (626a) corrin synthesis which was achieved by the photochemical cycloisomerization of the secocorrinoidic palladium complex involving a 1,16-hydrogen shift. Luminescence studies have decisively demonstrated the reasons for the effects of different transition metals on this reaction (209a).

Metalloporphyrins are interesting catalysts for the activation of C—H bonds in organic substrates (422a). The application of photochemistry to this problem also has great promise.

Apart from preparative applications, organometallic photochemistry should be of industrial use in "imaging" processes. The initiation of polymerization reactions is being extensively investigated (29, 417a, 552a). The potentially useful association of color changes with changes in electrical properties in organometallic photoreactions has not yet found major application in data fixation.

In the area of photobiology the effect of the coordination of heavy metals on DNA should be mentioned as well as the use of model compounds. The different effects of Ag^+ and Hg^{2+} on the photoinactivation of DNA (and on thymine dimerization) pose interesting questions about the role of transition metals in radiation damage.

The discovery that transition metal complexes, such as cis-$Pt(NH_3)_2Cl_2$, have useful properties as antitumor drugs will stimulate research on the mechanism of interaction of such complexes with model compounds of important cell components ($559a$). In the case of $Pt(NH_3)_2Cl_2$, only the cis and not the trans complex is effective; the photochemical cis-trans photoisomerization may be useful in this connection.

The wide range of sources for this article makes it appropriate to quote a saying of Edward John Phelps found in a recent book on photochemistry (534): "The man who makes no mistakes usually does not make anything." This principle has allowed us to infringe an old Latin proverb: *O si tacuisses, philosophus manisisses!*

ACKNOWLEDGMENTS

The writing of this article and the work from the authors' laboratory reported in it would not have been possible without substantial support from a variety of sources. For financial support of our work we are indebted to the Deutsche Forschungsgemeinschaft, to the Belgian National Science Foundation, and to the Alexander von Humboldt-Stiftung for a fellowship for L. H. G. Leenders, and to The Royal Society, London for a European Fellowship for Robin N. Perutz.

The authors would like to express their special gratitude to Mrs. Margaret Y. Hancox for being so efficient, enduring, and patient with the typing of the manuscript, to Mrs. Ilse Schneider and Dr. F.-W. Grevels for preparing all the drawings, and to Miss Rosi Wagner for all her help with organizing the references, tables, etc. For their help with collecting, checking, and listing of the references the authors are indebted to Dr. Pierre Kirsch, Mrs. Janet Leggo, Mrs. Cindy Dahlhoff, Mrs. Marion Budzwait, Dr. H.-W. Frühauf, Dr. and Mrs. Randy Bock, Mrs. Marianne Knöchel, and Mr. Peter Ritterskamp.

Last but not least, many helpful discussions within the worldwide family of "organometallic photochemists" have substantially contributed to our efforts. Professor Dr. H. J. Eméleus, F.R.S. the Editor of this serial, deserves a special tribute for his patience and understanding with the authors.

REFERENCES

1. Adamczyk, A., and Wilkinson, F., *J. Chem. Soc., Faraday Trans. II* 68, 2031 (1972).

1a. Adamson, A. W., and Fleischauer, P. O., Eds., "Concepts of Inorganic Photochemistry." Wiley, New York, 1975.

2. Ainscough, E. W., Brodie, A. M., and Furness, A. R., *Chem. Commun.* p. 1357 (1971).

3. Ainscough, E. W., Brodie, A. M., and Furness, A. R., *J. Chem. Soc., Dalton Trans.* p. 2360 (1973).
4. Alexander, J. J., and Wojcicki, A., *Inorg. Chem.* **12**, 74 (1973).
5. Ali, L. H., Cox, A., and Kemp, T. J., *J. Chem. Soc., Dalton Trans.* p. 1475 (1973).
6. Allen, D. M., Cox, A., Kemp, T. J., and Ali, L. H., *J. Chem. Soc., Dalton Trans.* p. 1899 (1973).
7. Allison, D. A., Clardy, J., and Verkade, J. G., *Inorg. Chem.* **11**, 2804 (1972).
8. Allsopp, S. R., and Wilkinson, F., *Chem. Phys. Lett.* **19**, 535 (1973).
9. Alper, H., LePort, P. C., and Wolfe, S., *J. Amer. Chem. Soc.* **91**, 7553 (1969).
10. Alpert, B., Banerjee, R., and Lindquist, L., *Proc. Nat. Acad. Sci. U.S.* **71**, 558 (1974).
11. Alt, H., and Rausch, M. D., *J. Amer. Chem. Soc.* **96**, 5936 (1974).
12. Anderson, W. P., Blenderman, W. G., and Drews, K. A., *J. Organometal. Chem.* **42**, 139 (1972).
13. Angelici, R. J., and Busetto, L., *Inorg. Chem.* **7**, 1935 (1968).
14. Antonini, E., Anderson, N. M., and Brunori, M., *J. Biol. Chem.* **247**, 319 (1972).
15. Antonini, E., and Brunori, M., *Annu. Rev. Biochem.* **39**, 977 (1970).
16. Ashley-Smith, J., Howe, D. V., Johnson, B. F. G., Lewis, J., and Ryder, I. E., *J. Organometal. Chem.* **82**, 257 (1974).
17. Ashworth, J., Bamford, C. H., and Smith, E. G., *Pure Appl. Chem.* **30**, 25 (1972).
18. Attig, T. G., Reich-Rohrwig, P., and Wojcicki, A., *J. Organometal. Chem.* **51**, C21 (1973).
19. Aumann, R., *J. Organometal. Chem.* **77**, C33 (1974).
20. Aumann, R., *J. Amer. Chem. Soc.* **96**, 2631 (1974).
21. Aumann, R., *J. Organometal. Chem.* **76**, C32 (1974).
22. Baddley, W. H., Panattoni, C., Bandoli, G., Clemente, D. A., and Belluco, U., *J. Amer. Chem. Soc.* **93**, 5590 (1971).
23. Bagga, M. M., Ferguson, G., Jeffreys, J. A. D., Mansell, C. M., Pauson, P. L., Robertson, I. C., and Sime, J. G., *Chem. Commun.* p. 672 (1970).
24. Ballard, D. G. H., and van Lienden, P. W., *Chem. Commun.* p. 564 (1971).
25. Ballardini, R., Varani, G., Moggi, L., and Balzani, V., *J. Amer. Chem. Soc.* **96**, 7123 (1974).
26. Balzani, V., *Chim. Ind.* **55**, 711 (1973).
27. Balzani, V., and Carassiti, V., "Photochemistry of Coordination Compounds." Academic Press, New York, 1970.
28. Balzani, V., Moggi, L., Manfrin, M. F., Boletta, F., and Laurence, G. F., *Coord. Chem. Rev.* **15**, 321 (1975).
29. Bamford, C. H., *Pure Appl. Chem.* **34**, 173 (1973).
30. Bamford, C. H., Bingham, J. F., and Block, H., *Trans. Farad. Soc.* **66**, 2612 (1970).
31. Bamford, C. H., Burley, J. W., and Coldbeck, M., *J. Chem. Soc., Dalton Trans.* p. 1846 (1972).
32. Bamford, C. H., Crowe, P. A., Hobbs, J., and Wayne, R. P., *Proc. Roy. Soc., Ser. A* **292**, 153 (1966).
33. Bamford, C. H., Eastmond, G. C., and Fildes, F. J. T., *Chem. Commun.* p. 144 (1970).
34. Bamford, C. H., and Mahmud, M. U., *Chem. Commun.* p. 762 (1972).

35. Bamford, C. H., and Mullik, S. U., *J. Chem. Soc., Faraday Trans. I* p. 1127 (1973).

36. Bamford, C. H., and Paprotny, J., *Polymer* **13**, 208 (1972).

37. Bamford, C. H., and Paprotny, J., *Chem. Commun.* p. 140 (1971).

38. Barbeau, C., and Turcotte, J., *Can. J. Chem.* **48**, 3583 (1970).

39. Bathelt, W., Herberhold, M., and Fischer, E. O., *J. Organometal. Chem.* **21**, 395 (1970).

40. Beck, W., Weis, J. C., and Wieczorck, J., *J. Organometal. Chem.* **30**, 89 (1971).

40a. Becker, R. S., "Theory and Interpretation of Fluorescence and Phosphorescence," p. 192. Wiley (Interscience), New York, 1969.

41. Beer, D. C., and Todd, L. J., *J. Organometal. Chem.* **55**, 363 (1973).

42. Beer, D. C., and Todd, L. J., *J. Organometal. Chem.* **36**, 77 (1972).

43. Behrens, H., Uhlig, D., and Lindner, E., *Z. Anorg. Allg. Chem.* **394**, 8 (1972).

44. Benn, R., *Chem. Phys.* (1976), in press.

45. Bennett, M. J., and Simpson, K. A., *J. Amer. Chem. Soc.* **93**, 7156 (1971).

46. Berry, A., and Brown, T. L., *Inorg. Chem.* **11**, 1165 (1972).

47. Bertrand, R. D., Allison, D. A., and Verkade, J. G., *J. Amer. Chem. Soc.* **92**, 71 (1970).

48. Bigotto, A., Costa, G., Mestroni, G., Pellizer, G., Puxeddu, A., Reisenhofer, E., Stefani, L., and Tauzher, G., *Inorg. Chim. Acta Rev.* **4**, 41 (1970).

49. Black, J. D., Boylan, M. J., Braterman, P. S., and Wallace, W. J., *J. Organometal. Chem.* **63**, C21 (1973).

50. Black, J. D., and Braterman, P. S., *J. Organometal. Chem.* **63** C19, (1973).

51. Boag, J. W., *Photochem. Photobiol.* **8**, 565 (1968).

52. Bock, C. R., Meyer, T. J., and Whitten, D. G., *J. Amer. Chem. Soc.* **97**, 2909 (1975).

53. Bock, C. R., Meyer, T. J., and Whitten, D. G., *J. Amer. Chem. Soc.* **96**, 4710 (1974).

54. Bock, P. L., Boschetto, D. J., Rasmussen, J. R., Demers, J. P., and Whitesides, G. M., *J. Amer. Chem. Soc.* **96**, 2814 (1974).

55. Bock, R., and Koerner von Gustorf, E., *Advan. Photochem.* in press.

56. Bokiy, N. G., Gatilov, Yu. V., Struchkov, Yu. T., and Ustynyuk, M. A., *J. Organometal. Chem.* **54**, 213 (1973).

57. Bolletta, F., Gleria, M., and Balzani, V., *J. Phys. Chem.* **76**, 3934 (1972).

58. Bolletta, F., Gleria, M., and Balzani, V., *Mol. Photochem.* **4**, 205 (1972).

59. Bonaventura, C., Bonaventura, J., Antonini, E., Brunori, M., and Wyman, J., *Biochemistry* **12**, 3424 (1973).

60. Bond, A., and Green, M., *J. Chem. Soc., Dalton Trans.* p. 763 (1972).

61. Bond, A., and Green, M., *Chem. Commun.* p. 12 (1971).

62. Bond, A., Green, M., Lewis, B., and Lowrie, S. F. W., *Chem. Comm.* p. 1230 (1971).

63. Bond, A., Green, M., and Taylor, S. H., *J. Chem. Soc., Chem. Commun.* p. 112 (1973).

64. Bond, C. M., Lees, K. A., and Enever, R. P., *J. Pharm. Pharmacol.* **24** (Suppl.), 143P (1972).

65. Bor, G., *Chem. Commun.* p. 641 (1969).

66. Bor, G., *Inorg. Chim. Acta* **3**, 191 (1969).

67. Boylan, M. J., Braterman, P. S., and Fullarton, A., *J. Organometal. Chem.* **31**, C29 (1971).

68. Bozak, R. E., *in* "Advances in Photochemistry" (W. A. Noyes, Jr., G. S.

Hammond, and J. N. Pitts, Jr., eds.), Vol. 8, p. 227. Wiley (Interscience), New York, 1971.

69. Braterman, P. S., and Black, J. D., *J. Organometal. Chem.* **39**, C3 (1972).

70. Braterman, P. S., Davies, R. C., and Williams, R. J. P., *Advan. Chem. Phys.* **7**, 359 (1964).

71. Braterman, P. S., and Fullarton, A., *J. Organometal. Chem.* **31**, C27 (1971).

72. Breeze, P. A., and Turner, J. J., *J. Organometal. Chem.* **44**, C7 (1972).

72a. Breeze, P. A., Ph.D. Thesis, University of Cambridge, 1975.

73. Brockhaus, M., Staudacher, F., and Vahrenkamp, H., *Chem. Ber.* **105**, 3716 (1972).

74. Brodie, A. M., Johnson, B. F. G., Josty, P. L., and Lewis, J., *J. Chem. Soc., Dalton Trans.* p. 2031 (1972).

75. Brodie, A. M., Johnson, B. F. G., and Lewis, J., *J. Chem Soc., Dalton Trans.* p. 1997 (1973).

76. Brookes, A., Knox, S. A. R., and Stone, F. G. A., *J. Chem. Soc., A* p. 3469 (1971).

77. Brookhart, M., Dedmond, R. E., and Lewis, B. F., *J. Organometal. Chem.* **72**, 239 (1974).

78. Brown, D. G., *Progr. Inorg. Chem.* **18**, 177 (1973).

79. Brown, G. M., Hopf, F. R., Ferguson, J. A., Meyer, T. J., and Whitten, D. G., *J. Amer. Chem. Soc.* **95**, 5939 (1973).

80. Brown, J. M., and Mertis, K., *J. Chem. Soc., Perkin Trans. II* p. 1993 (1973).

81. Brown, K. L., and Ingraham, L. L., *J. Amer. Chem. Soc.* **96**, 7681 (1974).

82. Brown, M. L., Meyer, T. J., and Winterton, N., *Chem. Commun.* p. 309 (1971).

83. Brown, R. A., and Dobson, G. R., *Inorg. Chim. Acta* **6**, 65 (1972).

84. Brown, R. A., and Dobson, G. R., *J. Inorg. Nucl. Chem.* **33**, 892 (1971).

85. Bruce, M. I., Cooke, M., Green, M., and Westlake, D. J., *J. Chem. Soc., A* p. 987 (1969).

86. Bruce, M. I., Iqbal, M. Z., and Stone, F. G. A., *Chem. Commun.* p. 1325 (1970).

87. Brunner, H., *J. Organometal. Chem.* **16**, 119 (1969).

88. Brunner, H., and Herrmann, W. A., *Z. Naturforsch. B* **28**, 606 (1973).

89. Brunner, H., and Herrmann, W. A., *Chem. Ber.* **105**, 770 (1972).

90. Brunori, M., Bonaventura, J., Bonaventura, C., Antonini, E., and Wyman, J., *Proc. Nat. Acad. Sci. U.S.* **69**, 868 (1972).

91. Brunori, M., and Giacometti, G. M., *Proc. Nat. Acad. Sci. U.S.* **70**, 3141 (1973).

92. Buchkremer, J., Ph. D. Thesis, Bochum, 1973.

93. Buchkremer, J., Grevels, F. W., Jaenicke, O., Kirsch, P., Knoesel, R., Koerner von Gustorf, E. A., and Shields, J., *Int. Conf. Photochem. 7th, 1973, Abstr.* p. 601.

94. Bücher, T., and Kaspes, J., *Biochim. Biophys. Acta* **1**, 21 (1947).

95. Burdett, J. K., *J. Chem. Soc., Chem. Commun.* p. 763 (1973).

96. Burdett, J. K., *J. Chem. Soc., Faraday Trans. II* **70**, 1599 (1974).

96a. Burdett, J. K., Perutz, R. N., Poliakoff, M., and Turner, J. J., *J. Chem. Soc., Chem. Commun.* p. 157 (1975).

96b. Burdett, J. K., *Inorg. Chem.* **14**, 375 (1975).

96c. Burdett, J. K., Graham, M. A., Perutz, R. N., Poliakoff, M., Rest, A. J., Turner, J. J., and Turner, R. F., *J. Amer. Chem. Soc.* **97**, 4805 (1975).

97. Burt, R., Cooke, M., and Green, M., *J. Chem. Soc., A.* p. 2975 (1970).

98. Burt, R., Cooke, M., and Green, M., *J. Chem. Soc., A.* p. 2981 (1970).
99. Butler, I. S., and Fenster, A. E., *J. Organometal. Chem.* **51**, 307 (1973).
100. Butler, W. L., *Accounts Chem. Res.* **6**, 177 (1973).
101. Calderazzo, F., and L'Eplattenier, F., *Inorg. Chem.* **6**, 1220 (1967).
102. Callear, A. B., *Proc. Roy. Soc., Ser. A* **265**, 71 (1961).
103. Callear, A. B., *Proc. Roy, Soc., Ser. A* **265**, 88 (1961).
104. Callear, A. B., and Oldman, R. J., *Trans. Faraday Soc.* **63**, 2888 (1967).
105. Callear, A. B., and Oldman, R. J., *Nature (London)* **210**, 730 (1966).
106. Callis, J. B., Knowles, J. M., and Gouterman, M. *J. Phys. Chem.* **77**, 154 (1973).
107. Calvert, J. G., and Pitts, J. N., Jr., "Photochemistry." Wiley, New York, 1966.
108. Carty, A. J., Efraty, A., Ng, T. W., and Birchall, T., *Inorg. Chem.* **9**, 1263 (1970).
109. Carty, A. J., Ng, T. W., Carter, W., and Palenik, G. J., *Chem. Commun.* p. 1101 (1969).
110. Chapman, O. L., Pacansky, J., and Wojtkoswki, P. W., *Chem. Commun.* p. 681 (1973).
111. Chatt, J., and Duncanson, L. A., *J. Chem. Soc.* p. 2939 (1953).
112. Chia, L. S., Cullen, W. R., and Harbourne, D. A., *Can. J. Chem.* **50**, 2182 (1972).
113. Chiaroni, A., and Pascard-Billy, C., *Bull. Soc. Chim. Fr.* p. 781 (1973).
114. Chow, Y. L., Fossey, J., and Perry, R. A., *Chem. Commun.* p. 501 (1972).
115. Clark, H. C., and Hauw, T. L., *J. Organometal. Chem.* **42**, 429 (1972).
116. Clark, R. J., Abraham, M. R., and Busch, M. A., *J. Organometal. Chem.* **35**, C33 (1972).
117. Clarke, D. A., Grigg, R., Johnson, A. W., and Pinnock, H. A., *Chem. Commun.* p. 309 (1967).
118. Cleland, A. J., Fieldhouse, S. A., Freeland, B. H., and O'Brien, R. J., *J. Organometal. Chem.* **32**, C15 (1971).
119. Coates, G. E., and Wade, K., "Organometallic Compounds," Vol. I. Methuen, London, 1967.
120. Collman, J. P., Hoyano, J. K., and Murphy, D. W., *J. Amer. Chem. Soc.* **95**, 3424 (1973).
121. Conder, H. L., and Darensbourg, M. Y., *J. Organometal. Chem.* **67**, 93 (1974).
122. Cooke, M., Green, M., and Kuc, T. A., *J. Chem. Soc., A* p. 1200 (1971).
123. Corriu, R. J. P., and Douglas, W. E., *J. Organometal. Chem.* **51**, C3 (1973).
124. Costa, G., Mestroni, G., and Pellizer, G., *J. Organometal. Chem.* **15**, 187 (1968).
125. Costa, G., and Mestroni, G., *Tetrahedron Lett.* p. 1781 (1967).
126. Cotton, F. A., and Day, V. W., *Chem. Commun.* p. 415 (1974).
127. Cotton, F. A., Demming, A. J., Jorty, P. L., and Ullah, S. S., *J. Amer. Chem. Soc.* **93**, 4624 (1971).
128. Cotton, F. A., Edwards, W. T., Rauch, F. C., Graham, M. A., Perutz, R. N., and Turner, J. J., *J. Coord. Chem.* **2**, 247 (1973).
129. Coville, N. J., and Butler, I. S., *J. Organometal. Chem.* **64**, 101 (1974).
130. Coville, N. J., and Butler, I. S., *J. Organometal. Chem.* **57**, 355 (1973).
131. Cox, A., and Kemp, T. J., "Introductory Photochemistry." McGraw-Hill, New York, 1971.
132. Crichton, O., Poliakoff, M., Rest, A. J., and Turner, J. J., *J. Chem. Soc., Dalton Trans.* p. 1321 (1973).

133. Crichton, O., and Rest, A. J., *J. Chem. Soc., Chem. Commun.* p. 407 (1973).
134. Crichton, O., and Rest, A. J., *Int. Conf. Organometal. Chem., 6th, 1973,* Abstr. p. 238.
135. Crow, J. P., Cullen, W. R., and Hou, F. L., *Inorg. Chem.* **11**, 2125 (1972).
136. Crowfoot-Hodgkin, D., Pickworth, J., Robertson, J. H., Trueblood, K. N., Prosen, R. J., and White, J. G., *Nature (London)* **175**, 325 (1955).
137. Cullen, W. R., and Harbourne, D. A., *Inorg. Chem.* **9**, 1839 (1970).
138. Cullen, W. R., and Harbourne, D. A., *Can. J. Chem.* **47**, 3371 (1969).
139. Cullen, W. R., Harbourne, D. A., Liengme, B. V., and Sams, J. R., *Inorg. Chem.* **8**, 1464 (1969).
140. Cullen, W. R., Harbourne, D. A., Liengme, B. V., and Sams, J. R., *Inorg. Chem.* **9**, 702 (1970).
141. Cullen, W. R., and Mihichuk, L., *Can. J. Chem.* **51**, 936 (1973).
142. Cullen, W. R., Sams, J. R., and Thompson, J. A., *Inorg. Chem.* **10**, 843 (1971).
143. Cundall, R. B., and Gilbert, A., "Photochemistry." Nelson, London, 1970.
144. Curtis, M. D., and Job, R. C., *J. Amer. Chem. Soc.* **94**, 2153 (1972).
145. Darensbourg, D. J., Darensbourg, M. Y., and Dennenberg, R. J., *J. Amer. Chem. Soc.* **93**, 2807 (1971).
146. Darensbourg, D. J., *Inorg. Nucl. Chem. Lett.* **8**, 529 (1972).
147. Dartiguenave, M., Dartiguenave, Y., and Gray, H. B., *Bull. Soc. Chim. Fr.* p. 4223 (1969).
148. Dauben, W. G., and Lorber, M. E., *Org. Mass. Spectr.* **3**, 211 (1970).
149. Davidson, J. L., and Sharp, D. W. A., *J. Chem. Soc., Dalton Trans.* p. 1957 (1973).
150. Davidson, J. L., and Sharp, D. W. A., *J. Chem. Soc., Dalton Trans.* p. 107 (1972).
151. Davison, A., and Ellis, J. E., *J. Organometal. Chem.* **31**, 239 (1971).
152. Davison, A., Krusell, W. C., and Michaelson, R. C., *J. Organometal. Chem.* **72** C7 (1974).
153. Davison, A., and Martinez, N., *J. Organometal. Chem.* **74**, C17 (1974).
154. Davison, A., Traficante, D. D., and Wreford, S. S., *J. Amer. Chem. Soc.* **96**, 2802 (1974).
155. Dawans, F., Dewailly, J., Meunier-Piret, J., and Piret, P., *J. Organometal. Chem.* **76**, 53 (1974).
156. de Beer, J. A., and Haines, R. J., *J. Organometal. Chem.* **36**, 287 (1972).
157. Deberitz, J., and Nöth, H., *J. Organometal. Chem.* **61**, 271 (1973).
158. Deberitz, J., and Nöth, H., *J. Organometal. Chem.* **49**, 453 (1973).
159. Deberitz, J., and Nöth, H., *Chem. Ber.* **103**, 2541 (1970).
160. Deckelmann, K., and Werner, H., *Helv. Chim. Acta* **53**, 139 (1970).
161. Deeming, A. J., Ullah, S. S., Domingos, A. J. P., Johnson, B. F. G., and Lewis, J., *J. Chem. Soc., Dalton Trans.* p. 2093 (1974).
162. DeFilippo, D., Devillanova, F., Preti, C., Trogu, E. F., and Viglino, P., *Inorg. Chim. Acta* **6**, 23 (1972).
163. DeFilippo, D. Lai, A., Trogu, E. F., Verani, G., and Preti, C., *J. Inorg. Nucl. Chem.* **36**, 73 (1974).
164. Deyrup, J. A., and Betkouski, M., *J. Org. Chem.* **37**, 3561 (1972).
165. Tom Dieck, H., and Renk, I. W., *Chem. Ber.* **104**, 110 (1971).
166. Dineen, J. A., and Pauson, P. L., *J. Organometal. Chem.* **71**, 91 (1974).
167. Dobbie, R. C., and Mason, P. R., *J. Chem. Soc., Dalton Trans.* p. 1124 (1973).

168. Dobbie, R. C., Mason, P. R., and Porter, R. J., *Chem. Commun.* p. 612 (1972).
169. Dodd, D., and Johnson, M. D., *Organometal. Chem. Rev.* 52, 1, (1973).
170. Doleiden, F. H., Fahrenholtz, S. R., Lamola, A. A., and Trozzolo, A. M., *Photochem. Photobiol.* 20, 519 (1974).
171. Dolphin, D., and Felton, R. H., *Accounts Chem. Res.* 7, 26 (1974).
172. Downs, A. J., and Peake, S. C., in "Molecular Spectroscopy, Vol. I. Specialist Periodical Report" (R. F. Barrow, D. A. Long, and D. J. Millen, eds.), The Chemical Society, London, 1973.
173. Duong, K. N. V., Ahond, A., Merienne, C., and Gaudemer, A., *J. Organometal. Chem.* 55, 375 (1973).
174. Duong, K. N. V., Fontaine, C., Giannotti, C., and Gaudemer, A., *Tetrahedron Lett.* p. 1187 (1971).
175. Dustin, D. F., Dunks, G. B., and Hawthorne, M. F., *J. Amer. Chem. Soc.* 95, 1109 (1973).
176. Efraty, A., Potenza, J., Sandhu, S. S., Jr., Johnson, R., Mastropaolo, M., Bystrek, R., Denney, D. Z., and Herber, R. H., *J. Organometal. Chem.* 70, C24 (1974).
176a. Ehrl, W., and Vahrenkamp, H., *J. Organometal. Chem.* 63, 389 (1973).
177. Ehrl, W., and Vahrenkamp, H., *Chem. Ber.* 106, 2563 (1973).
178. Ehrl, W., and Vahrenkamp, H., *Int. Conf. Organomet. Chem., 6th, 1973,* Abstr. p. 48.
179. Ehrl, W., and Vahrenkamp, H., *Chem. Ber.* 105, 1471 (1972).
180. Ehrl, W., and Vahrenkamp, H., *Chem. Ber.* 104, 3261 (1971).
181. Ehrl, W., and Vahrenkamp, H., *Chem. Ber.* 103, 3563 (1970).
182. Ellermann, J., and Uller, W., *Z. Naturforsch. B* 25, 1353 (1970).
183. Elmitt, K., Green, M. G. H., Forder, R. A., Jefferson, I., and Prout, K., *Chem. Commun.* p. 747 (1974).
184. Evans, M. C. W., Telfer, A., and Lord, A. V., *Biochim. Biophys. Acta* 267, 530 (1972).
185. Fenster, A. E., and Butler, I. S., *Inorg. Chem.* 13, 915 (1974).
186. Fenster, A. E., and Butler, I. S., *Can. J. Chem.* 50, 589 (1972).
187. Fieldhouse, S. A., Fullam, B. W., Neilson, G. W., and Symons, M. C. R., *J. Chem. Soc., Dalton Trans.* p. 567 (1974).
188. Fields, R., Germain, M. M. (in part), Haszeldine, R. N., and Wiggans, P. W., *J. Chem. Soc., A* p. 1964 (1970).
189. Fields, R., Germain, M. M. (in part), Haszeldine, R. N., and Wiggans, P. W., *J. Chem. Soc., A* p. 1969 (1970).
190. Fields, R., Godwin, G. L., and Haszeldine, R. N., *J. Organometal. Chem.* 26, C70 (1971).
191. Fierz-David, H. E., and Blangey, D. L., "Grundlegende Operationen der Farben chemie." Springer-Verlag, Wien, 1952.
192. Fischer, E. O., Bathelt, W., and Müller, J., *Chem. Ber.* 103, 1815 (1970).
193. Fischer, E. O., Bock, H. J., Kreiter, C. G., Lynch, J., Müller, J., and Winkler E., *Chem. Ber.* 105, 162 (1972).
194. Fischer, E. O., and Beck, H. J., *Angew. Chem.* 82, 44 (1970).
195. Fischer, E. O., and Fischer, H., *Chem. Ber.* 107, 657 (1974).
196. Fischer, E. O., and Herberhold, M., *Experientia, Suppl.* 9, 259 (1964).
197. Fischer, E. O., and Herrmann, W. A., *Chem. Ber.* 105, 286 (1972).
198. Fischer, E. O., and Louis, E., *J. Organometal. Chem.* 18, P26 (1969).
199. Fischer, E. O., and Schneider, R. J. J., *Angew. Chem.* 79, 537 (1967).

200. Fischer, H., Fischer, E. O., and Kreissl, F. R., *J. Organometal. Chem.* **64,** C41 (1974).

201. Fischler, I., Hildenbrand, K., and Koerner von Gustorf, E., *Angew. Chem.* **14,** 54 (1975).

202. Fischler, I, Kelly, J. M., Kirsch, P., and Koerner von Gustorf, E., *Mol. Photochem.* **5,** 497 (1973).

203. Fischler, I., and Koerner von Gustorf, E. A., *Naturwissenschaften* **62,** 63 (1975).

204. Font, J., Barton, S. C., and Strausz, O. P., *Chem. Commun.* p. 499 (1970).

205. Fontaine, C., Duong, K. N. V., Merienne, C., Gaudemer, A., and Giannotti C., *J. Organometal. Chem.* **38,** 167 (1972).

206. Foust, A. S., Hoyano, J. K., and Graham, W. A. G., *J. Organometal. Chem.* **32,** C65 (1971).

207. Franz, D. A., Miller, V. R., and Grimes, R. N., *J. Amer. Chem. Soc.* **94,** 412, (1972).

208. Fujita, M., Aoki, K., and Asahi, Y., *Takeda Kenkyusho Ho* **31,** 516 (1972).

209. Gaines, D. F., and Hildebrandt, S. J., *J. Amer. Chem. Soc.* **96,** 5574 (1974).

209a. Gardiner, M., and Thomson, A. J., *J. Chem. Soc., Dalton Trans.* p. 820 (1974).

210. Garnovskii, A. D., Kolobova, N. E., Osiov, O. A., Anisimov, K. N., Zlotina, I. B., Mitina, G. K., and Kolodyazhnyi, Y. V., *J. Gen. Chem. USSR* **42,** 918 (1972).

211. Garrou, P. E., and Hartwell, G. E., *J. Organometal. Chem.* **55,** 331 (1973).

212. Geoffroy, G. L., Gray, H. B., and Hammond, G. S., *J. Amer. Chem. Soc.* **96,** 5565 (1974).

213. George, A. D., and George, T. A., *Inorg. Chem.* **11,** 892 (1972).

214. Giannotti, C., and Bolton, J. R., *J. Organometal. Chem.* **80,** 379 (1974).

215. Giannotti, C., and Fontaine, C., *J. Organometal. Chem.* **52,** C41 (1973).

216. Giannotti, C., Fontaine, C., and Septe, B., *J. Organometal. Chem.* **71,** 107 (1974).

217. Giannotti, C., Fontaine, C., Septe, B., and Doue, D., *J. Organometal. Chem.* **39,** C74 (1972).

218. Giannotti, C., Fontaine, C., and Gaudemer, A., *J. Organometal. Chem.* **39,** 381 (1972).

219. Giannotti, C., Gaudemer, A., and Fontaine, C., *Tetrahedron Lett,* p. 3209 (1970).

220. Giannotti, C., and Green, M. L. H., *J. Chem. Soc., Chem. Commun.* p. 1114 (1972).

221. Giannotti, C., Septe, B., and Benlian, D., *J. Organometal. Chem.* **39,** C5 (1972).

222. Gibson, Q. H., *Biochem. J.* **71,** 293 (1959).

223. Gibson, Q. H., and Ainsworth, S., *Nature (London)* **180,** 1416 (1957).

224. Gibson, Q. H., and Antonini, E., *J. Biol. Chem.* **242,** 4678 (1967).

225. Gilbert, A., Kelly, J. M., Budzwait, M., and Koerner von Gustorf, E. A., *Z. Naturforsch.* Submitted for publication.

226. Gilbert, A., Kelly, J. M., and Koerner von Gustorf, E. A., *Mol. Photochem* **6,** 225 (1974).

227. Gillespie, R. J., "Molecular Geometry." van Nostrand-Reinhold, London, 1971.

228. Ginzburg, A. G., Nemirovskaya, I. B., Setkina, V. N., and Kursanov, D. M., *Dokl. Akad. Nauk. SSSR* **208,** 144 (1973).

229. Graham, M. A., Perutz, R. N., Poliakoff, M., and Turner, J. J., *J. Organometal. Chem.* **34**, C34 (1972).
230. Graham, M. A., Poliakoff, M., and Turner, J. J., *J. Chem. Soc., A* p. 2939 (1971).
231. Graham, M. A., Rest, A. J., and Turner, J. J., *J. Organometal. Chem.* **24**, C54 (1970).
232. Gray, C. J., "Enzyme-Catalyzed Reactions," p. 72. van Nostrand Reinhold Comp., London, 1971.
233. Green, M., and Lewis, B., *Chem. Commun.* p. 114 (1973).
234. Green, M. L. H., and Smith, M. J., *J. Chem. Soc., A.* p. 3220 (1971).
235. F.-W. Grevels, U. Feldhoff, J. Leitich, and C. Krüger, *J. Organometal. Chem.* (1976) in press.
236. Grevels, F. W., and Koerner von Gustorf, E. A., *Liebigs Ann. Chem.* p. 547 (1975).
237. Grevels, F. W., Schulz, D., and Koerner von Gustorf, E. A., *Angew. Chem.* **86**, 558 (1974).
238. Grevels, F. W., Schulz, D., and Koerner von Gustorf, E. A., unpublished results.
239. Grubbs, R. H., and Brunck, T. K., *J. Amer. Chem. Soc.* **94**, 2538 (1972).
240. Grubbs, R. H., Pancoast, T. A., and Grey, R. A., *Tetrahedron Lett,* p. 2425 (1974).
241. Hackett, P., and Manning, A. R., *J. Chem. Soc., Dalton Trans.* p. 2434 (1972).
242. Haines, R. J., Du Preez, A. L., and Nolte, C. R., *J. Organometal. Chem.* **55**, 199 (1973).
243. Haines, R. J., and Du Preez, A. L., *Inorg. Chem.* **11**, 330, (1972).
244. Haines, R. J., and Nolte, C. R., *J. Organometal. Chem.* **36**, 163 (1972).
245. Haines, R. J., Nolte, C. R., Greatrex, R., and Greenwood, N. N., *J. Organometal. Chem.* **26**, C45 (1971).
246. Haldane, J., and Lorrain-Smith, J., *J. Physiol. (London)* **20**, 497 (1896).
247. Hallam, H. E., *in* "Vibrational Spectroscopy of Trapped Species." Wiley, New York, 1973.
248. Hallock, S. A., and Wojcicki, A., *J. Organometal. Chem.* **54**, C27 (1973).
249. Halpern, J., *Accounts Chem. Res.* **3**, 386 (1970).
250. Hammond, A. L., Metz, W. D., and Maugh, T. H., "Energie für die Zukunft." Umschau Verlag, Frankfurt/Main, 1974.
251. Hart-Davis, A. J., White, C., and Mawby, R. J., *Inorg. Chim. Acta* **4**, 431 (1970).
252. Hawkins, C. J., "Absolute Configuration of Metal Complexes," p. 148. Wiley (Interscience), New York, 1971.
253. Heimbach, P., *Angew. Chem.* **85**, 1035 (1973).
254. Herberhold, M., and Brabetz, H., *Z. Naturforsch. B* **26**, 656 (1971).
255. Herberhold, M., and Jablonski, C. R., *Inorg. Chim. Acta* **7**, 241 (1973).
256. Herberhold, M., and Golla, W., *J. Organometal. Chem.* **26**, C27 (1971).
257. Herberhold, M., and Jablonski, C., *J. Organometal. Chem.* **14**, 457 (1968).
258. Herberhold, M., Leonhard, K., Golla, W., and Alt, H., *Int. Conf. Organometal. Chem. 6th, 1973,* Abstr. 117.
259. Herberhold, M., and Razavi, A., *Angew. Chem.* **84**, 1150 (1972).
260. Herberhold, M., and Razavi, A., *J. Organometal. Chem.* **67**, 81 (1974).
261. Herberich, G. E., and Müller, H., *Angew. Chem.* **83**, 1020 (1971).
262. Herman, A., and Wing, R. M., *J. Organometal. Chem.* **63**, 441 (1973).

263. Hill, B., Math, K., Pillsbury, D., Voecks, G., and Jennings, W., *Mol. Photochem.* **5**, 195 (1973).

263a. Hoard, J. L. *in* "Hemes and Hemoproteins," (B. Chance, R. W. Estabrook, and T. Yonetani, eds.) pp. 9. Academic Press, New York, 1966.

264. Hogenkamp, H. P. C., and Barker, H. A., *Fed. Proc.* **21**, 470 (1962).

265. Hopf, F. R., O'Brien, T. P., Scheidt, W. R., and Whitten, D. G., *J. Amer. Chem. Soc.* **97**, 277 (1975).

266. Hoyano, J. K., Elder, M., and Graham, W. A. G., *J. Amer. Chem. Soc.* **91**, 4568 (1969).

267. Hoyano, J. K., and Graham, W. A. G., *Inorg. Chem.* **11**, 1265 (1972).

268. Hubert, A. J., Georis, A., Warin, R., and Teyssié, P., *J. Chem. Soc., Perkin Trans. II* p. 366 (1972).

269. Hubert, A. J., Goebels, G., Moniotte, Ph., Warin, R., and Teyssié, P., *Int. Conf. Organometal. Chem., 6th, 1973*, Abstr. p. 187.

270. Hubert, A. J., Moniotte, P., Goebels, G., Warin, R., and Teyssié, P. *J. Chem. Soc., Perkin Trans. II* p. 1954 (1973).

271. Hyde, C. L., and Darensbourg, D. J., *Inorg. Chim. Acta* **7**, 145 (1973).

272. Inglis, T., Kilner, M., and Reynoldson, T., *J. Chem. Soc. Chem. Commun.* p. 774 (1972).

273. International Union of Pure and Applied Chemistry, *Appendices on Tentative Nomenclature, Symbols, Units, and Standards, Number 31, Nomenclature on Organic Chemistry: Section D, August 1973.*

274. Jablonski, C. R., and Sorensen, T. S., *Can. J. Chem.* **52**, 2085 (1974).

275. Jaenicke, O., Ph. D. Thesis, University of Vienna, 1973.

276. Jaouen, G., *Tetrahedron Lett.* p. 5159 (1973).

277. Jaouen, G., and Dabard, R., *J. Organometal. Chem.* **61**, C36 (1973).

278. Jaouen, G., and Dabard, R., *J. Organometal. Chem.* **72**, 377 (1974).

279. Jaouen, G., Meyer, A., and Simonneaux, G., *Tetrahedron Lett.* p. 5163 (1973).

280. Jeffery, J., and Mawby, R. J., *J. Organometal. Chem.* **40**, C42 (1972).

281. Jennings, W., and Hill, B., *J. Amer. Chem. Soc.* **92**, 3199 (1970).

282. Jensen, F. R., and Kiskis, R. C., *J. Organometal. Chem.* **49**, C46 (1973).

283. Jetz, W., and Graham, W. A. G., *Inorg. Chem.* **10**, 4 (1971).

284. Jetz, W., and Graham, W. A. G., *J. Amer. Chem. Soc.* **91**, 3375 (1969).

285. Job, R. C., *Diss. Abstr. B.* **32**, 6274-B (1972).

286. Job, R. C., and Curtis, M. D., *Inorg. Chem.* **12**, 2514 (1973).

287. Job, R. C., and Curtis, M. D., *Inorg. Chem.* **12**, 2510 (1973).

288. Johnson, B. F. G., Lewis, J., Matheson, T. W., Ryder, I. E., and Twigg, M. V., *J. Chem. Soc., Chem. Commun.* p. 269 (1974).

289. Johnson, B. F. G., Lewis, J., and Twigg, M. V., *J. Organometal. Chem.* **67**, C75 (1974).

290. Jun, M. J., and Müller, F. J., *Monatsh. Chem.* **103**, 1213 (1972).

291. Kaizu, Y., Fujita, I., and Kobayashi, H., *Z. Phys. Chem. (Frankfurt am Main)* **79**, 298 (1972).

292. Kaska, W. C., Mitchell, D. K., and Reichelderfer, R. F., *J. Organometal. Chem.* **47**, 391 (1973).

293. Katz, J. J., *in* "Inorganic Biochemistry" (G. L. Eichhorn, ed.), Vol. II, p. 1022. Elsevier, Amsterdam, 1973.

294. Ke, B., *Photochem. Photobiol.* **20**, 542 (1974).

295. Keable, H. R., Kilner, M., and Robertson, E. E., *J. Chem. Soc., Dalton Trans.* p. 639 (1974).

296. Keeley, D. F., and Johnson, R. E., J. Inorg. Nucl. Chem. 11, 33 (1959).
297. Keeton, D. P., and Basolo, F., Inorg. Chim. Acta 6, 33 (1972).
298. Kelly, J. M., Bent, D. V., Hermann, H., Schulte-Frohlinde, D., and Koerner von Gustorf, E., J. Organometal. Chem. 69, 259 (1974).
299. Kelly, J. M., Harrit, N., Hermann, H., and Koerner von Gustorf, E., unpublished results.
300. Kelly, J. M., Hermann, H., and Koerner von Gustorf, E., J. Chem. Soc., Chem. Commun. p. 105 (1973).
301. Kelly, J. M., and Koerner von Gustorf, E. A., unpublished results.
302. Kerber, R. C., and Koener von Gustorf, E. A., J. Organometal. Chem. (1976) in press.
303. Kettle, S. F. A., J. Chem. Soc., A p. 420 (1966).
304. Khatami, A. I., Ginzburg, A. G., Nefedova, M. N., Setkina, V. N., and Kwisanov, D. N., J. Gen. Chem. USSR 42, 2654 (1972).
305. King, R. B., and Efraty, A., J. Organometal. Chem. 23, 527 (1970).
306. King, R. B., and Efraty, A., J. Organometal. Chem. 20, 264 (1969).
307. King, R. B., Efraty, A., and Douglas, W. M., J. Organometal. Chem. 60, 125 (1973).
308. King, R. B., and Hodges, K. C., J. Amer. Chem. Soc. 96, 1263 (1974).
309. King, R. B., Kapoor, P. N., and Kapoor, R. N., Inorg. Chem. 10, 1841 (1971).
310. King, R. B., Kapoor, R. N., and Pannell, K. H., J. Organometal. Chem. 20, 187 (1969).
311. King, R. B., Kapoor, R. N., Saran, M. S., and Kapoor, P. N., Inorg. Chem. 10, 1851 (1971).
312. King, R. B., and Korenowski, T. F., Inorg. Chem. 10, 1188 (1971).
313. King, R. B., and Saran, M. S., Inorg. Chem. 10, 1861 (1971).
314. King, R. B., Zipperer, W. C., and Ishaq, M., Inorg. Chem. 11, 1361 (1972).
315. Kirsch, P., Buchkremer, J., Jaenicke, O., Knoesel, R., Rumin, R., Shields, J., and Koerner von Gustorf, E., Meeting Photochem. Group, German Chem. Soc., Göttingen, 1973, Abstr. 5.
316. Kirsch, P., Kerber, R. C., Rumin, R., and Koerner von Gustorf, E. A., to be published.
317. Kisch, H., J. Organometal. Chem. 30, C25 (1971).
318. Kling, O., Nikolaiski, E., and Schläfer, H. L., Ber. Bunsenges. Phys. Chem. 67, 883 (1963).
319. Knox, G. R., and Pryde, A., J. Organometal. Chem. 74, 105 (1974).
320. Knox, S. A. R., and Stone, F. G. A., J. Chem. Soc., A p. 2874 (1971).
321. Knox, S. A. R., and Stone, F. G. A., J. Chem. Soc., A p. 3147 (1970).
322. Knox, S. A. R., Mitchell, C. M., and Stone, F. G. A., J. Organometal. Chem. 16, P67, (1969).
323. Kooti, M., and Nixon, J. F., J. Organometal. Chem. 76, C29 (1974).
324. Köpf, H., and Räthlein, K. H., Angew. Chem. 81, 1000 (1969).
325. Koerner von Gustorf, E. A., Buchkremer, J., Budzwait, M., Fischler, I., Foulger, B., Grevels, F. W., Harrit, N., Kelly, J. M., Schulz, D., and Wagner, R., Symp. Metals Org. Chem., 1974, C7.
326. Koerner von Gustorf, E., Buchkremer, J., Pfajfer, Z., and Grevels, F. W., DBP 2105627 (21.12.1972, angemeldet 6.2.1971).
327. Koerner von Gustorf, E., Buchkremer, J., Pfajfer, Z., and Grevels, F. W., Angew. Chem. 83, 249 (1971).

328. Koerner von Gustorf, E. A., Budzwait, M., and Fischler, I., *J. Organometal. Chem.* **105**, 325 (1976).

329. Koerner von Gustorf, E., Fischler, I., Leitich, J., and Dreeskamp, H., *Angew. Chem.* **84**, 1143 (1972).

330. Koerner von Gustorf, E., Fischler, I., and Wagner, R., *J. Organometal. Chem.*, **112**, 155 (1976).

331. Koerner von Gustorf, E. A., Grevels, F. W., and Fischler, J., "The Organic Chemistry of Iron." Academic Press, New York, 1976. To be published.

332. Koerner von Gustorf, E., and Grevels, F. W., *Fortschr. Chem. Forsch.* **13**, 366 (1969).

333. Koerner von Gustorf, E., Grevels, F. W., Krüger, C., Olbrich, G., Mark, F., Schulz, D., and Wagner, R., *Z. Naturforsch. A* **27**, 392 (1972).

334. Koerner von Gustorf, E., Henry, M. C., and DiPietro, C., *Z. Naturforsch. B* **21**, 42 (1966).

335. Koerner von Gustorf, E. A., Hess, D., and Snatzke, G., to be published.

336. Koerner von Gustorf, E., and Hogan, J. C., *Tetrahedron Lett*, **28**, 3191 (1968).

337. Koerner von Gustorf, E., Hogan, J. C., and Wagner, R., *Z. Naturforsch. B* **27**, 140 (1972).

338. Koerner von Gustorf, E., Jaenicke, O., and Polansky, O. E., *Z. Naturforsch. B* **27**, 575 (1972).

339. Koerner von Gustorf, E. A., Jaenicke, O., Wolfbeis, O., and Eady, C. R., *Angew. Chem.* **87**, 300 (1975).

340. Koerner von Gustorf, E., Pfajfer, Z., and Grevels, F. W., *Z. Naturforsch. B* **26**, 66 (1971).

341. Koerner von Gustorf, E., and Wagner, R., *Angew. Chem.* **83**, 968 (1971).

342. Kolobova, N. E., Skripkin, V. V., and Anisimov, K. N., *Bull. Acad. Sci. USSR, Chem. Sci.* p. 2095 (1970).

343. Kolshorn, H., Meier, H., and Müller, E., *Tetrahedron Lett.* p. 1589 (1972).

344. Kolshorn, H., Meier, H., and Müller, E., *Tetrahedron Lett.* p. 1469 (1971).

345. Kramer, A. V., Labinger, J. A., Bradley, J. S., and Osborn, J. A., *J. Amer. Chem. Soc.* **96**, 7145 (1974); Kramer, A. V., and Osborn, J. A., *ibid.* **96**, 7832 (1974).

346. Kruck, T., and Hempel, H. U., *Angew. Chem.* **86**, 233, (1974).

347. Kruck, T., and Knoll, L., *Z. Naturforsch. B* **28**, 34 (1973).

348. Kruck, T., and Knoll, L., *Chem. Ber.* **106**, 3578 (1973).

349. Kruck, T., Knoll, L., and Laufenberg, J., *Chem. Ber.* **106**, 697 (1973).

350. Kruck, T., and Knoll, L., *Chem. Ber.* **105**, 3783 (1972).

351. Kruck, T., and Krause, V., *Z. Naturforsch. B* **27**, 302 (1972).

352. Kruck, T., and Prasch, A., *Z. Anorg. Allg. Chem.* **371**, 1 (1969).

353. Kruck, T., Sylvester, G., and Kunau, I. P., *Z. Naturforsch. B* **28**, 38 (1973).

354. Kündig, E. P., and Ozin, G. A., *J. Amer. Chem. Soc.* **96**, 3820 (1974).

355. Kummer, D., and Furrer, J., *Z. Naturforsch. B* **26**, 162 (1971).

356. Kursanov, D. N., Setkina, V. N., Ginzburg, A. G., Nefedova, M. N., and Khatami, A. I., *Bull. Acad. Sci. USSR, Ser. Chem.* p. 2276 (1970).

357. Laing, M., Ashworth, T., Sommerville, P., Singleton, E., and Reimann, R., *Chem. Commun.* p. 1251 (1972).

358. Lamola, A. A., *in* "Energy Transfer and Organic Photochemistry" (A. Weissberger, ed.), Vol. 14, p. 17. Wiley (Interscience), New York, 1969.

359. Lamola, A. A., ed., "Creation and Detection of the Excited State," Vol. 1, Parts A and B. Dekker, New York, 1971.

360. Landesberg, J. M., and Sieczkowski, J., *J. Amer. Chem. Soc.* **93**, 972 (1971).
361. Landesberg, J. M., and Sieczkowski, J., *J. Amer. Chem. Soc.* **91**, 2120 (1969).
362. Laurence, G. S., and Balzani, V., *Inorg. Chem.* **13**, 2976 (1974).
363. Law, P. Y., Brown, D. G., Lien, E. L., Babior, B. M., and Wood, J. M., *Biochemistry* **10**, 3428 (1971).
364. Law, P. Y., and Wood, J. M., *Biochim. Biophys. Acta* **331**, 451 (1973).
365. Lee, K. M., and Hester, R. E., *J. Organometal. Chem.* **57**, 169 (1973).
366. LeMoigne, F., and Dabard, R., *J. Organometal. Chem.* **60**, C14 (1973).
367. Levenson, R. A., Gray, H. B., and Ceasar, G. P., *J. Amer. Chem. Soc.* **92**, 3653 (1970).
368. Lever, A. B. P., "Inorganic Electronic Spectroscopy." Elsevier, Amsterdam, 1968.
369. Lindner, E., and Meier, W. P., *J. Organometal. Chem.* **67**, 277 (1974).
370. Lokshin, B. V., Zdanovich, V. I., Baranetskaya, N. K., Setkina, V. N., and Kursanov, D. N., *J. Organometal. Chem.* **37**, 331 (1972).
371. Magomedov, G. K. I., Syrkin, V. G., Morozova, L. V., and Frenkel, A. S., *J. Gen. Chem. USSR* **43**, 1930 (1973).
372. Mantzaris, J., and Weissberger, E., *J. Amer. Chem. Soc.* **96**, 1880 (1974).
373. Mantzaris, J., and Weissberger, E., *J. Amer. Chem. Soc.* **96**, 1873 (1974).
374. Manuel, T. A., *J. Org. Chem.* **27**, 3941 (1962).
375. Mark, G., and Mark, F., *Z. Naturforsch.* **A29**, 610 (1974).
376. Mathieu, R., Lenzi, M., and Poilblanc, R., *Inorg. Chem.* **9**, 2030 (1970).
377. Mathieu, R., and Poilblanc, R., *Inorg. Chem.* **11**, 1858 (1972).
378. McIntyre, J. A., *J. Phys. Chem.* **74**, 2403 (1970).
379. McPhail, A. T., Knox, G. R., Robertson, C. G., and Sim, G. A., *J. Chem. Soc.*, A p. 205 (1971).
380. Menzel, E. R., *Chem. Phys. Lett.* **26**, 45 (1974).
381. Merienne, C., Giannotti, C., and Gaudemer, A., *J. Organometal. Chem.* **54**, 281 (1973).
382. Meyer, B., "Low Temperature Spectroscopy." Elsevier, Amsterdam, 1971.
383. Miller, J. R., and Myers, D. H., *Inorg. Chim. Acta* **5**, 215 (1970).
384. Moelwyn-Hughes, J. T., Garner, A. W. B., and Gordon, N., *J. Organometal. Chem.* **26**, 373 (1971).
385. Moelwyn-Hughes, J. T., Garner, A. W. B., and Howard, A. S., *J. Chem. Soc.*, A p. 2370 (1971).
386. Moggi, L., and Balzani, V., *Mol. Photochem.* **4**, 559 (1972).
387. Moggi, L., Varani, G., Sabbatini, N., and Balzani, V., *Mol. Photochem.* **3**, 141 (1971).
388. Moritani, I., and Hosokawa, T., *Kagaku No Ryoiki Zokan* **93**, 183 (1970).
389. Moss, J. R., and Graham, W. A. G., *Chem. Commun.* p. 835 (1970).
390. Müller, J., Fenderl, K., and Mertschenk, B., *Chem. Ber.* **104**, 700 (1971).
391. Murahashi, S. I., Mizoguchi, T., Hosokawa, T., Moritani, I., Kai, Y., Kohara, M., Yasuoka, N., Kasai, N., *Chem. Commun.* p. 563 (1974).
392. Musco, A., Palumbo, R., and Paiaro, G., *Inorg. Chim. Acta* **5**, 157 (1971).
393. Musher, J. I., and Agosta, W. C., *J. Amer. Chem. Soc.* **96**, 1320 (1974).
394. Nasielski, J., and Denisoff, O., *Int. Conf. Organometal. Chem., 6th, 1973*, Abstr. p. 146.
395. Nasielski, J., Kirsch, P., and Wilputte-Steinert, L., *J. Organometal. Chem.* **27**, C13 (1971).

396. Nasielski, J., Kirsch, P., and Wilputte-Steinert, L., *J. Organometal. Chem.* **29**, 269 (1971).

397. Nesmeyanov, A. N., Chapovskii, Yu. A., Polovyanyuk, I. V., and Makarova, L. G., *Bull. Acad. Sci. USSR* p. 1536 (1968).

398. Nesmeyanov, A. N., Kolobova, N. E., Anisimov, K. N., and Skripkin, V. V., *Bull. Acad. Sci. USSR* p. 2698 (1969).

399. Nesmeyanov, A. N., Kolobova, N. E., Skripkin, V. V., Anisimov, K. N., and Fedorov, L. A., *Dokl. Chem.*, **195**, 816 (1970).

400. Nesmeyanov, A. N., Kolobova, N. E., Skripkin, V. V., Anisimov, Y. S., and Sizoi, V. F., *Dokl. Akad. Nauk SSSR* (Engl.) **213**, 902 (1973).

401. Nesmeyanov, A. N., Kursanov, D. N., Setkina, V. N., Vilchevskaya, V. D., Baranetskaya, N. K., Krylova, A. I., and Glushchenko, L. A., *Dokl. Chem.* **199**, 719 (1971).

402. Nesmeyanov, A. N., Makarova, L. G., Ustynyuk. N. A., and Bogatyreva, L. V., *J. Organometal. Chem* **46**, 105 (1972).

403. Nesmeyanov, A. N., Makarova, L. G., and Vinogradova, V. N., *Bull. Acad. Sci. USSR, Chem. Sci.* p. 1541 (1972).

404. Nesmeyanov, A. N., Makarova, L. G., and Vinogradova, V. N., *Bull. Acad. Sci. USSR, Chem. Sci.* p. 1869 (1971).

405. Nesmeyanov, A. N., Rybin, L. V., Rybinskaya, M. I., Gubenko, N. T., Letseva, I. F., and Ustynyak, Yu. A., *Bull. Acad. Sci., USSR, Chem. Sci.* p. 1145 (1969).

406. Nesmeyanov, A. N., Rybinskaya, M. I., Rybin, L. V., Kaganovich, V. S., and Petrovskii, P. V., *J. Organometal. Chem.* **31**, 257 (1971).

407. Nesmeyanov, A. N., Rybinskaya, M. I., Rybin, L. V., Azutyunyan, A. V., Kuzmina, L. G., and Stuichkov, Y. T., *J. Organometal. Chem.* **73**, 365 (1974).

408. Nesmeyanov, A. N., Polovyanyuk, J. V., Lokshim, B. V., Chapovskii, Yu. A., and Makarov, L. G., *J. Gen. Chem.* **37**, 1911 (1967).

409. Newlands, M. J., and Ogilvie, J. F., *Can. J. Chem.* **49**, 343 (1971).

410. Noack, K., and Calderazzo, F., *J. Organometal. Chem.* **10**, 101 (1967).

411. Noack, K., and Ruch, M., *J. Organometal. Chem.* **17**, 309 (1969).

412. Noble, R. W., Brunori, M., Wyman, J., and Antonini, E., *Biochemistry* **6**, 1216 (1967).

413. Norrish, R. G. W., *Angew. Chem.* **80**, 868 (1968).

414. Novak, J. R., and Windsor, M. W., *Proc. Roy. Soc., Ser. A* **308**, 95 (1968).

415. Öfele, K., and Herberhold, M., *Z. Naturforsch. B* **28**, 306 (1973).

416. Öfele, K., and Herberhold, M., *Angew. Chem.* **82**, 775 (1970).

417. Ogilvie, J. F., *Chem. Commun.* p. 323 (1970).

417a. Okimoto, T., Inaki, Y., and Takemoto, K., *J. Macromol. Sci. Chem., A* **7**, 1313 (1973).

418. Oliver, A. J., and Graham, W. A. G., *Inorg. Chem.* **10**, 1 (1971).

419. Oliver, A. J., and Graham, W. A. G., *Inorg. Chem.* **9**, 2578 (1970).

420. Osborne, H. G., and Stiddard, M. H. B., *J. Chem. Soc.* p. 634 (1964).

421. Pailes, W. H., and Hogenkamp, H. P. C., *Biochemistry* **7**, 4160 (1968).

422. Pankowski, M., and Bigorgne, M., *J. Organometal. Chem.* **19**, 393 (1969).

422a. Paulson, D. R., Ullman, R., Sloane, R. B., and Closs, G. L., *J. Chem. Soc., Chem. Commun.* p. 186 (1974).

423. Pauson, P. L., and Segal, J. A., *J. Organometal. Chem.* **63**, C13 (1973).

424. Pearson, R. G., and Muir, W. R., *J. Amer. Chem. Soc.* **92**, 5519 (1970).

425. Perutz, M. F., Heidner, E. J., Ladner, J. E., Beetlestone, C. H., and Slade, E. F., *Biochemistry* **13**, 2187 (1974).

426. Perutz, M. F., *Nature (London)* **237**, 495 (1972).

427. Perutz, M. F., *Nature (London)* **228**, 726 (1970).

428. Perutz, R. N., Ph. D. Thesis, Cambridge (1974).

429. Perutz, R. N., and Turner, J. J., *J. Amer. Chem. Soc.* **97**, 4791 (1975).

430. Perutz, R. N., and Turner, J. J., *Inorg. Chem.* **14**, 262 (1975).

431. Petersen, R. L., and Watters, K. L., *Inorg. Chem.* **12**, 3009 (1973).

432. Preti, C., and Filippo, D. De, *J. Chem. Soc. A* p. 1901 (1970).

433. Photochemistry, A Specialist Periodical Report, The Chemical Society, London.

434. Pitts, J. N., Jr., Wilkinson, F., and Hammond, G. S., *Advan. Photochem.* **1**, 1 (1963).

435. Platbrood, G., and Wilputte-Steinert, L., *Int. Conf. Organomet. Chem., 6th, 1973, Abstr. 49.*

436. Platbrood, G., and Wilputte-Steinert, L., *J. Organometal. Chem.* **70**, 393 (1974).

437. Platbrood, G., and Wilputte-Steinert, L., *J. Organometal. Chem.* **70**, 407 (1974).

438. Platbrood, G., and Wilputte-Steinert, L., *J. Organometal. Chem.* **85**, 199 (1975).

439. Platbrood, G., and Wilputte-Steinert, L., *Tetrahedron Lett*, p. 2507 (1974).

440. Poliakoff, M., personal communication.

441. Poliakoff, M., *J. Chem. Soc., Dalton Trans.* p. 210 (1974).

442. Poliakoff, M., and Turner, J. J., *J. Chem. Soc., Dalton Trans.* p. 1351 (1973).

443. Poliakoff, M., and Turner, J. J., *Int. Conf. Organometal. Chem., 6th, 1973, Abstr.* p. 45.

444. Poliakoff, M., and Turner, J. J., *J. Chem. Soc. A* p. 2403 (1971).

445. Poliakoff, M., and Turner, J. J., *J. Chem. Soc., Dalton Trans.* p. 2276 (1974).

446. Poliakoff, M., and Turner, J. J., *J. Chem. Soc., Faraday Trans. II* **70**, 93 (1973).

447. Pomeroy, R. K., Gay, R. S., Evans, G. O., and Graham, W. A. G., *J. Amer. Chem. Soc.* **94**, 272 (1972).

448. Porter, G., and West, M. A., *in* "Techniques of Chemistry," Vol. 6, Part II, p. 367. Wiley, New York, 1974.

449. Porter, G., *Angew. Chem.* **80**, 882 (1968).

450. Pratt, J. M., "Inorganic Chemistry of Vitamin B_{12}." Academic Press, New York, 1972.

451. Pratt, J. M., and Craig, P. J., *Advan. Organometal. Chem.* **11**, 331 (1973).

452. Pratt, J. M., and Whitear, B. R. D., *J. Chem. Soc. A* p. 252 (1971).

453. Prince, R. H., and Stotter, D. A., *J. Inorg. Nucl. Chem.* **35**, 321 (1973).

454. Quinkert, G., *Angew. Chem.* **84**, 1157 (1972).

455. Rabinowitch, E., and Govindjee, "Photosynthesis." Wiley, New York, 1969.

456. Rahman, T. S., and Knox, R. S., *Phys. Status Solidi B* **58**, 715 (1973).

456a. Rahn, R. O., Setlow, J. K., and Landry, L. C., *Photochem. Photobiol.* **18**, 39 (1973).

457. Ramsey, B. G., "Electronic Transitions in Organometalloids." Academic Press, New York, 1969.

457a. Razuvaev, G. A., and Domrachev, G. A., *Tetrahedron* **19**, 341 (1963).

458. Reardon, E. J., Jr., and Brookhart, M., *J. Amer. Chem. Soc.* **95**, 4311 (1973).

459. Reed, T. E., and Hendricker, D. G., *J. Coord. Chem.* **2**, 83 (1972).

460. Rees, C. W., and von Angerer, E., *Chem. Commun.* p. 420 (1972).
461. Reeves, P., Henery, J., and Pettit, R., *J. Amer. Chem. Soc.* **91**, 5888 (1969).
462. Rehder, D., *J. Organometal. Chem.* **37**, 303 (1972).
463. Reid, K. I. G., and Paul., J. C., *Chem. Commun.* 1106 (1970).
464. Reimann, R. H., and Singleton, E., *J. Organometal. Chem.* **38**, 113 (1972).
465. Renk, I. W., and tom Dieck, H., *Chem. Ber.* **105**, 1403 (1972).
466. Rest, A. J., *J. Organometal. Chem.* **40**, C76 (1972).
467. Rest, A. J., *Chem. Commun.* p. 345 (1970).
468. Rest, A. J., *J. Organometal. Chem.* **25**, C30 (1970).
469. Rest, A. J., and Turner, J. J., *Chem. Commun.* p. 1026 (1969).
470. Rest, A. J., and Turner, J. J., *Chem. Commun.* p. 375 (1969).
471. Rest, A. J., and Turner, J. J., *Int. Conf. Organometal Chem. 4th, 1966* Abstr. N6.
472. Rimington, C., *Proc. Roy. Soc., Ser. B* **127**, 106 (1939).
472a. Robbins, D. J., and Thomson, A. J., *Mol. Phys.* **25**, 1103 (1973).
473. Rosenblum, M., North, B., Wells, D., and Giering, W. P., *J. Amer. Chem. Soc.* **94**, 1239 (1972).
474. Roustan, J. L., Charrier, C., Mérour, J. Y., Bénaim, J., and Grancotti, C., *J. Organometal. Chem.* **38**, C 37 (1972).
475. Roy, J. K., Caroll, F. A., and Whitten, D. G., *J. Amer. Chem. Soc.* **96**, 6349 (1974).
476. Roy, J. K., and Whitten, D. G., *J. Amer. Chem. Soc.* **94**, 7162 (1972).
477. Ruff, J. K., *Inorg. Chem.* **10**, 409 (1971).
478. Ruff, J. K., *Inorg. Chem.* **8**, 86 (1969).
479. Ruff, J. K., *Inorg. Chem.* **6**, 1502 (1967).
480. Ruff, J. K., White, R. P., Jr., and Dahl, L. F., *J. Amer. Chem. Soc.* **93**, 2159 (1971).
481. Rybinskaya, M. I., Rybin, L. V., Gubenko, N. T., and Nesmeyanov, A. N., *J. Gen. Chem. USSR* **41**, 2041 (1971).
482. Sabherwal, I. H., and Burg, A. B., *Chem. Commun.* p. 853 (1969).
483. Salomon, R. G., and Kochi, J. K., *Tetrahedron Lett.* p. 2529 (1973).
484. Salomon, R. G., and Kochi, J. K., *J. Amer. Chem. Soc.* **96**, 1137 (1974).
485. Salomon, R. G., Folting, K., Streib, W. E., and Kochi, J. K., *J. Amer. Chem. Soc.* **96**, 1145 (1974).
486. Sandhu, S. S., and Mehta, A. K., *Inorg. Nucl. Chem. Lett.* **7**, 891 (1971).
487. Sandhu, S. S., and Mehta, A. K., *J. Organometal. Chem.* **77**, 45 (1974).
487a. Sastri, V. S., *Inorg. Chim. Acta* **7**, 381 (1973).
488. Scheer, H., Norris, J. R., Druyan, M. E., and Katz, J. J., to be published.
489. Scheer, H., Svec, W. A., Cope, B. T., Studier, M. H., Scott, R. G., and Katz, J. J., *J. Amer. Chem. Soc.* **96**, 3714 (1974).
490. Schenk, G. O., "Jahrbuch der Max-Planck-Gesellschaft zur Förderung der Wissenschaften," e.V., p. 161. 1960.
491. Schenk, G. O., Gollnick, K., and Neumüller, O. -A., *Liebigs Ann. Chem.* **603**, 46 (1957).
492. Schenck, G. O., Koerner von Gustorf, E. A., and Jun, M.-J., *Tetrahedron Lett.* p. 1059 (1962).
493. Schlientz, W. J., Lavender, Y., Welcman, N., King, R. B., and Ruff, J. K., *J. Organometal. Chem.* **33**, 357 (1971).
494. Schmid, G., and Kempny, H.-P., *Angew. Chem.* **85**, 720 (1973).
495. Schmidt, M., and Schenck, W. A., *Naturwissenschaften* **58**, 96 (1971).

180   ERNST A. KOERNER VON GUSTORF ET AL.

496. Schneider, M., and Weiss, E., *J. Organometal. Chem.* **73**, C7 (1974).
497. Schoenberg, A. R., and Anderson, W. P., *Inorg. Chem.* **11**, 85 (1972).
498. Scholes, G., Grahem, C. R., and Brookhart, M., *J. Amer. Chem. Soc.* **96**, 5665 (1974).
499. Schrauzer, G. N., *Accounts. Chem. Res.* **1**, 97 (1968).
500. Schrauzer, G. N., *Pure Appl. Chem.* **33**, 545 (1973).
501. Schrauzer, G. N., and Kisch, H., *J. Amer. Chem. Soc.* **95**, 2501 (1973).
502. Schrauzer, G. N., Lee, L. P., and Sibert, J. W., *J. Amer. Chem. Soc.* **92**, 2997 (1970).
503. Schrauzer, G. N., Michaely, W. J., and Holland, R. J., *J. Amer. Chem. Soc.* **95**, 2024 (1973).
504. Schrauzer, G. N., Sibert, J. W., and Windgassen, R. J., *J. Amer. Chem. Soc.* **90**, 6681 (1968).
505. Schrauzer, G. N., and Windgassen, R. J., *J. Amer. Chem. Soc.* **88**, 3738 (1966).
506. Schrauzer, G. N., and Windgassen, R. J., *J. Amer. Chem. Soc.* **89**, 3607 (1967).
507. Schroeder, M. A., and Wrighton, M. S., *J. Organometal. Chem.* **74**, C29 (1974).
508. Schumann, H., and Breunig, H. J., *J. Organometal. Chem.* **76**, 225 (1974).
509. Schumann, H., and Breunig, H. J., *J. Organometal. Chem.* **27**, C28 (1971).
510. Schumann, H., Breunig, H. J., and Frank, U., *J. Organometal. Chem.* **60**, 279 (1973).
511. Schumann, H., and Kuhlmey, J., *J. Organometal. Chem.* **42**, C57 (1972).
512. Schumann, H., Mohtachemi, R., Kroth, H.-J., and Frank, U., *Chem. Ber.* **106**, 2049 (1973).
513. Schumann, H., Mohtachemi, R., Kroth, H.-J., and Franck, U., *Chem. Ber.* **106**, 1555 (1973).
514. Schumann, H., Pfeifer, G., and Röser, H., *J. Organometal. Chem.* **44**, C10 (1972).
515. Schumann, H., Stelzer, O., Kuhlmey, J., and Niederreuther, U., *Chem. Ber.* **104**, 993 (1971).
516. Schumann, H., Stelzer, O., Kuhlmey, J., and Niederreuther, U., *J. Organometal. Chem.* **28**, 105 (1971).
517. Schumann, H., Stelzer, O., and Niederreuther, U., *J. Organometal. Chem.* **16**, P 64 (1969).
518. Schumann, H., Stelzer, O., Niederreuther, U., and Rösch, L., *Chem. Ber.* **103**, 2350 (1970).
519. Schumann, H., Stelzer, O., Weiss, R., Mohtachemi, R., and Fischer, R., *Chem. Ber.* **106**, 48 (1973).
520. Schumann, H., and Weiss, R., *Angew. Chem.* **82**, 256 (1970).
521. Schwartz, J., *J. Chem. Soc., Chem. Commun.* p. 814 (1972).
522. Schwarz, F. P., Gouterman, M., Muliani, Z., and Dolphin, D. H., *Bioinorg. Chem.* **2**, 1 (1972).
523. Schwenzer, G., Darensbourg, M. Y., and Darensbourg, D. J., *Inorg. Chem.* **11**, 1967 (1972).
524. Seel, F., and Röschenthaler, G. V., *Angew. Chem.* **82**, 182 (1970).
525. Sellmann, D., *Angew. Chem.* **85**, 1123 (1973).
526. Sellmann, D., *J. Organometal. Chem.* **36**, C17 (1972).
527. Sellmann, D., *J. Organometal. Chem.* **44**, C46 (1972).
528. Sellmann, D., Brandl. A., and Endell, R., *J. Organometal. Chem.* **49**, C22 (1973).
529. Sellmann, D., and Maisel, G., *Z. Naturforsch. B* **27**, 718 (1972).
530. Sellmann, D., and Maisel, G., *Z. Naturforsch. B* **27**, 465 (1972).

531. Silverman, R. B., and Dolphin, D., J. Amer. Chem. Soc. 95, 1686 (1973).
532. Silverstein, H. T., Beer, D. C., and Todd, L. J., J. Organometal. Chem. 21, 139 (1970).
533. Simmons, E. L., and Wendlandt, W. W., Coord. Chem. Rev. 7, 11 (1971).
534. Simons, J. P., "Photochemistry and Spectroscopy." Wiley (Interscience), London, 1971.
535. Singleton, E., Moelwyn-Hughes, J. T., and Garner, A. W. B., J. Organometal. Chem. 21, 449 (1970).
536. Sneddon, L. G., Beer, D. C., and Grimes, R. N., J. Amer. Chem. Soc. 95, 6623 (1973).
537. Solar, W., Weigel, H., and Mark, F., unpublished (1972).
538. Sovocool, G. W., Hopf, F. R., and Whitten, D. G., J. Amer. Chem. Soc. 94, 4350 (1972).
539. Starzewski, K. A. O., Tom Dieck, H., Franz, K. D., and Hohmann, F., J. Organometal. Chem. 42, C35 (1972).
540. Stelzer, O., and Schmutzler, R., J. Chem. Soc., A p. 2867 (1971).
541. Stillman, M. J., and Thomson, A. J., J. Chem. Soc., Faraday Trans. II 70, 790 (1974); ibid. 70, 805 (1974).
542. Stolz, I. W., Dobson, G. R., and Sheline, R. K., J. Amer. Chem. Soc. 85, 1013 (1963).
543. Stolz, I. W., Dobson, G. R., and Sheline, R. K., J. Amer. Chem. Soc. 84, 3589 (1962).
544. Strausz, O. P., Barton, S. C., Duholke, W. K., Gunning, H. E., and Kebarle, P., Can. J. Chem. 49, 2048 (1971).
545. Strohmeier, W., Angew. Chem. 76, 873 (1964).
546. Strohmeier, W., J. Organometal. Chem. 60, C60 (1973).
547. Strohmeier, W., and Csontos, G., J. Organometal. Chem. 72, 277 (1974).
548. Strohmeier, W., and Hartmann, P., Z. Naturforsch. B 25, 550 (1970).
549. Strohmeier, W., and Hartmann, P., Z. Naturforsch. B 24, 939 (1969).
550. Strohmeier, W., and Müller, F. J., Chem. Ber. 102, 3613 (1969).
551. Strohmeier, W., and Müller, F. J., Chem. Ber. 102, 3608 (1969).
552. Strohmeier, W., and von Hobe, D., Z. Naturforsch. B 18 981 (1963).
552a. Sugita, K., Muroga, H., and Suzuki, S., Polym. J. 4, 351 (1973).
553. Swanson, B., and Satija, S. K., J. Chem. Soc., Chem. Commun. p. 40 (1973).
554. Szabo, A., and Karplus, M., Proc. Nat. Acad. Sci. U.S. 70, 673 (1973).
555. Takahashi, K., Iwanami, M., Tsai, A., Chang, P. L., Harlow, R. L., Harris, L. E., McCaskie, J. E., Pfluger, C. E., and Dittmer, D. C., J. Amer. Chem. Soc. 95, 6113 (1973).
556. Takahashi, K., Iwanami, M., Tsai, A., Chang, P. L., Stamos, I., Blidner, B. B., McCaskie, J. E., Harris, L. E., Harlow, R., Pfluger, C. E., and Dittmer, D.C., Int. Conf. Organometal. Chem. 6th, 1973, Abstr. p. 15.
557. Tan, L. Y., Winer, A. M., and Pimentel, G. C., J. Chem. Phys. 57, 4028 (1972).
558. Tang Wong, K. L., Thomas, J. L., and Brintzinger, H. H., J. Amer. Chem. Soc. 96, 3694 (1974).
559. Taylor, R. T., Smucker, L., Hanna, M. L., and Gill, J., Arch. Biochem. Biophys. 156, 521 (1973).
559a. Thomson, A. T., Williams, R. T. P., and Resova, S., Struct. Bonding (Berlin) 11, 1 (1972).
560. Traverso, O., Carassiti, V., Graziani, M., and Belluco, U., J. Organometal. Chem. 57, C22 (1973).

561. Trecker, D. J., Henry, J. P., and McKeon, J. E., *J. Amer. Chem. Soc.* **87,** 3261 (1965).

562. Treichel, P. M., and Benedict, J. J., *J. Organometal. Chem.* **17,** P37 (1969).

563. Treichel, P. M., Dean, W. K., and Douglas, W. M., *Inorg. Chem.* **11,** 1609 (1972).

564. Treichel, P. M., Dean, W. K., Douglas, W. M., *J. Organometal. Chem.* **42,** 145 (1972).

565. Tripathi, S. C., and Srivastava, S. C., *J. Organometal. Chem.* **23,** 193 (1970).

566. Tripathi, S. C., Srivastava, S. C., and Pandey, R. D., *J. Inorg. Nucl. Chem.* **35,** 457 (1973).

568. Turro, N. J., "Molecular Photochemistry." Benjamin, New York, 1965.

569. Tyerman, W. J. R., Kato, M., Kebarle, P., Masamune, S., Strausz, O. P., and Gunning, H. E., *Chem. Commun.* p. 497 (1967).

570. Udovich, C. A., and Clark, R. J., *Inorg. Chem.* **8,** 938 (1969).

571. Udovich, C. A., Clark, R. J., and Haas, H., *Inorg. Chem.* **8,** 1066 (1969).

572. Valentine, D., Valentine, G. G., and Sarver, B., "Excited State Chemistry" (J. M. Pitts, Jr., ed.), pp. 107–119. Gordon & Breach, New York, 1970.

573. Valentine, J. S., and Valentine, D., Jr., *J. Amer. Chem. Soc.* **92,** 5795 (1970).

574. Vernon, L. P., *Photochem. Photobiol.* **18,** 529 (1973).

575. Victor, R., and Ben-Shoshan, R., *Chem. Commun.* p. 93 (1974).

576. Victor, R., Ben-Shoshan, R., and Sarel, S., *Tetrahedron Lett.* p. 4211 (1973).

577. Victor, R., Ben-Shoshan, R., and Sarel, S., *J. Org. Chem.* **37,** 1930 (1972).

578. Victor, R., Ben-Shoshan, R., and Sarel, S., *Chem. Commun.* p. 1241 (1971).

579. Victor, R., Ben-Shoshan, R., and Sarel, S., *Tetrahedron Lett.* p. 4253 (1970).

580. Victor, R., Ben-Shoshan, R., and Sarel, S., *Tetrahedron Lett.* p. 4257 (1970).

581. Victor, R., Deutsch, J., and Sarel, S., *J. Organometal. Chem.* **71,** 65 (1974).

582. Voecks, G. E., Jennings, P. W., Smith, G. D., and Caughlan, C. N., *J. Org. Chem.* **37,** 1460 (1972).

583. Vogler, A., private communication.

584. Vogler, A., in "Concepts of Inorganic Photochemistry" (A. W. Adamson and P. D. Fleischauer, eds.), p. 269. Wiley, New York, 1975.

585. Vogler, A., private communication, cited by J. F. Endicott in Vogler [**584**].

586. Vogler, A., *Z. Naturforsch. B* **25,** 1069 (1970).

587. Volhardt, K. P. C., Bercaw, J. E., and Berman, R. G., *J. Amer. Chem. Soc.* **96,** 4998 (1974).

588. Wagner, F., *Annu. Rev. Biochem.* **35,** 405 (1966).

589. Walther, D., *Anorg. Allg. Chem.* **396,** 46 (1973).

590. Waltz, N. L., and Sutherland R. G., *Chem. Soc. Rev.* **1,** 241 (1972).

591. Wang, A. H. J., Paul, I. C., and Schrauzer, G. N., *Chem. Commun.* p. 736 (1972).

592. Warburg, O., "Heavy Metal Prosthetic Groups and Enzyme Action." Oxford Univ. Press (Clarendon), London and New York, 1949.

593. Warburg, O., and Negelein, E., *Biochem. Z.* **204,** 495 (1929).

594. Ward, J. S., and Pettit, R., *J. Amer. Chem. Soc.* **93,** 262 (1971).

595. Warren. J. D., Busch, M. A., and Clark, R. J., *Inorg. Chem.* **11,** 452 (1972).

596. Warren, J. D., and Clark, R. J., *Inorg. Chem.* **9,** 373 (1970).

597. Wayne, R. P., "Photochemistry." Butterworths, London, 1970.

598. Wegner, P. A., Guggenberger, L. J., and Muetterties, E. L., *J. Amer. Chem Soc.* **92,** 3473 (1970).

599. Weigel, H., Koerner von Gustorf, E. A., and Mark, F., *Meeting German Chem. Soc. Abstr.* **227,** *1975.*

600. Wege, D., and Wilkinson, S. P., *Chem. Commun.* p. 795 (1973).

601. Weis, J. C., and Beck, W., *J. Organometal. Chem.* **44**, 325 (1972).

602. Whitesides, T. H., and Budnik, R. A., *Chem. Commun.* p. 1514 (1971).

603. Whitesides, T. H., and Slaven, R. W., *J. Organometal. Chem.* **67**, 99 (1974).

604. Whitten, D. G., Wildes, P. D., and DeRosier, C. A., *J. Amer. Chem. Soc.* **94**, 7811 (1972).

605. Wikholm, G. S., and Todd, L. J., *J. Organometal. Chem.* **71**, 219 (1974).

606. Wilkinson, F., *J. Pure Appl. Chem.*, **41**, 661 (1975).

607. Wojcicki, A., *Advan. Organometal. Chem.* **11**, 87 (1973).

608. Wood, J. M., and Brown, D. G., *Struct. Bonding (Berlin)* **11**, 47 (1972).

609. Wrighton, M., *Chem. Rev.* **74**, 401 (1974).

610. Wrighton, M., *Inorg. Chem.* **13**, 905 (1974).

610a. Wrighton, M., and Markham, J., *J. Phys. Chem.* **77**, 3042 (1973).

611. Wrighton, M., and Bredesen, D., *J. Organometal. Chem.* **50**, C35 (1973).

612. Wrighton, M. S., Pdungsap, L., and Morse, D. L., *J. Phys. Chem.*, **79**, 66 (1975).

613. Wrighton, M. S., and Ginley, D. S., *J. Amer. Chem. Soc.*, **97**, 2065 (1975).

614. Wrighton, M. S., Ginley, D. S., Schroeder, M. A., and Morse, D. L., *IUPAC Symp. Photochem. 5th 1974. Plenary Lect., Pure Appl. Chem.* **41**, 671 (1975).

615. Wrighton, M., Gray, H. B., and Hammond, G. S., *Mol. Photochem.* **5**, 165 (1973).

616. Wrighton, M., Hammond, G. S., and Gray, H. B., *J. Amer. Chem. Soc.* **93**, 3285 (1971).

617. Wrighton, M., Hammond, G. S., and Gray, H. B., *J. Amer. Chem. Soc.* **92**, 6068 (1970).

618. Wrighton, M., Hammond, G. S., and Gray, H. B., *Mol. Photochem.* **5**, 179 (1973).

619. Wrighton, M., Hammond, G. S., and Gray, H. B., *Inorg. Chem.* **11**, 3122 (1972).

620. Wrighton, M., Hammond, G. S., and Gray, H. B., *J. Amer. Chem. Soc.* **93**, 6048 (1971).

621. Wrighton, M., Hammond, G. S., and Gray, H. B., *J. Amer. Chem. Soc.* **93**, 4336 (1971).

622. Wrighton, M., Hammond, G. S., and Gray, H. B., *J. Organometal. Chem.* **70**, 283 (1974).

623. Wrighton, M., and Morse, D. L., *J. Amer. Chem. Soc.* **96**, 998 (1974).

624. Wrighton, M., and Schroeder, M. A., *J. Amer. Chem. Soc.* **95**, 5764 (1973).

625. Wrighton, M., and Schroeder, M. A., *Int. Conf. Organometal. Chem.*, *6th, 1973*, Abstr. 47.

626. Wrighton, M. S., and Schroeder, M. A., *J. Amer. Chem. Soc.* **96**, 6235 (1974).

626a. Yamada, Y., Miljkovic, D., Wehrli, P., Golding, B., Löliger, P., Keese, R., Müller, K., and Eschenmoser, A., *Angew. Chem. Int. Engl. Ed.* **8**, 343 (1969).

626b. Yoneda, H., Nakashima, Y., and Sakaguchi, U., *Chem. Lett.* p. 1343 (1973).

627. Zarnegar, P. P., Bock, C. R., and Whitten, D. G., *J. Amer. Chem. Soc.* **95**, 4367 (1973).

628. Zarnegar, P. P., and Whitten, D. G., *J. Amer. Chem. Soc.* **93**, 3776 (1971).

629. Zink, J. I., *Inorg. Chem.* **12**, 1018 (1973).

630. Zink, J. I., *Mol. Photochem.* **5**, 151 (1973).

631. Zink, J. I., *J. Amer. Chem. Soc.* **96**, 4464 (1974).

632. Zollinger, H., "Chemie der Azofarbstoffe." Birkhäuser, Basel, 1958.

NITROGEN–SULFUR–FLUORINE IONS

R. MEWS

Institute of Inorganic Chemistry, University of Göttingen, Göttingen, West Germany

I. Introduction

Although the chemistry of nitrogen–sulfur–fluorine compounds was fully discussed in a recent article (58), the rapid development of this field renders a new review timely, especially as a fresh approach is adopted. Summaries of work on sulfur–nitrogen anions were published some time ago (12, 13). The literature that has been covered earlier (58) will be dealt with briefly here insofar as it is necessary for the general cohesion of this article, but the main emphasis is placed on more recent work.

The chemistry of nitrogen–sulfur–fluorine compounds is in the main concerned with covalent substances, the end products being also mostly covalent, but intermediates are often ionic in character. In this article it is planned to show which ions have been prepared and which are in general possible, and also to put this wide range of possibilities on a systematic basis. Known reactions are examined to determine if they occur through ionic intermediates and how these fit into the reaction scheme. In this connection, ways may sometimes be found to trap these ions. Nitrogen–sulfur–fluorine ions are defined primarily as compounds

185

containing these elements, but derivatives that can be prepared from
N—S—F compounds (e.g., NS^+) will also be considered. Insofar as it is
necessary to secure an overall picture, related compounds will also be
discussed (e.g., pure SF ions or perfluoroalkyl derivatives) although the
selection of literature in this connection is somewhat arbitrary. The
chemistry of N—S—F ions is still almost in its infancy. Whereas much
information is available on anions, especially for tetracoordinated
compounds of S(VI), little is known about the cations. This paucity of
information does, however, enable us to recognize what possibilities
there are, particularly for new syntheses.

II. Classification of Sulfur Compounds According to Their Coordination Numbers

Sulfur is one of the most interesting of the chemical elements
because it occurs with such different valencies and coordination
numbers and is able to form bonds varying in type from formal single
bonds to triple bonds. In addition, it readily enters into the formation
of chains and rings as, for example, in sulfur–nitrogen chemistry.

It is difficult to deal with this versatility, but the problem may be
simplified by classifying sulfur compounds according to their coordination numbers. Structural investigations available so far show the free
electron pair in sulfur compounds to be sterically active, and treatment
of these molecules in terms of the Gillespie and Nyholm rules (48) is
completely justified. Table I shows the various coordination possibilities
in sulfur fluorides and oxyfluorides.

Sulfur (VI) has no free pair of electrons, and coordination numbers
of 6, 5, 4, and 3 are found. Four-valent sulfur can reach a maximum
coordination number of 5, the position of the doubly bonded oxygen in
OSF_4 being taken in $\ominus SF_4$ by a free pair of electrons. Two doubly
bonded oxygens and singly bonded fluorine atoms lead to tetrahedral
coordination of S(VI) in SO_2F_2, the corresponding compound in the
S(IV) system being $\ominus SOF_2$, whereas in $F\overset{\ominus}{\underset{\ominus}{S}}F$ the last oxygen is also
replaced by a free pair of electrons. As in the transition from OSF_4
to $\ominus SF_4$, a corresponding gradation in chemical behavior in the series
SO_2F_2—$\ominus SOF_2$—$\overset{\ominus}{\underset{\ominus}{S}}F_2$ is found. The relationship between isocoordi-
nated compounds with different oxidation states is essentially greater
than that between differently coordinated representatives of the same
oxidation state.

TABLE I

COORDINATION NUMBERS IN NEUTRAL SULFUR FLUORIDES,
OXYFLUORIDES, AND OXIDES

Coordination No.	Ṡ(VI)	S(IV)	S(II)
6		—	—
5			—
4			
3			

Variation in coordination number is fairly restricted in sulfur ions and for S(VI) the anionic and cationic species encountered are almost exclusively tetra- or hexacoordinated. The same is true in the main for sulfur(IV), although here cations with lower coordination are possible.

III. Acyclic Nitrogen–Sulfur–Fluorine Anions

The negative charge is carried by nitrogen in nitrogen–sulfur–fluorine anions, whereas in cations the positive charge may be assigned to sulfur. As a result, the chemistry of the anions is more that of the nitrogen atom, whereas for the cations reactions occur at the central sulfur atom. Exceptions are the sulfur–fluorine anion $\ddot{C}SF_5^-$ and, perhaps, also $F\ddot{S}O_2^-$, for which attempts to transfer the species intact in ionic reactions result in decomposition to F^- and SF_4 or SO_2.

A. HEXACOORDINATED ANIONS

Hexacoordinated anions have been known for a long time in the chemistry of both sulfur(VI) and sulfur(IV) through the species OSF_5^- (135) and $\ddot{C}SF_5^-$ (154). The corresponding compound of bivalent sulfur,

TABLE II
Possible Hexacoordinated Nitrogen–Sulfur–Fluorine Anions

cannot be prepared both because of the great instability of SF_2 (*141*) and because addition of 2 fluoride ions to tetrahedrally coordinated SF_2 is thermodynamically unfavorable. No examples are known in sulfur chemistry in which it has been possible to add fluoride ions to a neutral tetrahedral molecule and form a stable anion. Thus there remain in nitrogen–sulfur–fluorine chemistry only the possibilities shown in Table II. It is theoretically possible in OSF_5^- or in SF_5^- to replace an atom of fluorine by an amino group or, in the case of S(VI), to introduce in place of $\overline{|O}^-$ — the isosteric $R\overline{N}^-$ — group.

1. *Pentafluorosulfuroxy Anion*, SF_5O^-, *and Its Derivatives*

The OSF_5^- ion has been assumed to be an intermediate in the catalytic fluorination of OSF_4 to SF_5OF (*39, 40, 135*) and the cesium salt may be prepared at room temperature in CH_3CN (*90*) or at elevated temperature without a solvent (*135, 23*):

$$OSF_4 + CsF \xrightarrow[\text{RT}]{CH_3CN} Cs^+ OSF_5^- \tag{1}$$

The salt is stable at room temperature, but OSF_4 is evolved almost quantitatively at higher temperatures. One fluorine atom in sulfuroxy tetrafluoride may be replaced by an amino or phenoxy group (*53, 159, 160, 161, 131*), and it has also proved possible to replace 2 fluorine atoms by phenoxy- or perfluoroalkyl groups. As for OSF_5^-, these derivatives should also yield hexacoordinated anions. It would, how-ever, be expected that replacement of fluorine atoms by bulky and less electronegative groups would reduce the ability of the central sulfur atom to accept fluoride ions, and it is, in fact, found that dialkylamino-sulfur oxytrifluoride is unable to add fluoride ions (94):

$$(CH_3)_2NSOF_3 + CsF \xrightarrow[RT]{CH_3CN} \text{no stable salt formed} \qquad (2)$$

Few reactions have so far been carried out with OSF_5^-. Compound SF_5OF is formed with fluorine (135), and SF_5OCl can be isolated following decomposition of the salt with FCl (140):

$$Cs^+ SF_5O^- + FX \longrightarrow SF_5OX + CsF \qquad (X = F, Cl) \qquad (3)$$

It has not so far proved possible to transfer the OSF_5^- ion to other systems by means of the cesium salt: Decomposition of the ion to F^- and OSF_4 occurs. Reactions analogous to those shown in Eq. (3) are hardly to be expected with derivatives of the type $\{R—SOF_3\}$ as they would involve cleavage of the S—R bond, e.g.,

$$CF_3—\overset{O}{\underset{}{S}}—F + F_2 \text{ (excess)} \xrightarrow{CsF} CF_4 + SF_5OF \quad (109) \qquad (4)$$

2. Pentafluorosulfanylamide Anions, $SF_5—\overline{N}—R^-$

a. Derivatives with Perfluoroalkyl Groups. Acids containing these anions may be prepared from perhaloazomethines and HF (154):

$$R—\overset{Cl}{\underset{|}{C}}=N—SF_5 + 2HF \xrightarrow{-HCl} R—\overset{F}{\underset{F}{C}}—\overset{H}{N}—SF_5 \qquad (R = CF_3, C_2F_5) \qquad (5)$$

The mercury derivative is produced from HgF_2 and the correspond-ing perfluorinated azomethine (154):

$$2CF_2=N—SF_5 + HgF_2 \longrightarrow Hg\left(N\diagdown^{SF_5}_{CF_3}\right)_2 \qquad (6)$$

Alkali metal derivatives have not actually been described, although the formation of dimers on heating the azomethine with KF shows that they must be formed as intermediates (154):

$$SF_5N=CF_2 + KF \xrightarrow{225°C} SF_5—\overline{N}—CF_3^- K^+ \xrightarrow{SF_5N=CF_2}$$

$$SF_5—N=\overset{F}{C}—N\diagup^{SF_5}_{\diagdown CF_3} + KF \qquad (7)$$

Partial isomerization is also observed in this reaction,

$$SF_5N{=}CF_2 \xrightarrow{\ \sim\ } F_4S{=}N{-}CF_3 \tag{8}$$

This too may be satisfactorily explained by assuming the formation of an ionic intermediate:

$$SF_5{-}N{=}CF_2 \underset{-KF}{\overset{+KF}{\rightleftarrows}} K^+ \; N \overset{SF_5}{\underset{CF_3^-}{\diagdown}} \underset{+KF}{\overset{-KF}{\rightleftarrows}} CF_3{-}N{=}SF_4 \tag{9}$$

Equilibrium (9) has not been investigated, but, in purely qualitative terms, it follows from these observations that F^- addition occurs more readily to the $C{=}N{-}$ than to the $S{=}N{-}$ double bond.

The mercury compound reacts with active halides (154), e.g.,

$$Hg\left(N \overset{CF_3}{\underset{SF_5}{\diagdown}}\right)_2 + 2C_6H_5COCl \longrightarrow 2C_6H_5CON \overset{CF_3}{\underset{SF_5}{\diagdown}} + HgCl_2 \tag{10}$$

The preparation of the corresponding N—Cl compound presumably occurs by a radical mechanism:

$$\overset{H}{CF_3N}{-}SF_5 + AgF_2 \longrightarrow \{CF_3{-}N{-}SF_5\} + HF + AgF$$
$$\downarrow Cl_2 \tag{11}$$
$$\overset{Cl}{CF_3}{-}N{-}SF_5$$

b. *Derivatives with Inorganic Groups.* If a proton in pentafluoro-sulfanylamine, SF_5NH_2, is replaced by an inorganic group rather than a perfluoroalkyl group, the resulting compound will naturally possess acid character. Compounds of this type mentioned in the literature include the following:

i. N-Chloropentafluorosulfanylamine (28),

$$SF_5NCl_2 + HOH \longrightarrow \overset{H}{SF_5NCl} + HOCl \tag{12}$$

No information on this compound is available apart from its mass spectrometric identification.

ii. N-Fluorosulfonylpentafluorosulfanylamine (73),

$$F_5S{-}N{=}SOF_2 + H_2O \xrightarrow[-HCl]{(C_6H_5)_4AsCl} [F_5S{-}\overset{=}{N}{-}SO_2F]^- [(C_6H_5)_4As]^+ + HF \tag{13}$$

It is obtained as the tetraphenylarsonium salt in quantitative yield by alkaline hydrolysis of SF_5NSOF_2. The hydrolysis mechanism will be considered in discussing the fluorosulfonylamine.

3. The $\overset{\ominus}{S}F_5^-$ Ion and Its Derivatives

Cesium fluoride reacts with $\bar{S}F_4$ as follows (154):

$$\overset{\ominus}{S}F_4 \text{ (excess)} + \text{CsF} \xrightarrow[250^\circ C]{125^\circ -} \text{Cs}^+ \ \overset{\ominus}{S}F_5^- \tag{14}$$

Products obtained in this way are, however, heavily contaminated with CsF (154) and so far it has not been possible to prepare the compound in a pure state. Exact information on the stability of the salt is also lacking [the dissociation pressure is greater than 1 atm at 150°C (154)]. It is known that $(CH_3)_4N^+ \ \overset{\ominus}{S}F_5^-$, which is prepared by a reaction analogous to that shown in Eq. (14), has only a small SF_4 dissociation pressure at room temperature (38, 155). Nuclear magnetic resonance studies have not been made, but the vibrational spectrum shows clearly that the central sulfur atom in the ion is octahedrally coordinated, the fluorine ligands forming a square pyramid (22, 38, 154).

Since the $\overset{\ominus}{S}F_5^-$ ion is only moderately stable, therefore stable derivatives of the type $R\overset{\ominus}{S}F_4^-$ are hardly to be expected. The SF_5^- ion is an intermediate in the preparation of SF_5Cl (154). Compound CF_3SF_4Cl is isolated in the reaction between $CF_3\bar{S}F_3$ and Cl_2 in presence of CsF (34) and this may be explained by postulating intermediate formation of $Cs^+ \ CF_3\overset{\ominus}{S}F_4^-$:

$$CF_3\bar{S}F_3 + \text{CsF} \longrightarrow [Cs^+ \ \overset{\ominus}{S}(CF_3)F_4^-]$$
$$\downarrow Cl_2 \tag{14a}$$
$$\text{CsCl} + CF_3SF_4Cl$$

In the corresponding reaction with $(CH_3)_2N\bar{S}F_3$, only decomposition products, such as SF_5Cl, are found. This does not necessarily mean that the $(CH_3)_2N\bar{S}F_4^-$ ion has not been formed, although it is not very likely. Since $\overset{\ominus}{S}F_4$ adds fluoride ions less readily than OSF_4, a similar reaction path would be expected for the amino derivative.

B. PENTACOORDINATED ANIONS

As would be expected, pentacoordinated anions have not been isolated, since F^- is split off and they go over to more stable tetra-coordinated neutral compounds (11). Exchange of the OH group in alcohols for fluorine with the aid of SF_4 is a reaction of preparative significance (*70*):

$$ROH + :SF_4 \xrightarrow{-HF} \{RO\overset{\odot}{-}SF_3\} \longrightarrow RF + :SOF_2 \qquad (15)$$

Aminosulfur trifluorides or bisaminosulfur difluorides are more selective fluorinating agents (*100*):

$$ROH + :S\underset{\underset{F}{|}}{\overset{\overset{F}{|}}{<}}\overset{X}{_{NR_2}} \xrightarrow{-HF} \left[:S\overset{X}{\underset{F}{<}}\overset{OR}{_{NR_2}}\right] \longrightarrow \left[R^+ :S\overset{X}{\underset{F}{<}}\overset{\overset{\odot}{O}|^-}{_{NR_2}}\right] \longrightarrow$$

$$(A)$$

$$RF + \overset{\overset{\odot}{\cdot}}{\underset{O}{S}}\overset{X}{_{NR_2}} \qquad (X = F, NR_2) \quad (16)$$

The $R_2NSOF_2^-$ ion should be a better leaving group than $(R_2N)_2SOF^-$ and, as a result, the intermediate (A) has more ionic character in the first case. The stronger carbonium ion character of R^+ is established experimentally by the greater proportion of rearranged products. The solvent dependence of these reactions supports the mechanism given (*100*).

C. TETRACOORDINATED ANIONS

Table III shows the theoretical possibilities for tetracoordinated nitrogen–sulfur–fluorine anions. In discussing these numerous possibilities systematically we will take as our starting point the monobasic fluorooxy acids and the neutral oxyfluorides.

On paper we would expect the OH^- groups in them to be replaceable by —$\overline{N}$—R and the doubly bound oxygen by $R\overline{N}=$ or the iso-
  H
electronic $\overline{N}^-=$ group. Although the possibilities for S(VI) are quite numerous, the free electron pair in S(IV) limits this variation very much, and for S(II) only a single possibility exists.

In Table III account has been taken of the fact that a singly bonded OH group adjacent to a doubly bonded nitrogen is unstable. Rearrangement always takes place to the energetically favored S=O— doubly bonded and —S—N— singly bonded system. A further rule

TABLE III

Tetrahedrally Coordinated Nitrogen–Sulfur–Fluorine Anions Derived from Fluorooxy Acids and Oxyfluorides

Acid	Oxidation state	Parent acid	Nitrogen derivatives	Parent oxifluoride	Nitrogen derivatives
Monobasic	VI	$F{-}\overset{O}{\underset{O}{S}}{-}OH$	$F{-}\overset{O}{\underset{O}{S}}{-}\overset{H}{N}{-}R$ (Fluorosulfonyl-amides) $F{-}\overset{O}{\underset{N{-}R}{S}}{=}\overset{H}{N}{-}R$ (Iminofluorosulfonyl-amides) $F{-}\overset{N{-}R}{\underset{N{-}R}{S}}{=}\overset{H}{N}{-}R$ (Bisiminofluoro-sulfonylamides)	OSF_2O	$HN{=}\overset{F}{\underset{F}{S}}{=}O$ (Sulfuroxy-difluorideimide) $HN{=}\overset{F}{\underset{F}{S}}{=}NR$ (Iminosulfur-difluorideimides)
	IV	$F{-}\overset{(:)}{\underset{O}{S}}{-}OH$	$F{-}\overset{(:)}{\underset{O}{S}}{-}\overset{H}{N}R$ (Fluorosulfinicamides) $F{-}\overset{(:)}{\underset{N{-}R}{S}}{=}\overset{H}{N}R$ (Iminofluorosulfinicamides)	$\ominus SF_2O$	$HN{=}\overset{F}{\underset{F}{\overset{..}{S}}}{\ominus}$ (Sulfurdifluorideimides)
	II	$F\overset{(:)}{\underset{(:)}{S}}{-}OH$	$F{-}\overset{(:)}{S}{-}\overset{H}{N}R$ (Fluorosulfenicamides)	—	—

continued

TABLE III—Continued

Acid	Oxidation state	Parent acid	Nitrogen derivatives		Parent oxifluoride	Nitrogen derivatives
Dibasic	VI	$F-\overset{O}{\underset{O}{S}}-OH$	$F-\overset{O}{S}-O-H \longrightarrow F-SO$ with NH below, NH_2 Fluorosulfonicamide	$FS-\overset{O}{\underset{NH}{\parallel}}-N-R \longrightarrow \overset{O}{FS}=NR$, NH_2 Iminofluorosulfonicamide	OSF_2O	$HN=SF_2=NH$ Bis(imino)-sulfurdifluorideimide
			$FS=\overset{NH}{\underset{N-R}{=}}\overset{H}{NR} \longrightarrow FS=\overset{NH_2}{\underset{NR}{NR}}$	Bisiminofluorosulfonicamide		
	IV	$F-\overset{(\cdot\cdot)}{\underset{O}{S}}-OH$	$FS-\overset{(\cdot\cdot)}{\underset{NH}{\parallel}}OH \longrightarrow F-S-NH_2$ with O Fluorosulfinicamide	$FS-\overset{(\cdot\cdot)}{\underset{NH}{\parallel}}NR \longrightarrow F-\overset{(\cdot\cdot)}{S}=NR$, NH_2 Iminofluorosulfinicamide	$\subset SF_2O$	—
Tribasic	VI	$\overset{O}{FS}-OH$ with O	$FS-\overset{NH}{\underset{NH}{\parallel}}OH \longrightarrow FS=O$ with NH_2 Amidosulfuroxyfluoride-imide	$F-\overset{NH}{\underset{NH_2}{\parallel}}S=NR$ Iminoamidosulfurfluoride-imide	OSF_2O	—

appears to be that imido groups go over to amides in so far as this is possible.

1. Fluorosulfonylamide Ions $FS\overset{O}{\underset{O}{—}}\overline{N}—R^-$

Since fluorosulfonic acid is one of the strongest known acids (75), the fluorosulfonylamides, in which the OH group of FSO_2OH is replaced by —NHR, should also show acidic properties. The acid strength in this case depends on the nature of R and, as would be expected, increases roughly in the order alkyl < carboxy < perfluoroalkyl < derivatives with inorganic groups (e.g., SO_2F or SO_2CF_3); the latter will be described in a special section.

Starting mostly from FSO_2NH_2 and FSO_2NCO, it is possible to prepare a large number of compounds of the type FSO_2NHR (58). e.g.,

$$\underset{O}{\overset{O}{FSNCO}} + RH \longrightarrow \underset{O}{\overset{O}{FS}}—\overset{H}{N}—\overset{O}{C}—R \quad (24,\ 126)$$

$$(R = OH,\ NRR')$$

$$\underset{O}{\overset{O}{FSNCO}} + RCOOH \xrightarrow{-CO_2} \underset{O}{\overset{O}{FS}}—\overset{H}{\underset{}{N}}—\overset{O}{C}—R \quad (121,\ 123)$$

$$(R = \text{alkyl, haloalkyl})$$

(17)

$$\underset{O}{\overset{O}{ClSN}}{=}C\overset{Cl}{\underset{Cl}{\diagdown}} + HF\ (\text{excess}) \xrightarrow{-HCl} \underset{O}{\overset{O}{FS}}—\overset{H}{N}—CF_3 \quad (112,\ 113) \tag{18}$$

$$\underset{O}{\overset{O}{FSN}}{=}PCl_3 + CF_3COOH \longrightarrow \underset{O}{\overset{O}{FS}}—\overset{H}{N}—\overset{O}{C}—CF_3 + POCl_3 \quad (123) \tag{19}$$

$$R_3SiCl + FSO_2NH_2 \xrightarrow{Et_3N} \underset{O}{\overset{O}{FS}}—\overset{H}{N}—SiR_3 + Et_3NHCl \quad (115) \tag{20}$$

$$FSO_2NH_2 + OSCl_2 \xrightarrow{reflux} FSO_2NSO + \underset{O}{\overset{O}{FS}}—\overset{H}{N}—\overset{H}{N}—\overset{O}{SF}$$

$$+ \text{other products} \quad (111) \quad (21)$$

The acidic character of the above compounds may be shown in part by the formation of the corresponding salts with Ph_4PCl or Ph_4AsCl in aqueous solution, e.g.,

$$\underset{O}{\overset{O}{FSNHR}} + Ph_4PCl \xrightarrow[-HCl]{HOH} Ph_4P^+ \underset{O}{\overset{O}{FSNR}}{}^- \quad (126,\ 112) \tag{22}$$

$$(R = CH_3O—\overset{O}{C}—,\ CF_3—)$$

Mercury salts are obtained by reaction with Hg_2CO_3 in benzene (15):

$$Hg_2CO_3 + 2F\overset{O}{\underset{O}{\overset{\|}{\underset{\|}{S}}}}NHR \longrightarrow Hg_2\left(N\overset{R}{\underset{SO_2F}{}}\right)_2 + H_2O + CO_2$$

$$[R = \overset{O}{\overset{\|}{-C}}-OC_2H_5, \ \overset{O}{\overset{\|}{-C}}-OCH_3, \ \overset{O}{\overset{\|}{-C}}-N(C_2H_5)_2] \qquad (23)$$

$$Hg_2CO_3 + F\overset{O}{\underset{O}{\overset{\|}{\underset{\|}{S}}}}NH-\overset{O}{\overset{\|}{C}}-NH\overset{O}{\underset{O}{\overset{\|}{\underset{\|}{S}}}}F \longrightarrow [-Hg-N-\overset{SO_2F}{\underset{O}{\overset{\|}{C}}}-N-Hg-]_n$$

A very simple way for preparing salts of the type $M^+N\overset{COF^-}{\underset{SO_2F}{}}$

is by reaction of FSO_2NCO with alkali fluorides (110):

$$MF + F\overset{O}{\underset{O}{\overset{\|}{\underset{\|}{S}}}}NCO \xrightarrow[CH_3CN]{25°C} M^+N\overset{COF^-}{\underset{SO_2F}{}} \qquad (M = Na, K, Cs) \qquad (24)$$

Transfer of the $N\overset{COF}{\underset{SO_2F}{}}$ group takes place in the reaction of the
alkali metal salts with $S_2O_6F_2$ (110):

$$M^+N\overset{COF^-}{\underset{SO_2F}{}} + S_2O_6F_2 \longrightarrow F\overset{O}{\underset{O}{\overset{\|}{\underset{\|}{S}}}}ON\overset{COF}{\underset{SO_2F}{}} + MOSO_2F \qquad (25)$$

The reaction should occur by a radical mechanism rather than by
an ionic.

The N–halogen derivatives $F\overset{O}{\underset{O}{\overset{\|}{\underset{\|}{S}}}}-NHX$ are particularly interesting.

Whereas $F\overset{O}{\underset{O}{\overset{\|}{\underset{\|}{S}}}}NHF$ results from direct fluorination of $F\overset{O}{\underset{O}{\overset{\|}{\underset{\|}{S}}}}NH_2$ (114),

$$\mathrm{F\overset{O}{\underset{O}{S}}NH_2 + F_2 \longrightarrow F\overset{O}{\underset{O}{S}}NHF + HF} \tag{26}$$

the corresponding N—Cl compound is obtained by hydrolysis of the dichloroamine (*117*),

$$\mathrm{F\overset{O}{\underset{O}{S}}NSO + 2ClF \longrightarrow F\overset{O}{\underset{O}{S}}NCl_2 + OSF_2} \tag{27}$$

$$\mathrm{F\overset{O}{\underset{O}{S}}NCl_2 + HOH \longrightarrow F\overset{O}{\underset{O}{S}}NHCl + HOCl} \tag{27a}$$

No salts of $\mathrm{F\overset{O}{\underset{O}{S}}NHF}$ have been described, but the chlorine derivative has been isolated in the form of its tetraphenylphosphonium or arsonium salts $\mathrm{Ph_4M^+N\begin{smallmatrix}\diagup Cl\\ \diagdown SO_2F\end{smallmatrix}}$ (M = P, As).

If a second sulfonyl group (FSO_2, CF_3SO_2) is introduced as the substituent R in the fluorosulfonylamide $\mathrm{F\overset{O}{\underset{O}{S}}NHR}$, there is a drastic increase in the acid strength [pK_a (FSO_2NHSO_2F) = 1.28 (*132*)].

The bis(fluorosulfonyl)amides are best prepared by reaction of $(ClSO_2)_2NH$ with AsF_3 (*136*) [other preparative methods are given in (Ref. *58*)]:

$$\mathrm{Cl\overset{O}{\underset{O}{S}}NH_2 + PCl_5 \xrightarrow{-2HCl} Cl\overset{O}{\underset{O}{S}}{-}N{=}PCl_3} \tag{28a}$$

$$\mathrm{Cl\overset{O}{\underset{O}{S}}N{=}PCl_3 + ClSO_3H \xrightarrow[-POCl_3]{} (Cl\overset{O}{\underset{O}{S}})_2NH} \tag{28b}$$

$$\mathrm{3(Cl\overset{O}{\underset{O}{S}})_2NH + 2AsF_3 \xrightarrow{-2AsCl_3} 3(F\overset{O}{\underset{O}{S}})_2NH} \tag{28c}$$

The compound $CF_3SO_2NHSO_2F$ is obtained by the following reaction (*122*):

$$\mathrm{CF_3\overset{O}{\underset{O}{S}}NPCl_3 + FSO_3H \xrightarrow[-POCl_3]{} CF_3\overset{O}{\underset{O}{S}}{-}\overset{H}{N}{-}\overset{O}{\underset{O}{S}}F} \tag{29}$$

The reverse route using CF_3SO_3H does not, however, give the desired product (122):

$$\overset{O}{\underset{O}{FSNPCl_3}} + 2CF_3SO_3H \longrightarrow \overset{O}{\underset{O}{FSNHPOCl_2}} + HCl + \overset{O\ O}{\underset{O\ O}{CF_3SOSCF_3}} \quad (29a)$$

Bissulfonylamides react with metallic oxides and carbonates to give the corresponding salts:

$$M_2CO_3 + 2HN(SO_2F)_2 \longrightarrow 2M^+N(SO_2F)_2{}^- + H_2O + CO_2 \quad (132, 15) \quad (30)$$
$$[M = Na, K, Cs, Hg(I)]$$

$$Ag_2O + 2HN\overset{\displaystyle /SO_2F}{\underset{\displaystyle \backslash SO_2R}{}} \xrightarrow{-H_2O} 2AgN\overset{\displaystyle /SO_2F}{\underset{\displaystyle \backslash SO_2R}{}} \quad (132, 122) \quad (30a)$$

$$(R = F, CF_3)$$

They cleave the tin–carbon bond in $Sn(CH_3)_4$ in the same way as do strong oxy acids (94):

$$Sn(CH_3)_4 + HN(SO_2F)_2 \longrightarrow (CH_3)_3SnN(SO_2F)_2 + CH_4 \quad (31)$$

An analogous behavior is also observed in the reaction with xenon difluoride, and this led recently to the synthesis of the first noble gas–nitrogen compound (84):

$$XeF_2 + HN(SO_2F)_2 \xrightarrow[\substack{0°C/4d. -HF}]{CF_2Cl_2} \underset{89.2\%}{FXeN(SO_2F)_2} \quad (32)$$

Bis(fluorosulfonyl)aminoxenon fluoride is a white solid that decomposes quantitatively at 70° according to (84):

$$2FXeN(SO_2F)_2 \longrightarrow Xe + XeF_2 + [(FSO_2)_2N]_2 \quad (33)$$

The existence of the xenon compound shows that the $FS\overset{O}{\underset{O}{—}}\overline{N}\overset{O}{\underset{O}{—}}SF^-$ ion is extraordinarily stable and has a high electronegativity. It resembles the halogens in its behavior, the radicals combining to form a dimer (133); with fluorine and chlorine the "interhalogen" compounds $(FSO_2)_2NF$ (136) and $(FSO_2)_2NCl$ are obtained (133):

$$2(FSO_2)_2NCl \xrightarrow{UV} [(FSO_2)_2N]_2 + Cl_2 \quad (34)$$

$$(FSO_2)_2NH + F_2 \longrightarrow (FSO_2)_2NF + HF \quad (35)$$

$$(FSO_2)_2N^-Ag^+ + Cl_2 \longrightarrow (FSO_2)_2NCl + AgCl \quad (36)$$

The N—Cl compound prepared from the $(FSO_2)_2\overline{N}^-$ ion provides a key to the synthesis of organobis(fluorosulfonyl)amides. It may be added by either a radical or a polar mechanism to multiple bonds. Insertion of methylenes into the N—Cl bond is likewise possible.

$$
(FSO_2)_2NCl \begin{cases} \xrightarrow{+\,CO} & \underset{Cl}{\overset{O}{\underset{\|}{C}}}{-}N(SO_2F)_2 \quad (133) \\[2mm] \xrightarrow{+\,XCN} & XCN \cdot ClN(SO_2F)_2 \quad (133) \\[2mm] \xrightarrow{+\,CF_3CF=CF_2} & CF_3{-}CFCl{-}CF_2N(SO_2F)_2 \quad (156) \\[2mm] \xrightarrow[-\,N_2]{+\,(CF_3)_2CN_2} & (CF_3)_2C(Cl)N(SO_2F)_2 \quad (156) \end{cases} \tag{37}
$$

Alkyl halides will also react directly with the silver salt (132):

$$
AgN(SO_2F)_2 + CH_3I \longrightarrow CH_3N(SO_2F)_2 + AgI \tag{38}
$$

In this way it was possible to introduce the $N(SO_2F)_2$ group into transition metal chemistry (46, 96):

$$
M(CO)_5Br + AgN(SO_2F)_2 \xrightarrow[-\,AgBr]{CH_2Cl_2} M(CO)_5N(SO_2F)_2 \tag{39}
$$
$$
(M = Mn, Re)
$$
$$
Mn(CO)_4\,(Ph_3P, As)Br + AgN(SO_2F)_2 \xrightarrow{-\,AgBr} Mn(CO)_4\,(Ph_3P, As)N(SO_2F)_2 \tag{39a}
$$
$$
C_5H_5Fe(CO)_2I + AgN(SO_2F)_2 \xrightarrow{-\,AgI} C_5H_5Fe(CO)_2N(SO_2F)_2 \tag{39b}
$$
$$
C_5H_5Cr(NO)_2Cl + AgN(SO_2F)_2 \xrightarrow{-\,AgCl} C_5H_5Cr(NO)_2N(SO_2F)_2 \tag{39c}
$$

The transition metal–nitrogen bond in these complexes is highly polar, and, in their reactions with ligands such as CH_3CN, Ph_3P, and Ph_3As, the first step is always exchange of the bis(fluorosulfonyl)amino group with formation of ionic compounds (46, 96):

$$
M(CO)_5N(SO_2F)_2 + CH_3CN \longrightarrow [M(CO)_5CH_3CN]^+N(SO_2F)_2{}^- \tag{40}
$$
$$
M(CO)_5N(SO_2F)_2 + Ph_3P \longrightarrow [M(CO)_5(Ph_3P)]^+N(SO_2F)_2{}^- \tag{40a}
$$
$$
(M = Mn, Re)
$$
$$
C_5H_5Fe(CO)_2N(SO_2F)_2 + L \longrightarrow [C_5H_5Fe(CO)_2L]^+N(SO_2F)_2{}^- \tag{40b}
$$
$$
(L = CH_3CN, Ph_3P)
$$

Closely related to the acids discussed above is compound SS-bis-(sulfur oxydifluorideimidosulfonyl)amide (17):

$$
\underset{SO_2Cl}{\overset{SO_2Cl}{HN{\Big\langle}}} + Hg(NSOF_2)_2 \longrightarrow HN{\Big\langle}\begin{matrix} \overset{O}{\underset{O}{S}}{-}N{=}S{\overset{O}{\underset{F}{\diagup}}}^F_F \\[4mm] \overset{O}{\underset{O}{S}}{-}N{=}S{\overset{O}{\underset{F}{\diagdown}}}^F_F \end{matrix} + HgCl_2 \tag{41}
$$

In its reactions it behaves like the other bis(sulfonyl) amides (17):

$$HN(SO_2NSOF_2)_2 \begin{cases} \xrightarrow{+AgCF_3COO} Ag^+ N(SO_2NSOF_2)_2^- \\ \xrightarrow{+Li-n-C_4H_9} Li^+ N(SO_2NSOF_2)_2^- \\ \xrightarrow[\text{in } H_2O]{+Ph_4MCl} Ph_4M^+ N(SO_2NSOF_2)_2^- \end{cases} \quad (42)$$

(M = P, As)

a. *Iminofluorosulfonylamide Ions,* $FS\overset{\displaystyle O}{\underset{\displaystyle NR}{\overline{\parallel\!\!\!-N\!-R'^-}}}$. In the imino-fluorosulfonylamides, an oxygen atom of the fluorosulfonylamide is formally replaced by an isosteric RN= group. They may be prepared by the following methods:

i. Decomposition of sulfur oxydifluoride imides with primary amines (119),

$$\underset{O}{\overset{O}{FS}}\!-\!N\!=\!\underset{F}{\overset{O,\,F}{S}} + 2CH_3NH_2 \longrightarrow \underset{O}{\overset{O}{FS}}\!-\!N\!=\!\underset{F}{\overset{O\ H}{S}}\!-\!NCH_3 + CH_3NH_3F \quad (43)$$

ii. By hydrolysis of bis(imino)sulfur difluorides (127),

$$\underset{O}{\overset{O\quad F\quad O}{FS\!-\!S\!=\!N\!-\!SF}} + OH^- \xrightarrow{-F^-} \left\{ \underset{O}{\overset{O}{FS}}\!-\!N\!\overset{H\!-\!O}{\underset{F}{\overset{\nwarrow\,\searrow}{=\!S}}}\!=\!N\!-\!\underset{O}{\overset{O}{SF}} \right\}$$

$$\longrightarrow \underset{O}{\overset{O\ H}{FS\!-\!N}}\!-\!\overset{O}{\underset{F}{\overset{\parallel}{S}}}\!=\!N\!-\!\underset{O}{\overset{O}{SF}} \quad (44)$$

The hydrolysis product is isolated as the tetraphenylphosphonium salt. Replacement of one oxygen in the fluorosulfonylamides by imino groups reduces the acidity, but it is possible by changing R and R' in

$$\underset{O}{\overset{F}{\diagdown}}\!S\!\underset{\underset{H}{NR'}}{\overset{NR}{\diagup}}$$

to vary the acid strength at will.

b. *Bis(imino)fluorosulfonylamide Ions,* $F\overset{\displaystyle NR}{\underset{\displaystyle NR'}{\overline{\parallel\!\!\!-S\!-\overline{N}\!-R''^-}}}$. Although the reactions of sulfur difluoride diimides with secondary amines have been described (*118*),

$$RN{=}\underset{F}{\overset{F}{S}}{=}NR + 2R'_2NH \longrightarrow RN{=}\underset{NR'_2}{\overset{F}{S}}{=}NR + R'_2NH_2F \qquad (45)$$

there are no reports on the reactions with primary amines, which would give the desired type of compound.

c. Sulfur Oxydifluorideimide Ion, $O{=}\underset{F}{\overset{F}{S}}{=}\overline{\overline{N}}^-$. Several methods are available for preparing the free acid:

$$OSF_4 + 3NH_3 \longrightarrow O{=}\underset{F}{\overset{F}{S}}{=}NH + 2NH_4F \quad (108) \qquad (46)$$

$$Hg(NSOF_2)_2 + 2HCl \longrightarrow 2O{=}\underset{F}{\overset{F}{S}}{=}NH + HgCl_2 \quad (149) \qquad (46a)$$

Sulfur oxydifluoride imide has also been detected as a product of the hydrolysis of NSF$_3$ (*124*). This reaction is especially interesting as it illustrates a general principle in sulfur–nitrogen chemistry. It has been found in all hydrolysis reactions so far investigated that proton migration occurs if a sulfur–oxygen double bond can form at the expense of a sulfur–nitrogen multiple bond:

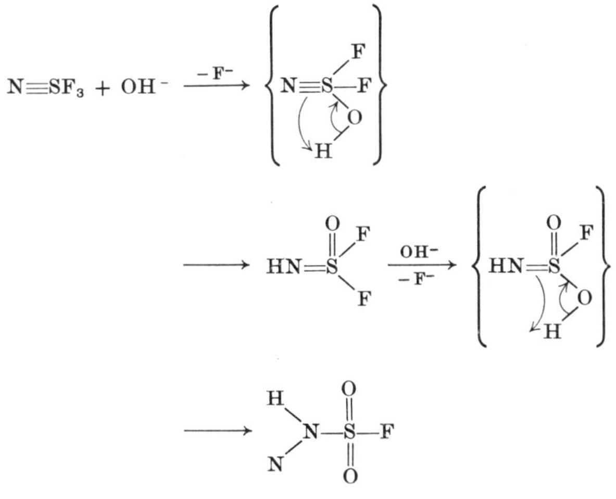

This reaction principle is not restricted to linear thiazenes but is also encountered in S—N heterocycles (Section V, A, 2).

Salts of sulfur oxydifluoride imide are obtained by cleavage of the

CN bond in COFNSOF$_2$ or of the SiN bond in R$_3$SiNSOF$_2$ (*59, 64, 134, 148*):

$$COFNSOF_2 + CsF(HgF_2) \xrightarrow{-COF_2} Cs^+ NSOF_2^- [Hg(NSOF_2)_2] \tag{48}$$

$$2R_3SiNSOF_2 + HgF_2 \xrightarrow{-2R_3SiF} Hg(NSOF_2)_2 \tag{48a}$$

The method that makes use of the acid character of the imide is, however, particularly simple (*43*):

$$HN\!\!=\!\!S\!\!=\!\!O \begin{cases} + M_2CO_3 \xrightarrow{ether} 2MNSOF_2 + H_2O + CO_2 \\ + Ag_2O \longrightarrow 2AgNSOF_2 + H_2O \\ + 2NH_3 \longrightarrow 2NH_4^+ NSOF_2^- \ (146) \end{cases} \tag{49}$$

(with F, F on the sulfur)

Decomposition of the cesium, mercury, or silver salts with halogens leads to the corresponding N–halogen compounds (*43, 59, 134, 149*):

$$M(NSOF_2)_n + nX_2 \longrightarrow nXNSOF_2 + MX_n \tag{50}$$
$$(n = 1,2; \ M = Cs, Ag, Hg; \ X = F, Cl, Br, I)$$

These N–halogen compounds provide a route to organosulfur oxydifluoride imides:

$$XNSOF_2 \begin{cases} + R\!\!-\!\!C\!\!=\!\!C \xrightarrow{UV} R\!\!-\!\!\overset{X}{\underset{}{C}}\!\!-\!\!C\!\!-\!\!NSOF_2 + R\!\!-\!\!\overset{}{C}\!\!-\!\!\overset{}{\underset{NSOF_2}{C}}\!\!-\!\!X \quad (78, 95, 156) \\ \\ + (CF_3)_2CN_2 \xrightarrow{UV} \overset{F_3C}{\underset{F_3C}{>}}C\overset{X}{\underset{NSOF_2}{<}} \quad (156) \end{cases} \tag{51}$$

(X = Cl)

The mercury salt is very suitable for the preparation of inorganic sulfur oxydifluoride imides, and the NSOF$_2$ ion can be transferred to a large number of elements by reaction with the corresponding halides (*18, 72, 76, 129, 153*):

$$MCl_n + (n/2)Hg(NSOF_2)_2 \xrightarrow[CH_2Cl_2]{40°} M(NSOF_2)_n + (n/2)HgCl_2 \tag{52}$$

M =	B	Sb	Sb	OSe	OCNSO$_2^-$	CH$_3$O—S(O)—	Cl$_3$PN—SO$_2$—
n =	3	3	"5"	2	1	1	1

M =	P	As	(CH$_3$)$_{4-n}$Si	OS	FSO$_2$NS	(CH$_3$)$_3$Sn
n =	3	3	1–4	2	2	1

An interesting rearrangement is observed in the reaction with COFNSCl$_2$ (*18*):

$$COFNSCl_2 + Hg(NSOF_2)_2 \longrightarrow HgCl_2 + \left\{ COF-N=S\underset{NSOF_2}{\overset{NSOF_2}{\diagup}} \right\}$$

$$\downarrow \sim \qquad (53)$$

$$\overset{O}{\underset{\parallel}{F_2S}}=N-\overset{O}{\underset{\parallel}{C}}-N=\overset{F}{\underset{\mid}{S}}-N=\overset{O}{\underset{\parallel}{SF_2}}$$

In many cases $AgNSOF_2$ has proved to be a better reagent, because AgCl is less volatile and more insoluble than $HgCl_2$ and, thus, no separation problems arise. Alkyltin, phosphoryl, and thiophosphoryl derivatives have been prepared in this way, the composition of the products depending on the stoichiometric ratio taken (44):

$$AgNSOF_2 \begin{cases} + (CH_3)_2SnX_2 \longrightarrow & \begin{array}{l} R_2Sn(NSOF_2)_2 \\ R_2SnClNSOF_2 \\ R_2SnBrNSOF_2 \end{array} \\ + OPCl_3 \longrightarrow & OP(NSOF_2)_3, OPCl(NSOF_2)_2, OPCl_2(NSOF_2) \quad (54) \\ + OPFCl_2 \longrightarrow & OPF(NSOF_2)_2, OPFCl(NSOF_2)_2 \\ + SPCl_3 \longrightarrow & SP(NSOF_2)_3, SPCl(NSOF_2)_2, SPF(NSOF_2)_2, \\ & SPCl_2NSOF_2 \end{cases}$$

The silver salt has also been used to introduce the $NSOF_2$ group into transition metal chemistry (97):

$$Re(CO)_5Br + AgNSOF_2 \xrightarrow[CH_2Cl_2]{RT} Re(CO)_5NSOF_2 + AgBr \qquad (55)$$

$$\downarrow 90°C, \textit{n}\text{-heptane}$$

$$\tfrac{1}{2}[Re(CO)_4NSOF_2]_2 + 2CO$$

$$2Mn(CO)_5Br + 2AgNSOF_2 \xrightarrow[CH_2Cl_2]{RT} [Mn(CO)_4NSOF_2]_2 + 2AgBr + 2CO \quad (55a)$$

Compounds $C_5H_5Fe(CO)_2NSOF_2$ and $C_5H_5Cr(NO)_2NSOF_2$ may be similarly prepared from the corresponding iodide or chloride (45).

Structural investigations show that bridging in the dimeric Re and Mn compounds occurs only through the nitrogen atom of the $NSOF_2$ group (20). Bond distances and angles in the geminal $NSOF_2$ group of $Re(CO)_5NSOF_2$ hardly deviate from the values found for the bridging ligand itself (Figs. 1 and 2).

Dimers also occur with substituted Mn tetracarbonyl halogenides (45):

$$2Mn(CO)_4(Ph_3X)Br + 2AgNSOF_2 \longrightarrow [Mn(CO)_3(Ph_3X)NSOF_2]_2 + 2AgBr + 2CO$$
$$(X = P, As, Sb) \qquad (56)$$

Mn – N 2.116 (6)
Mn – N' 2.108(6)
N – S 1.434(6)
S – O 1.402(8)
S – F 1.556(7)
S – F' 1.535(7)

N–Mn–N 76.5°
Mn–N–Mn 103.6°
F – S – F 90.6°

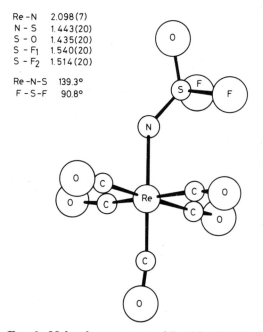

FIG. 1. Molecular structure of $[Mn(CO)_4NSOF_2]_2$.

Re – N 2.098(7)
N – S 1.443(20)
S – O 1.435(20)
S – F_1 1.540(20)
S – F_2 1.514(20)

Re –N–S 139.3°
F – S – F 90.8°

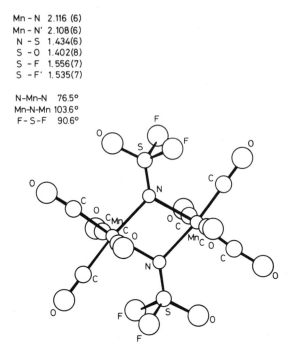

FIG. 2. Molecular structure of $Re(CO)_5NSOF_2$.

The bridge structure remains unchanged in the reaction of $[Mn(CO)_4NSOF_2]_2$ with nitrogen bases (94):

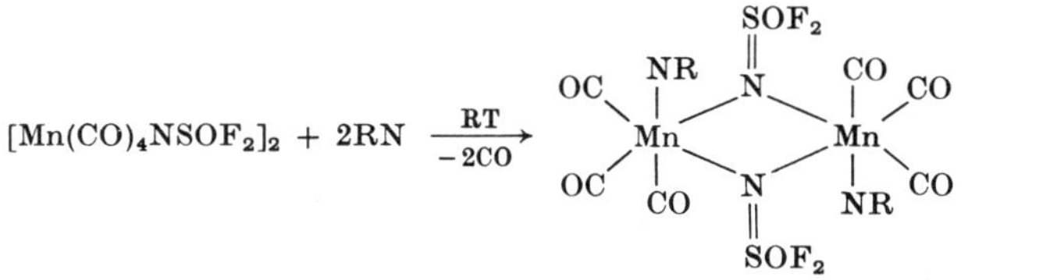

$$[Mn(CO)_4NSOF_2]_2 + 2RN \xrightarrow[-2CO]{RT}$$

(57)

(NR = pyr, CH$_3$CN)

The rhenium compound does not react under these conditions. With Ph$_3$P and Ph$_3$As, bridge cleavage occurs (94):

$$[M(CO)_4NSOF_2]_2 + 4Ph_3X \longrightarrow M(CO)_3(Ph_3X)_2NSOF_2 + 2CO \qquad (58)$$

(M = Mn, Re; X = P, As)

d. Iminosulfur Difluorideimide Ions, $RN{=}S{=}N^-$ (with two F on S). One representative of this class of compound is prepared by the hydrolysis of

$$N{\equiv}SF_2{-}N{=}S{=}O \quad (54)$$

(with two F on S)

the acid being isolated as the tetraphenylarsonium salt,

$$FSN{=}S{=}\bar{N}^- \; Ph_4As^+$$

(with O, O on first S and F, F on second S)

$$|N{\equiv}S{-}N{=}S{=}O + OH^- \xrightarrow{-F^-} \left\{ \begin{array}{c} \end{array} \right\} \longrightarrow H{-}N{=}S{=}N{-}SO_2F$$

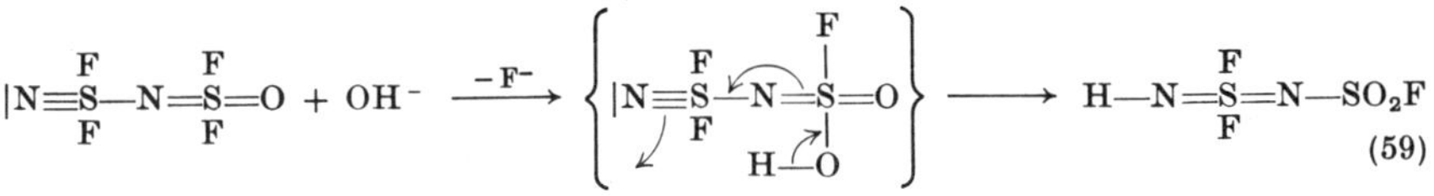

(59)

Structural investigations (Fig. 3) show clearly that there are three different S—N distances, the bond to the geminal nitrogen being the shortest (1.439 Å), followed by that between the bridge nitrogen atom and the FSO$_2$ group (1.517 Å). These S–N bond distances suggest that the structure of this special iminosulfur difluoride is better described as difluorothiazyl fluorosulfonylamide

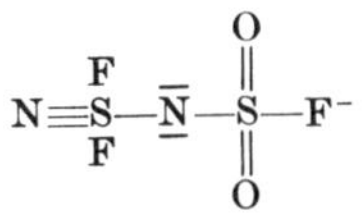

$$N{=}S{-}\bar{N}{-}\overset{O}{\underset{O}{\overset{\|}{\underset{\|}{S}}}}{-}F^-$$

(the $N{\equiv}SF_2^-$ group being isoelectronic to FSO_2^-) This ion should be structurally closely related to $(FSO_2)_2N^-$.

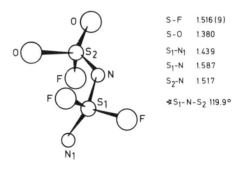

S–F	1.516 (9)
S–O	1.380
S_1–N_1	1.439
S_1–N	1.587
S_2–N	1.517
$\not< S_1$–N–S_2	119.9°

FIG. 3. Molecular structure of $NSF_2NSO_2F^-$ (*19*).

2. Fluorosulfinamide, $F\overset{(\cdot\cdot)}{\underset{\parallel}{\overset{}{S}}}\text{—}\overline{\overline{N}}\text{—}R^-$, and Iminofluorosulfinamide Ions,

$$F\overset{(\cdot\cdot)}{\underset{\parallel}{S}}\text{—}\overline{\overline{N}}R^-$$
$$NR$$

If the systematic treatment is continued for S(IV) in the same way as for S(VI), the corresponding starting point is fluorosulfinic acid

$$\underset{(\cdot\cdot)}{\overset{O}{\diagdown}}S\underset{OH}{\overset{F}{\diagup}}$$ or thionyl fluoride $$\underset{(\cdot\cdot)}{\overset{O}{\diagdown}}S\underset{F}{\overset{F}{\diagup}}$$. In this case the theoretical

possibilities for the formation of tetrahedrally coordinated anions are limited since the free pair of electrons can occupy the position of the double-bonded oxygen and also that of the imino group. As numerous examples show, sulfur(IV) has some tendency to assume a coordination number of 3. Fluorosulfinic acid itself decomposes to HF and OSO (*145*), and similar reactions are to be expected for the sulfur(IV)–nitrogen system:

$$OSF_2 + RNH_2 \xrightarrow{-HF} \left\{ F\overset{O}{\underset{(\cdot\cdot)}{\overset{\parallel}{S}}}\text{—}\overset{H}{N}\text{—}R \right\} \xrightarrow[-\text{base H}^+]{+\text{ base}}$$

$$\left\{ F\text{—}\overset{O}{\underset{(\cdot\cdot)}{\overset{\parallel}{S}}}\text{—}\overline{N}\text{—}R^- \right\} \xrightarrow{-F^-} \underset{O}{\overset{(\cdot\cdot)}{\diagup}}S\diagdown_{NR} \qquad (69) \quad (60)$$

$$R\!-\!N\!=\!SF_2 + R'NH_2 \xrightarrow{-HF} \left\{ \begin{array}{c} N\!-\!R \\ \| \quad H \\ F\overset{..}{S}\!-\!N\!-\!R' \end{array} \right\} \xrightarrow[-\text{base H}^+]{+\text{base}}$$

$$\left\{ \begin{array}{c} N\!-\!R \\ \| \\ F\!-\!\overset{..}{S}\!-\!N'\!-\!R'- \end{array} \right\} \xrightarrow{-F^-} \begin{array}{c} \overset{..}{S} \\ RN \diagdown \quad \diagup NR' \end{array} \qquad (32) \quad (61)$$

On the other hand, salts of fluorosulfinic acid may be prepared by addition of F^- to SO_2 (*144, 145*):

$$MF + \underset{O}{\overset{..}{\underset{\diagdown}{S}}}{\diagup}_{O} \longrightarrow M^+O\!=\!\overset{..}{\underset{F}{S}}\!=\!O^- \qquad (62)$$

If one or both of the oxygen atoms in SO_2 are replaced by the isosteric $RN=$ group, the F^- acceptor property of the central sulfur atom is reduced. Reaction analogous to that in Eq. (62) might be successful only if R = perfluoroalkyl or an inorganic group.

a. Sulfur Difluorideimide Ion, $\overset{F}{\underset{F}{\diameter S}}\!=\!\overline{\underline{N}}^-$. The free acid $HN\!=\!\overset{F}{\underset{F}{S\diameter}}$ is unstable. Earlier unsuccessful attempts to prepare it have recently been confirmed:

$$NH_3 + SF_4 \xrightarrow{-NH_4F} \{HN\!=\!\underline{S}F_2\} \xrightarrow[-NH_4F]{NH_3} N\!\equiv\!\overline{\underline{S}}\!-\!F \quad (58) \qquad (63)$$

$$Ph_3P\!=\!NH + SF_4 \xrightarrow{-Ph_3PF_2} \{HN\!=\!\underline{S}F_2\} \xrightarrow{-HF} N\!\equiv\!\overline{\underline{S}}\!-\!F \quad (2) \qquad (63a)$$

the NSF is decomposed in liquid HF (*94*):

$$|N\!\equiv\!\overline{\underline{S}}F \xrightarrow{HF} \left\{ H\!-\!N\!=\!\overset{..}{\underset{F}{S}}\!-\!F \right\} \xrightarrow{HF} \left\{ H_2N\!-\!\overset{F}{\underset{F}{S}}\!-\!F \right\} \xrightarrow{HF} NH_4F + \diameter SF_4$$

$$(63b)$$

Salts of sulfur difluorideimide may, however, be prepared (i) by addition of F^- to NSF (*50, 134*),

$$N\!\equiv\!S\overset{F}{\underset{\diameter}{\diagup}} + CsF \longrightarrow Cs^+\overline{\underline{N}}\!=\!\overset{..}{\underline{S}}F_2^- \qquad (64)$$

or (ii) by cleavage of the C—N bond in acylsulfur difluorideimides with CsF or HgF_2 (*52, 60, 134*),

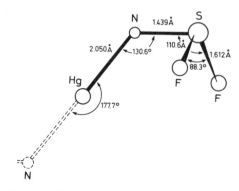

FIG. 4. Molecular structure of $Hg(NSF_2)_2$.

$$\underset{(R = F, CF_3)}{R-\overset{O}{\overset{\|}{C}}-NSF_2 + CsF} \longrightarrow Cs^+\overline{N}=\underset{=}{S}F_2{}^- + Cs^+RCF_2O^- \qquad (64a)$$

$$2R\overset{O}{\overset{\|}{C}}NSF_2 + HgF_2 \longrightarrow Hg(N\underset{=}{S}F_2)_2 + 2RCOF \qquad (64b)$$

The structure of the mercury salt has been determined (80) (Fig. 4).

Reaction of the mercury salt with chlorides has so far resulted in chlorine–fluorine exchange and the splitting off of NSF, but not in transfer of the NSF_2 group. With halogens, the corresponding N–halogen compounds are obtained:

$$\underset{(X = Cl, Br)}{CsNSF_2 + X_2} \longrightarrow CsX + XNSF_2 \quad (134) \qquad (65)$$

$$\underset{(X = F, Cl, Br, I)}{Hg(NSF_2)_2 + 2X_2} \longrightarrow HgX_2 + 2XNSF_2 \quad (60, 61) \qquad (65a)$$

These compounds open up a valuable route to organosulfur difluoride-imides, as has been demonstrated with $ClNSF_2$ and $BrNSF_2$ (58, 156):

$$XNSF_2 \begin{cases} + \underset{/}{\overset{\backslash}{C}}{=}\underset{\backslash}{\overset{/}{C}} \xrightarrow[\text{or } \Delta]{UV} -\overset{X}{\underset{|}{C}}-\underset{|}{\overset{|}{C}}-NSF_2 \\[2em] + (CF_3)_2CN_2 \xrightarrow{UV} (CF_3)_2C\overset{\textstyle X}{\underset{\textstyle NSF_2}{\diagdown}} \end{cases} \qquad (66)$$

3. Fluorosulfenamide Ions, $F\overset{(:)}{\underset{(:)}{S}}-\overline{N}-R^-$

The only possible type of a tetrahedrally coordinated S(II) anion, $F-\overset{(:)}{\underset{(:)}{S}}-\overline{N}R^-$, has not so far been synthesized and, from the known instability of sulfur(II)–fluorine compounds such a preparation appears to be very unlikely. Attempts to add fluorine ions to the $C=N$ bond in $(CF_3)_2C=N-\overset{(\cdot)}{\underset{(\cdot)}{S}}F$ (158) resulted in slow decomposition of the sulfenyl fluoride (157):

$$
\underset{F_3C}{\overset{F_3C}{>}}C=N-\overset{(\cdot)}{\underset{(\cdot)}{S}}-F
\begin{cases}
+ \text{CsF} \longrightarrow (CF_3)_2CFNSF_2 + (CF_3)_2C=N-S-N=C(CF_3)_2 \\
\qquad\qquad\qquad\quad + (CF_3)_2C=N-S-S-N=C(CF_3)_2 \\
+ \text{HgF}_2 \xrightarrow{-\text{Hg}_2\text{F}_2} (CF_3)_2CFN=SF_2 \qquad\qquad\qquad\qquad (67)
\end{cases}
$$

4. Anions of Di- and Tribasic Acids

As Table III shows, all of the theoretically possible dibasic acids [except $F_2S(NH)_2$] occur as amines. In addition to $F\overset{O}{\underset{O}{S}}NH_2$, only $\overset{O}{\underset{O}{FS}}-N=\overset{O}{\underset{F}{S}}-NH_2$ has so far been prepared (127):

$$
\overset{O}{\underset{O}{FSN}}=\overset{F}{\underset{F}{S}}=O + 2NH_3 \longrightarrow \overset{O}{\underset{O}{FS}}-N=\overset{O}{\underset{F}{S}}-NH_2 + NH_4F \qquad (68)
$$

The reaction of sulfur difluoridediimides with ammonia to give the corresponding bisimino derivatives, $R-N=\overset{\overset{\displaystyle N-R}{\|}}{\underset{\underset{\displaystyle F}{|}}{S}}-NH_2$, has not been described, nor is SS-difluorosulfur diimide or its salts known. Work on $F\overset{(\cdot)}{\underset{O}{S}}NH_2$ (69) should be checked as this compound would be expected to decompose very rapidly to HNSO and HF. Interaction of $HN=SOF_2$ with NH_3 gives the ammonium salt $NH_4^+NSOF_2^-$ (146) rather than the amino derivative.

Aminosulfonyl compounds exhibit acidic properties, as is shown, for example, in the case of $CF_3SO_2NH_2$ (14):

$$
CF_3SO_2NH_2 + 2Ag^+ + 3NH_3 \longrightarrow CF_3SO_2NAg_2 \cdot NH_3 + 2NH_4^+
$$

$$
200°C \Big\downarrow -NH_3 \qquad\qquad (69)
$$

$$
CF_3\overset{O}{\underset{O}{S}}-\bar{N}|Ag_2
$$

Compound FSO_2NH_2, on the other hand, is decomposed by alkali. The desired silver salt cannot be isolated by reaction with CF_3COOAg.

At 140° to 160°C, the sulfur-containing heterocycle

is obtained in 10–15% yield (in addition to CF_3COF, CF_3CN and CO_2) (*46*). This compound is formed in better yield from $ClSO_2NPCl_3$ and CF_3COOH (*116*).

IV. Acyclic Nitrogen–Sulfur–Fluorine Cations

The coordination number 4 dominates the chemistry of nitrogen–sulfur–fluorine cations. Neutral pentacoordinated compounds always split off fluorine ions on treatment with Lewis acids to give tetracoordinated cations. By contrast, it is not possible to prepare threefold coordinated cations from tetracoordinated neutral S(VI) compounds with the aid of Lewis acids. In the case of S(IV), threefold coordinated cations may be obtained by selecting suitable ligands and, in $N{\equiv}S{\odot}^+$, we even find formal twofold coordination.

A. TETRACOORDINATED CATIONS

Since the time that $\odot SF_3^+$ (*9, 10, 142, 143*) and $O{=}SF_3^+$ cations (*142, 143*) were first prepared by splitting off F^- from the neutral pentacoordinated compounds,

$$OSF_4 + MF_5(BF_3) \longrightarrow OSF_3^+ MF_6^- (BF_4)^- \qquad (70)$$
$$\odot SF_4 + MF_5(BF_3) \longrightarrow \odot SF_3^+ MF_6^- (BF_4)^- \qquad (70a)$$
$$(M = Sb, As)$$

a series of investigations on them have been carried out. Infrared, Raman (*3, 16, 42, 143*) and NMR spectra (*3, 8*) established that these compounds were indeed salts of the type shown, but only recently have the structures been determined by X-ray analysis (Fig. 5).

It was possible to prepare the first mixed halogen derivative (Fig. 6) by oxidative addition of a chlorine cation to OSF_2 (*82, 83*):

$$O{=}\overset{F}{\underset{F}{S}}{\odot} + Cl_2F^+AsF_6^- \longrightarrow OSF_2Cl^+AsF_6^- + ClF \qquad (71)$$

Analogous compounds are unknown in S(IV) chemistry. In addition to $\odot SF_3^+$ only $\odot SCl_3^+$ has been described (*79*):

$$SCl_2 + 2AsF_3 + 2Cl_2 \longrightarrow \odot SCl_3^+AsF_6^- + AsCl_3 \qquad (72)$$

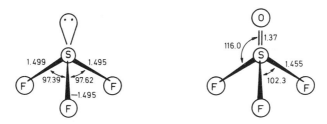

FIG. 5. Structure of $\overset{+}{\text{S}}\text{F}_3^+$ (47) and OSF_3^+ cations (81).

Addition of NOCl at low temperatures to a solution of $\bar{\text{S}}\text{F}_3^+\text{AsF}_6^-$ in SO_2, followed by excess of AsF_5 after completion of the reaction, gives a mixture of $\text{NO}^+\text{AsF}_6^-$, $\bar{\text{S}}\text{F}_3^+\text{AsF}_6^-$, and $\bar{\text{S}}\text{Cl}_3^+\text{AsF}_6^-$ (94):

$$\overset{..}{\text{S}}\text{F}_3^+\text{AsF}_6^- + \text{NOCl} \xrightarrow[\text{SO}_2\text{ liq.}]{-60°\text{C}} \text{NO}^+\text{AsF}_6^- + \{\overset{..}{\text{S}}\text{F}_3\text{Cl}\}$$

$$\bigg\downarrow \text{AsF}_5 \qquad (73)$$

$$\overset{..}{\text{S}}\text{F}_3^+\text{AsF}_6^- + \overset{..}{\text{S}}\text{Cl}_3^+\text{AsF}_6^- \longleftarrow \{\overset{..}{\overset{+}{\text{S}}}\text{F}_2\text{Cl AsF}_6^-\}$$

Whether the exchange occurs at the stage where the neutral pentacoordinated compound is present or whether the tetracoordinated cation disproportionates is not yet known.

Although replacement of fluorine in the cations by other halogens was possible in only one exceptional case, the existence of a series of known perhaloalkyl (130, 137–139), aryl (150), amino (37, 53, 159, 160), and aryloxy (33, 161, 131) derivatives of OSF_4 or $\overset{..}{\text{S}}\text{F}_4$ shows that derivatives of the OSF_3^+ and $\overset{..}{\text{S}}\text{F}_3$ cations can certainly be made. In the case of S(VI) it is also possible to replace the doubly bonded oxygen by an RN= group and, thus, obtain iminosulfur trifluoride cations, $\text{RN}{=}\overset{+}{\text{S}}\text{F}_3$.

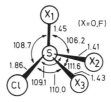

FIG. 6. Structure of the OSF_2Cl^+ cation.

1. Derivatives of the OSF_3^+ Cation

Derivatives of the OSF_3^+ cation may be prepared in three ways.

a. By cleavage of fluoride ions (i) from pentacoordinated neutral compounds [as in Eq. (70)],

$$R\overset{O}{\overset{\|}{-}}SF_3 + MF_5 \longrightarrow R\overset{O}{\overset{\|}{\underset{+}{S}}}F_2\ MF_6^- \tag{74}$$

(ii) from a neighboring group to an already four-coordinated sulfur,

$$\begin{array}{c}\diagdown\\A-N=S\\\diagup\ |\\F\end{array}\overset{O}{\overset{\|}{}}\diagup^F_{\diagdown F} + MF_5 \longrightarrow \begin{array}{c}\diagdown\\A=N-\underset{+}{S}F_2\end{array}\overset{O}{\overset{\|}{}}\ MF_6^- \tag{75}$$

or

$$\begin{array}{c}\diagdown\\A-N=S\\\diagup\ |\\X\end{array}\overset{O}{\overset{\|}{}}\diagup^F_{\diagdown F} + AgMF_6 \xrightarrow{-AgX} \begin{array}{c}\diagdown\\A=N-\underset{+}{S}F_2\end{array}\overset{O}{\overset{\|}{}}\ AsF_6^- \tag{75a}$$

$$(X = Cl,\ Br)$$

b. By addition of cations to tetracoordinated neutral compounds:

$$R-N=S\overset{O}{\overset{\|}{}}\diagup^F_{\diagdown F} + X^+MF_6^- \longrightarrow RN-\underset{+}{S}\overset{X}{\underset{\diagdown F}{\overset{\|}{}}}\overset{O}{\diagup^F}\ MF_6^- \tag{76}$$

The type of reaction shown in Eq. (74) is possible for all derivatives of OSF_4. It would be expected that replacement of one or several fluorine atoms by other groups would increase the ability to donate fluorine ions because of the greater $+I$ effect. This is particularly marked for amino derivatives (94):

$$(CH_3)_2NSOF_3 + MF_5 \longrightarrow (CH_3)_2NSOF_2^+MF_6^- \tag{77}$$
$$(M = As,\ white\ solid,\ m.p.\ 223°C,\ or\ Sb\ white\ solid,\ m.p.\ 229°C)$$

These unexpectedly high-melting salts are thermally extraordinarily stable and the vibrational and NMR spectra show discrete anions and cations to be present both in the solid state and in solution.

Fluorformylsulfur oxydifluorideimide reacts according to Eq. (75), losing a fluoride ion of the fluorformyl group (94):

$$COFN=SF_2\overset{O}{\overset{\|}{}} + MF_5 \longrightarrow OCN\underset{+}{S}F_2\overset{O}{\overset{\|}{}}\ MF_6^- \tag{78}$$
$$(M = As,\ Sb)$$

No stable salt is obtained with BF_3. Properties and reactions of the isocyanatosulfur oxydifluoride cation are discussed in connection with the corresponding derivatives of SF_4.

The protonation of CH_3NSOF_2 is an example of reaction type b [Eq. (76)]:

$$CH_3NSOF_2 + AgAsF_6 + HCl \xrightarrow[-AgCl]{SO_2 \text{ liq.}} CH_3\overset{H}{N}SOF_2{}^+AsF_6{}^- \quad (79)$$

The interaction of cationic transition metal–SO_2 complexes (93) with this compound (46) may possibly be a further example:

$$M(CO)_5Br + AgAsF_6 + CH_3NSOF_2 \xrightarrow[-AgBr]{SO_2 \text{ liq.}}$$

$$[M(CO)_5SO_2]^+AsF_6{}^- + CH_3NSOF_2 \xrightarrow{-SO_2} [M(CO)_5CH_3NSOF_2]^+AsF_6{}^- \quad (79a)$$
$$(M = Mn, Re)$$

$$C_5H_5Fe(CO)_2I + AgAsF_6 + CH_3NSOF_2 \xrightarrow[-AgI]{SO_2 \text{ liq.}}$$

$$[C_5H_5Fe(CO)_2CH_3NSOF_2]^+AsF_6{}^- \quad (79b)$$

So far it has been impossible to determine from spectroscopic data if CH_3NSOF_2 in these cations is coordinated to the transition metal through nitrogen or oxygen:

Coordination through nitrogen is more probable from what is so far known of the chemistry of such substances.

2. Iminosulfur Trifluoride Cations, $RN{=}SF_3{}^+$

These cations may be prepared in three different ways. (a) cleavage of fluoride ions from the pentacoordinated sulfur tetrafluorideimides [as in Eq. (70a)],

$$R{-}N{=}SF_4 + MF_5 \longrightarrow RN{=}SF_3{}^+MF_6{}^- \quad (80)$$

(b) fluorination of sulfur difluoride imides in presence of Lewis acids,

$$R{-}N{=}SF_2 + F_2 + MF_5 \longrightarrow RN{=}SF_3{}^+MF_6{}^- \quad (81)$$

and (c) addition of cations to the nitrogen of NSF_3,

$$NSF_3 + X^+ MF_6{}^- \longrightarrow X{-}N{=}SF_3{}^+ MF_6{}^- \quad (82)$$

Reactions of the types shown in Eqs. (80) and (81) have not yet been fully investigated. On irradiation of sulfur difluoride imides with OF_2, sulfur oxydifluorideimides and sulfur difluoride diimides are obtained (*151*):

$$3R\text{—}NSF_2 + OF_2 \xrightarrow{\text{UV}} RNSOF_2 + RNSF_2NR + SF_4 \tag{83}$$

Formation of sulfur tetrafluorideimides is sometimes observed in these reactions. $(CF_3)_2CFNSF_4$ reacts with AsF_5 according to Eq. (80):

$$(CF_3)_2CFNSF_4 + AsF_5 \longrightarrow (CF_3)_2CF\text{—}N\overset{+}{=}SF_3AsF_6^- \ (154) \tag{84}$$

Equation (81) brings together both of the last reactions (oxidation and cation formation).

In the reaction of thiazyl trifluoride, attack by electrophiles on nitrogen plays a major role in addition to the exchange of fluorine atoms on sulfur by nucleophiles. Recent investigations have shown that nitrogen carries a substantial negative charge (*31*, *77*), and this has been used in preparative work involving reactions with Lewis acids (*58*, *56*):

$$NSF_3 \begin{cases} + \ BF_3 \longrightarrow F_3B\text{—}NSF_3 \\ + \ MF_5 \longrightarrow F_5M\text{—}NSF_3 \quad (M = As, Sb) \end{cases} \tag{85}$$

Derivatives of NSF_3 behave similarly: No cations are formed, and, as would be expected, sulfur(VI) retains tetracoordination.

Coordinatively unsaturated transition metal complexes react as Lewis acids and add electrophillically (*89*, *98*):

$$M(CO)_5Br + Ag^+AsF_6^- \xrightarrow[-\text{AgBr}]{\text{NSF}_3 \text{ liq.}} \{M(CO)_5{}^+AsF_6{}^-\}$$
$$\downarrow NSF_3 \tag{86}$$
$$[M(CO)_5NSF_3]^+AsF_6{}^-$$

$$(M = Mn, Re)$$

[In the meantime it has been shown that "pentacarbonyl hexafluoro-arsenates" in the solid state have the structure $(CO)_5M\text{-}F\text{---}AsF_5$; in their reactions, however, they behave as coordinatively unsaturated cations (*93*).]

Reactions are better carried out in liquid SO_2, only a small excess of NSF_3 being necessary (*89*), e.g.,

$$C_5H_5Fe(CO)_2I + AgAsF_6 \xrightarrow[-\text{AgI}]{\text{SO}_2 \text{ liq.}} [C_5H_5Fe(CO)_2SO_2]^+AsF_6{}^- \xrightarrow[-\text{SO}_2]{+\text{NSF}_3}$$
$$[C_5H_5Fe(CO)_2NSF_3]^+AsF_6{}^- \tag{86a}$$

Addition reactions with transition metal cations give an indication of the mechanism of additions to the $N\equiv S$ bond in NSF_3:

$$NSF_3 \begin{cases} + \ 2HF \longrightarrow SF_5NH_2 \quad (25) \\ + \ ClF \longrightarrow SF_5NCl_2 \quad (28) \end{cases} \tag{87}$$

Both reagents are highly polar $\left(\overset{\delta+}{H}-\overset{\delta-}{F};\ \overset{\delta+}{Cl}-\overset{\delta-}{F}\right)$, and in the first step the cation is added to nitrogen, following which this cation is neutralized by F^-:

$$N\equiv SF_3 + \overset{\delta+}{X}-\overset{\delta-}{F} \longrightarrow \left\{XN\overset{+}{=}SF_3\,|\bar{F}|^-\right\} \longrightarrow \{X-N\equiv SF_4\} \qquad (88)$$

The tetrafluorideimide adds a further X—F molecule. It seems more probable that a hexacoordinated anion is formed rather than a pentacoordinated cation.

$$\{X-N\equiv SF_4\} + \overset{\delta+}{X}-\overset{\delta-}{F} \longrightarrow \left\{X\bar{N}-SF_5{}^- \atop X^+\right\} \longrightarrow \overset{X}{\underset{X}{\diagdown}}N-SF_5 \qquad (88a)$$

Compound SF_5NSF_2 is isolated in 3% yield from the reaction of NSF_3 with SF_4 in presence of BF_3 (27). Addition of the $\bar{S}F_3{}^+$ cation to nitrogen has been proposed as the primary step in the reaction mechanism:

$$(SF_4 + BF_3) \longrightarrow \textcircled{:}\,SF_3{}^+BF_4{}^- + NSF_3 \longrightarrow \left\{F_3\overset{\textcircled{..}}{S}-N\equiv SF_3{}^+\ BF_4{}^-\right\} \longrightarrow$$

$$BF_3 + \left\{F_3\overset{\textcircled{..}}{S}-N\equiv SF_4\right\} \overset{\sim}{\longrightarrow} F_2\bar{S}\equiv N-SF_5 \qquad (89)$$

The primary cation that would be expected is unstable under the reaction conditions. It takes F^- from the anion to form neutral $F_3\bar{S}-N\equiv SF_4$, a compound with 2 pentacoordinated sulfur atoms. Disproportionation then leads to $SF_5-N\equiv\bar{S}F_2$, with 1 hexa- and 1 tetracoordinated sulfur. Equations (87)–(89) show that, outside of transition metal chemistry, there is evidence that iminosulfur trifluoride cations may be prepared by electrophilic attack on NSF_3. If the transition metal halide in Eq. (86) is replaced by HCl, it is possible to protonate the nitrogen (94):

$$NSF_3 + HCl + AgAsF_6 \xrightarrow[-\,AgCl]{SO_2\ liq.} [HN\overset{+}{=}SF_3]\,AsF_6{}^- \qquad (90)$$

Isolation of pure $HNSF_3{}^+AsF_6{}^-$ was not possible, because some complexation of the silver occurs, and because two further stable compounds ($AsF_5\cdot NSF_3$ and $SF_5NH_2\cdot AsF_5$) are formed in the system $NSF_3/AsF_5/HF$.

3. Derivatives of the CSF_3^+ Cation

The CSF_3^+ ion is the analog in the chemistry of S(IV) of cations $O{=}SF_3^+$ or $RN{=}SF_3^+$ in that of S(VI). The only derivative described in the literature is $CF_3SF_2^+$ (79a); amino complexes may be prepared (a) by reactions of amino derivatives of CSF_4 with Lewis acids, (b) by cleavage of fluoride ions from a group bonded to tetra-coordinated sulfur, and (c) by addition of cations to the nitrogen of a sulfur difluorideimide group.

As for OSF_4, substitution of one or more fluorine atoms in CSF_4 by other groups, especially amino groups, brings about a very big increase in the ability of the molecule to act as a fluoride ion donor (94):

$$(CH_3)_2NSF_3 \begin{cases} + MF_5 \longrightarrow (CH_3)_2NSF_2^+MF_6^- \\ \qquad [M = As\ (m.p.\ 214°C)\ or\ Sb\ (m.p.\ 108°C,\ dec.)] \\ + BF_3 \longrightarrow (CH_3)_2NSF_2^+BF_4^-\ (m.p.\ 159°C) \end{cases} \quad (91)$$

In all cases, stable colorless high-melting solids are produced, which are readily soluble in SO_2 and are indefinitely stable in the solid state in dry glass vessels. Yields in the preparations are quantitative.

Reaction type (b) is illustrated by the N-fluoroformyl derivatives:

$$\begin{cases} + MF_5 \longrightarrow OCNSF_2\,MF_6^- \quad (M = As,\ Sb) \\ + BF_3 \longrightarrow OCNSF_2\,BF_4^- \end{cases} \quad (92)$$

Whereas the hexafluoroarsenate and the antimonate are very stable, the tetrafluoroborate decomposes with loss of BF_3. On the basis of their vibrational spectra, the above compounds are not to be considered as acylium sulfur difluorideimide salts ($OCN{=}SF_2^+$).

Reaction (92) represents a general route to isocyanatosulfur cations:

$$+ MF_5 \longrightarrow OCN{-}\overset{+}{S}Cl_2MF_6^- \quad (91) \qquad (92a)$$

(M = As, Sb)

$$+ AsF_5 \longrightarrow OCN{-}\overset{+}{S}(F)(CF_3)\ AsF_6^- \quad (152) \qquad (92b)$$

Additions to the $S{=}N$ double bond of polar reagents are possible for the sulfur difluoride imides in the same way as for the triple bond in NSF_3, except that it is accompanied by cleavage of SF_4:

$$R-N\overset{(\cdot\cdot)}{=}SF_2 + 2\overset{\delta+}{Cl^+}-\overset{\delta-}{F} \longrightarrow RNCl_2 + |SF_4 \quad (35, 36) \qquad (93)$$

$$R-N\overset{(\cdot\cdot)}{=}SF_2 + 3\overset{\delta+}{H}-\overset{\delta-}{F} \longrightarrow RNH_3F + |SF_4 \quad (94) \qquad (93a)$$

It must be assumed again here that the primary step involves addition of a cation to nitrogen. Under similar conditions, as was described for $RN\overset{O}{\overset{||}{=}}SF_2$, it should be possible to trap the tetracoordinated nitrogen–sulfur cations.

4. Reactions of Tetracoordinated Cations

The characteristic of the cations is naturally their positive charge and they are, therefore, predestined for electrophilic attack or neutralization reactions with anions. Three possible types of reaction arise.

a. Addition to Double Bonds or Reactions Resembling Friedel-Crafts Acylation. Reactions of this sort have not been carried out in nitrogen–sulfur–fluorine chemistry. It has so far been necessary to proceed in the reverse way, increasing the nucleophilic character of the unsaturated reaction partner by anion formation, rather than increasing the electrophilic character of the N—S—F compound by cation formation, e.g.,

$$CF_3-CF=CF_2 + CsF + |SF_4 \longrightarrow \{(CF_3)_2\bar{C}F^-Cs^+\} + |SF_4 \quad (130) \qquad (94)$$
$$\longrightarrow (CF_3)_2CF\bar{S}F_3 + CsF$$

$$N\equiv C-NSOF_2 + CsF + |SF_4 \longrightarrow \left\{\bar{N}\overset{F}{\underset{}{=}}C-N=SOF_2{}^-Cs^+\right\} + |SF_4 \quad (94a)$$
$$\longrightarrow \left\{F_3\bar{S}-N\overset{F}{\underset{}{=}}C-NSOF_2\right\} + CsF$$

$$\Big\downarrow^{\sim}$$

$$F_2\bar{S}=N-\overset{F}{\underset{F}{C}}-NSOF_2$$

b. Reactions with Nucleophiles. The electrophilic character is so strongly enhanced by the positive charge on the central sulfur atom that even weak nucleophiles can be induced to react. The reaction of $\text{C}SF_3{}^+BF_4{}^-$ with NSF_3 has already been described [Eq. (89)], and is of this type. $O=SF_4$ is appreciably less reactive than $\text{C}SF_4$. Thus, whereas $Si(NCO)_4$ reacts with $\text{C}SF_4$ even at room temperature, it is necessary with OSF_4 to use BF_3 as a catalyst and to heat to $235°–240°C$ (*134*):

$$\{OSF_4 + BF_3\} \longrightarrow O\overset{+}{S}F_3BF_4{}^- + Si(NCO)_4 \xrightarrow{-SiF_4}$$

$$\{O=\overset{+}{S}F_2NCO\,BF_4{}^-\} \xrightarrow{-BF_3} \overset{O}{\underset{F}{\overset{\diagdown}{C}}}-N=SOF_2 \qquad (95)$$

c. Reactions with Anions.   Although $\mathbb{C}SF_3{}^+$ and $O{=}SF_3{}^+$ cations have been known for more than 15 years, there is no work in which their reactions as cations is described, e.g., reactions in which they interact with anions to form covalent molecules.

Using liquid SO_2 as solvent, both $\mathbb{C}SF_3{}^+$ and $OSF_3{}^+$ will react with NCO^- to yield the corresponding fluoroformylsulfur imides (*94*):

$$\mathbb{C}SF_3{}^+ + NCO^- \xrightarrow{\text{SO}_2 \text{ liq.}} \left\{ \begin{matrix} F \\[-2pt] \overset{|}{\underset{\underset{F}{|}}{\mathbb{C}{=}S}} {\overset{F}{\diagup}} \\ N{=}CO \end{matrix} \right\} \xrightarrow{\sim} \underset{F}{\overset{O}{\diagdown}} C{-}N{=}\bar{S}F_2 \qquad (96)$$

$$O{=}SF_3{}^+ + NCO^- \xrightarrow{\text{SO}_2 \text{ liq.}} \left\{ \begin{matrix} F \\[-2pt] \overset{|}{\underset{\underset{F}{|}}{O{=}S}} {\overset{F}{\diagup}} \\ N{=}CO \end{matrix} \right\} \xrightarrow{\sim} \underset{F}{\overset{O}{\diagdown}} C{-}N{-}SOF_2 \qquad (96a)$$

These ionic reactions occur at very low temperatures under milder conditions than usual.

Using as the reaction principle the splitting off of a single fluorine followed by addition of an anion to form a new neutral compound, it is possible to exchange selected and, under mild conditions, single fluorine atoms. This will be illustrated by taking the haloformylsulfur imides as an example.

N-Fluoroformylsulfur difluorideimide may be prepared very readily from $Si(NCO)_4$ and $\bar{S}F_4$. With PCl_5 it is then possible first to exchange the fluorine atoms on sulfur and then that on carbon:

$$Si(NCO)_4 + 4\bar{S}F_4 \longrightarrow 4COFN\bar{S}F_2 + SiF_4 \quad (26) \qquad (97)$$

$$COFN\bar{S}F_2 + PCl_5 \xrightarrow{0°} COFN\bar{S}Cl_2 + P(Cl,F)_5 \quad (128) \qquad (97a)$$

$$COFN\bar{S}Cl_2 + PCl_5 \xrightarrow{\text{RT}} COClN\bar{S}Cl_2 + P(Cl,F)_5 \quad (128) \qquad (97b)$$

In the reaction of $\bar{S}F_2NCO^+AsF_6{}^-$ with $NOCl$, on the other hand, the first step is not a nucleophilic attack of Cl^- on sulfur, as the spectroscopic data would lead us to expect, but attack on the carbon of the isocyanate group (*92*):

$$\mathbb{C}SF_2NCO^+AsF_6{}^- + NOCl \xrightarrow[- NO^+AsF_6{}^-]{\text{SO}_2 \text{ liq.}} \underset{Cl}{\overset{O}{\diagdown}} C{-}N{=}\bar{S}{\overset{F}{\underset{F}{\diagup}}} \xrightarrow{\sim} \underset{F}{\overset{O}{\diagdown}} C{-}N{=}\bar{S}{\overset{Cl}{\underset{F}{\diagup}}} \qquad (98)$$

The resulting $COClN\bar{S}F_2$ is unstable and rearranges to $COFN\bar{S}ClF$ which, in turn, dismutes at room temperature in an analogous way to $O{=}\bar{S}{\overset{Cl}{\underset{F}{\diagup}}}$, giving $COFN\bar{S}F_2$ and $COFN\bar{S}Cl_2$. Compound

$OSF_2NCO^+AsF_6^-$ reacts to give the more stable $COClNSOF_2$, and rearrangement to $COFNSOClF$ is not observed:

$$OSF_2NCO^+ AsF_6^- + NOCl \xrightarrow[-NOAsF_6]{SO_2 \text{ liq.}} COClNSOF_2 \qquad (98a)$$

For exchange reactions of this sort, nitrosyl salts are in many cases particularly suitable as they are more soluble in SO_2 than the alkali salts. The nitrosyl compounds are often readily accessible by reaction of $NOCl$ with the corresponding silver salts.

B. THREEFOLD COORDINATED CATIONS

Whereas coordination number 3 occurs relatively frequently in the chemistry of quadrivalent sulfur (e.g., $O{=}\bar{S}{=}O$, $R{-}N{=}\bar{S}{=}O$, $R{-}N{=}\bar{S}{=}N{-}R$), it is seldom found for sulfur(VI). Apart from the monomer of SO_3, only very recently have some aza derivatives become known: $(R_3SiN{=})_3S$ (66); $(R_3SiN{=})_{3-x}(CF_3CON{=})_xS$ ($x = 1, 2$) (74); $(R_3SiN{=})_2S{=}O$ (49). This showed for the first time that the silylimino group is very suitable for stabilizing low coordination. In the meantime this principle has been used with great success in phosphorus chemistry (107), and the iminoaminophosphane, $R_3SiN{=}P{-}N$ $(SiR_3)_2$ (105), and bis(imino)aminophosphorane, $(R_3SiN{=})_2P{-}N(SiR_3)_2$ (106), have proved to be key substances in a very large number of interesting reactions.

1. Threefold Coordinated Cations of Sulfur(VI)

Preparation of these cations is successful only in exceptional cases (87). The following reaction gave the desired sulfonylium salts, the hexafluoroantimonate of which is stable at room temperature (88):

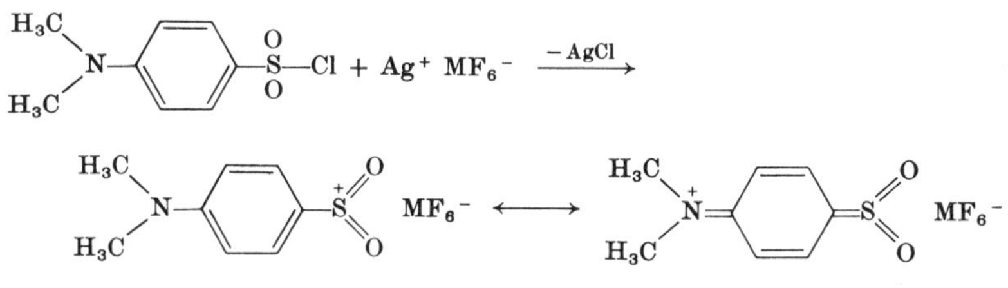

The extraordinary stability of the p-amino compound was explained by supposing that the sulfur is able to transfer its positive charge to the aromatic group by forming quinoid limiting structures. If other sulfonyl compounds are used in reaction (99), either chlorine–fluorine exchange or rearrangement to donor–acceptor complexes of the type $R{-}\underset{F}{\overset{O}{S}}{=}O{-}{-}{-}MF_5$

takes place (88). Complexes of this sort are also encountered in the reaction of aminosulfonyl compounds, R_2NSF, with SbF_5 or AsF_5, and it is believed that the fluoride ion donor function is not greatly strengthened by the inductive effect of the amino group.

When one oxygen of sulfonyl fluoride, $O{=}S{=}O$, is replaced by a methylimino group, stable adducts are formed with Lewis acids (94):

$$R{-}N{=}S \underset{F}{\overset{O}{\big/}} {}^{F} + MF_6 \longrightarrow RNSF{\cdot}MF_5 \quad (M = As, Sb) \quad (100)$$

It is not possible to make any definite prediction as to the nature of the coordination (cf. Section IV, A, 1).

2. Threefold Coordinated Cations of Sulfur(IV)

Whereas $O\overline{S}F_2$ does not form adducts of any sort with Lewis acids, replacement of 1 fluorine atom by a dialkylamino group leads to 1:1 adducts which, from NMR measurements, are formulated as salts (94):

$$R_2N{-}\underset{(\cdot)}{\overset{O}{S}}F + MF_5 \longrightarrow [R_2N{-}\underset{(\cdot)}{\overset{+}{S}}{=}O]\,MF_6{}^- \quad (101)$$

This class of compound is of interest because, as in the case of tetracoordinated S(IV) cations, the central sulfur atom does not violate the octet rule. The oxygen–sulfur bond may be described as a pure $p_\pi - p_\pi$ bond.

A second way of facilitating fluoride ion cleavage from the tetracoordinated fluoride is by replacement of the oxygen by an imino group (94):

$$CH_3N\overline{S}F_2 + AsF_5 \longrightarrow \text{decomposition} \quad (102)$$
$$CF_3N\overline{S}F_2 + AsF_5 \longrightarrow \text{no reaction} \quad (102a)$$

The primary adduct of $CH_3N\overline{S}F_2$ and AsF_5 decomposes at room temperature; the polyfluoroalkyl derivatives do not react. On substitution of 1 of the fluorine atoms by an amino group, salt formation again occurs, however (94):

$$R_2N{-}\underset{(\cdot)}{\overset{F}{S}}{=}NR_f + MF_5 \longrightarrow R_2N{-}\underset{(\cdot)}{\overset{+}{S}}{=}N{-}R_f\,MF_6{}^- \quad (102b)$$

Thus sulfur–nitrogen double bond may also be described as a pure $p_\pi - p_\pi$ bond.

In their bonding and structure the aminooxy- and aminoimino-

TABLE IV
RELATIONSHIP BETWEEN THREEFOLD COORDINATED
NEUTRAL AND CATIONIC SULFUR(IV) SPECIES

sulfur(IV) cations are related to the thionyl imides and diimides (Table IV). Formally the diimides (IIIa) and thionylimides (III) contain two double bonds, and in the fluorides (I or Ia) there is one single and one double bond. In cations II or IIa, the positive charge on the central atom leads to strengthening of both bonds. Shortening of the single bond will occur in any case. Whether this is also detectable for the double bonds relative to thionyl imides or diimides must be settled by structural investigations (cf. Section VI).

C. DOUBLY COORDINATED CATIONS

From the discussion of threefold coordinated cations it is to be expected that it would be almost impossible to prepare doubly coordinated cations of sulfur(VI), for example "$N\equiv\overset{+}{S}=O$." The fluoride corresponding to this cation, "NSOF," would be expected as an intermediate in the decomposition of some sulfur oxydifluoride imides (72), e.g.,

(103)

but it stabilizes itself at once by increasing its coordination number.

Replacement of the doubly bonded oxygen by a free electron pair leads to $N\equiv\overset{\ominus}{S}$—F, the analogous substance in the chemistry of sulfur(IV). This exists as a monomer and either polymerizes to the trimer or adds fluoride ions (increase in the coordination number). It can, however, also lose F^- and go over to the thiazyl cation (58):

$$(104)$$

1. Thiazyl Cation, $N\equiv\overset{+}{S}\odot$

The thiazyl cation may be prepared from $N\overline{S}F$ and MF_5 at reduced pressure in the gas phase (55, 56) or, on a preparative scale, in liquid SO_2 (94):

$$N\equiv\overline{S}F + MF_5 \longrightarrow N\equiv\overset{+}{S}\odot MF_6^- \qquad (M = As, Sb) \qquad (105)$$

As in all of these reactions which result in salts or adduct formation, yields are almost quantitative. On vacuum sublimation the hexafluoro arsenate is obtained as colorless waxy solid which, in contrast to the very unstable $N\overline{S}F$, may be kept without decomposition for years in dry glass vessels. Thiazyl tetrafluoroborate (62, 63) is stable only at low temperatures, decomposing completely at room temperature to $N\overline{S}F$ and BF_3.

So far there have been no structural investigations on thiazyl salts as it has not been possible to obtain single crystals. From the vibrational spectra it is apparent that there is a considerable strengthening of the N—S bond in relation to NSF (55, 56).

Thiazyl salts are suitable for introducing N—S units into other molecules. Synthetic possibilities include reaction with anions, with nucleophiles, and with compounds having polar bonds. Their use in transition metal chemistry has also been suggested.

a. Reactions with Anions (94):

$$N\overline{S}^+AsF_6^- + NO^+CF_3SO_3^- \xrightarrow[-NO^+AsF_6^-]{SO_2 \text{ liq.}} N\overline{S}^+CF_3SO_3^- \qquad (m.p. 120°C) \quad (106)$$

The colorless and extremely hygroscopic salt may be separated from

$NO^+AsF_6^-$ by sublimation (30°C/0.01 torr). An ionic structure is assigned on the basis of the Raman spectrum.

b. Reactions with Nucleophiles (94):

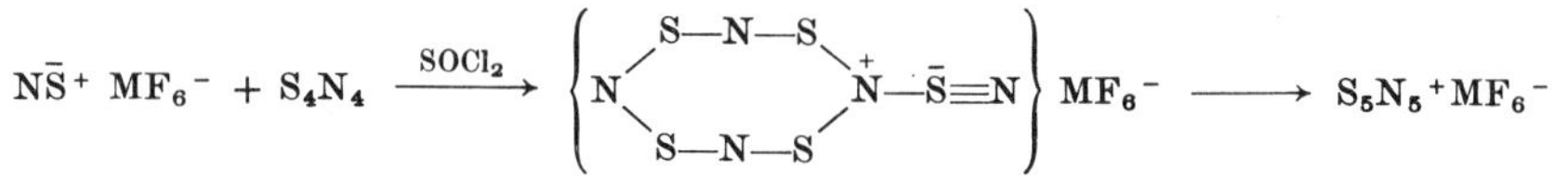

$$[M = \text{As (yellow solid, m.p. } 267°) \text{ or Sb (yellow solid, m.p. } 175°)] \tag{107}$$

The $N\bar{S}^+$ cation has been postulated (4) as an intermediate in the reaction of $N_3S_3Cl_3$ with S_4N_4 and $AlCl_3$, $FeCl_3$, or $SbCl_5$ in $SOCl_2$ as a solvent in the ratio 1:3:3. This mechanism seems to be supported by reaction (107). However, it is not possible to isolate $N\bar{S}^+SbCl_6^-$ from $(NSCl)_3$ and an excess, for example, of $SbCl_5$ in $SOCl_2$. Therefore, instead of the formation of $N\bar{S}^+$ it is possible that cyclic cations (cf. Section V, B) and their decomposition products are attacked nucleophilically by S_4N_4 in order to give the S_5N_5 salts.

The formulation of the intermediate in Eq. (107) is quite arbitrary because cycloaddition products of $N\equiv\overset{+}{S}\!\!\circlearrowleft$ to S_4N_4 or linear cationic thiazenes would also be possible.

c. Reactions with Polar Bonds (94):

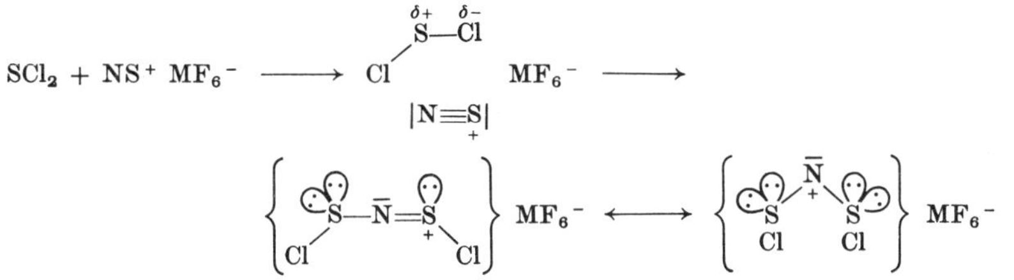

The $N(\bar{S}Cl)_2^+$ cation was first prepared from NSF_3 and BCl_3 as the tetrachloroborate (57). The more accessible tetrachloroaluminate is prepared (67) as follows:

$$S_3N_3Cl_3 + 3\bar{S}Cl_2 + 3AlCl_3 \longrightarrow 3[N(\bar{S}Cl)_2]^+ AlCl_4^- \tag{108a}$$

d. Applications in Transition Metal Chemistry.

Reference has been made in the literature to the analogy between the $N\bar{S}^+$ cation and the lighter homolog NO^+ (21, 29), but preparation of the first thiazyl (thionitrosyl) complex was carried out in a completely different way (21):

$$[MoN(S_2CNR_2)_3] + S_8 \xrightarrow[\text{reflux}]{CH_3CN} [MoNS (S_2CNR_2)_3] \tag{109}$$

TABLE V
POSSIBLE NEUTRAL AND CATIONIC SPECIES IN ACYCLIC S(IV)–N CHEMISTRY

$N{\equiv}\overset{+}{S}{:}$

$N{\equiv}S\overset{{:}^{-}}{\diagdown}$ F

$\overset{\displaystyle S^{-}}{\underset{R-N^{+}\diagup\diagdown F}{}}$

$R-N{=}\overset{{:}^{-}}{\underset{\diagdown F}{S}}-F$	$R-N\overset{\displaystyle S^{-}}{\diagup\diagdown}O$	$R-N\overset{\displaystyle S^{-}}{\diagup\diagdown}N-R$
$\overset{R}{\underset{R'}{\diagdown}}N-\overset{{:}^{-}}{\underset{F}{S^{+}}}-F$	$\overset{R}{\underset{R'}{\diagdown}}N-\overset{{:}^{-}}{\underset{O}{S^{+}}}$	$\overset{R}{\underset{R'}{\diagdown}}N-\overset{{:}^{-}}{S^{+}}{=}N-R$
$\overset{R}{\underset{R'}{\diagdown}}N-\overset{F\,F}{\underset{F}{S}}{:}^{-}$	$\overset{R}{\underset{R'}{\diagdown}}N-\overset{{:}^{-}}{\underset{O}{S}}-F$	$\overset{R}{\underset{R'}{\diagdown}}N-\overset{{:}^{-}}{\underset{N-R}{S}}-F$

Since the NO^{+} cation can be introduced directly into transition metal complexes through its salts (*30*), it should be possible for $N\overline{S}^{+}$ (*21*, *29*).

Table V shows the various possibilities for bonding sulfur(IV) and nitrogen. They range from a formal single bond to a triple bond strengthened by a positive charge. By variation of the group R and by forming cations and anions, it is possible to build up a continuous spectrum of bond types. In the chemistry of the anions, only $N\overline{S}F_{2}^{-}$ and $N\overline{S}O^{-}$ are so far known but, in the case of the cations, the last missing type of binding $\left(RN{=}\overset{+}{S}\overset{\diagup F}{\diagdown{:}^{-}}\right)$ has possibly been identified in $[Re(CO)_{5}$ NSF]$^{+}$ (*89*).

V. Cyclic Nitrogen–Sulfur–Fluorine Ions

So far as coordination numbers are concerned, the chemistry of sulfur–nitrogen–halogen heterocycles (Table VI) is considerably simpler than that of the acyclic compounds, as sulfur is exclusively tetrahedrally coordinated. Other coordination numbers are rarely encountered among

TABLE VI

RELATIONSHIP BETWEEN CYCLIC AND ACYCLIC
N—S—F COMPOUNDS

(Ia)   (Ib)   (Ic)

(IIa)   (IIb)   (IIc)

the cyclic compounds of sulfur(VI), exceptions being the recently prepared disubstituted derivatives of SF_6 and OSF_4 (1, 168). In the sulfanuryl (Ia) or thiazyl fluorides (IIa), there are steric relations similar to those in the acyclic fluorosulfonylamides (Ib) and fluorosulfinylamides (IIb) or their imino derivatives (Ic, IIc).

These heterocycles may be thought of as built up from acyclic units. Acyclic and cyclic systems are also very similar in their reactivity toward nucleophilic reagents. Aminosulfonyl fluorides may be prepared in aqueous solution from the sulfonyl chlorides by chlorine–fluorine exchange. Sulfinic acid amides react vigorously with water. Sulfanuryl fluorides are hydrolyzed only slowly, but thiazyl halides react spontaneously with destruction of the ring system.

Sterically, there is no great difference between a thiazyl, $—N{=}\overset{\displaystyle (\cdot\cdot)}{\underset{\displaystyle X}{S}}—$,

and a sulfanuryl group, $—N{=}\overset{\displaystyle O}{\underset{\displaystyle X}{\overset{\|}{S}}}—$, therefore they are interchangeable

at will. The more sulfanuryl groups there are in a mixed ring, the greater is its stability.

Substitution reactions have been carried out successfully only on $(NSOF)_3$ (58, 101) and $(NSOF)_2(N\overline{S}F)$ (85) [substitution taking place at S(IV)]. With pure thiazyl halides under the same conditions, one observes ring cleavage.

A. Cyclic Anions

Substitution reactions with sulfanuryl halides have been described in the literature (58, 101), but only recently have careful systematic investigations led to a deeper understanding of them. From $(NSOF)_3$ and secondary amines, it is possible to obtain mono-, di-, or trisubstituted products by choice of conditions (solvent, stoichiometric proportions, temperature, reaction time) (58, 101, 164, 166). By reaction with primary amines, hydrolysis, alcoholysis, or reduction, three different types of anion may be obtained: $(NSOF)_x(NSONR)_{3-x}^{(3-x)-}$ (166), $(NSOF)_2(NSO_2)^-$ (162), and $(NSOF)_2(NSO)^-$ (162).

1. Aminosulfanuryl Fluoride Ions, $(NSOF)_x(NSO\overline{N}R)_{3-x}^{(3-x)-}$

In the reaction of $(NSOF)_3$ with primary amines, only 1 fluorine atom is exchanged initially (166):

$$(NSOF)_3 + 3RNH_2 \longrightarrow (NSOF)_2(NSONHR)\cdot RNH_2 + RNH_3F \qquad (110)$$

The substitution product is obtained as the amine adduct and is dissociated to ions in aqueous solution, giving difficultly soluble salts with Ph_4AsCl and Ph_4PCl:

$$(NSOF)_2(NSONHR)\cdot RNH_2 \xrightarrow{H_2O} (NSOF)_2(NSO\overline{N}R)^- + RNH_3^+ \xrightarrow[-RNH_3Cl]{Ph_4MCl}$$

$$Ph_4M^+ (NSOF)_2(NSO\overline{N}R)^- \qquad (111)$$

The free acid $(NSOF)_2(NSONHR)$ is produced by passing HCl into a suspension of the mixture from reaction (110) in ether:

$$(NSOF)_2(NSONHR)\cdot RNH_2 + HCl \longrightarrow (NSOF)_2(NSONHR) + RNH_3Cl \qquad (112)$$

Exchange of further fluorine atoms takes place only with great difficulty: The sulfanuryl ring in the amine adducts is present in the anionic form and nucleophilic exchange reactions with anions occur only under very vigorous conditions. Further exchange with primary amines has, indeed, been successfully carried out but results only in the trisubstituted product (164):

$$(NSOF)_2(NSONHR)\cdot RNH_2 + 3RNH_2 \longrightarrow (NSONHR)_3 + 2RNH_3F \qquad (113)$$

This is explained by the fact that the disubstituted product is less acidic and an amine adduct would have hardly any ionic character. As a result, exchange will occur substantially more readily than for the monosubstituted derivative. $(NSONHR)_3$ is devoid of acid character and amine adducts are not observed (164). If, on the other hand, one starts from the N,N-dialkyl derivatives the second fluorine atom can be exchanged and the product isolated (*164*):

$$(NSOF)_2(NSON(CH_3)_2) + 3RNH_2 \longrightarrow$$

$$(NSOF)(NSONR_2)(NSONHR) \cdot RNH_2 + RNH_3F \quad (114)$$

2. Sulfanuric Fluoride–Sulfimide Ions, $(NSOF)_{3-x}(NSO_2)_x{}^{x-}$

Although the trisulfimide is formed by hydrolysis of $(NSOCl)_3$ (*71*),

$$(NSOCl)_3 + 3H_2O + Ag^+ \text{ (excess)} \xrightarrow[-3AgCl]{} (AgNSO_2)_3 \cdot 3H_2O \quad (115)$$

it was reported in one of the first investigations on $(NSOF)_3$ that only one SF bond is cleaved on hydrolysis, the ring system remaining intact (*147*). This primary product may be isolated as a salt by using large cations (*162*):

$$(NSOF)_3 + HOH + Ph_4P^+ \longrightarrow (NSOF)_2(NSO_2)^- Ph_4P^+ + HF + H^+ \quad (116)$$

A more elegant route to this anion is afforded by reaction with methanol in presence of $(CH_3)_3N$ (*162*):

$$(117)$$

As with the acyclic derivatives, it seems that there is again a shift of the double bond to oxygen.

The tetramethylammonium salt is soluble in water without decomposition. Using an ion exchange resin, it is possible to isolate the free acid $(NSOF)_2(NHSO_2)$ as a hydrate. Reaction with silver carbonate yields the corresponding salt, which, in turn, may be used as a starting material for the preparation of mixed sulfanuryl fluoride–sulfimide derivatives (*163, 165*):

$$(118)$$

Alkylation occurs exclusively at the most nucleophilic site, whereas with trimethylchlorosilane or -stannane the thermodynamically favored oxygen derivative results. The first optically active inorganic ring was obtained with these alkysulfimide–sulfanuryl fluoride derivatives (*164*).

The foregoing reactions can also be carried out with sulfanuryl fluoride derivatives (*165*):

$$\tag{119}$$

As would be expected, the amino derivatives are less acidic than $(NSOF)_2(SO_2NH)$, and they are less dissociated in aqueous solution. Since covalent compounds undergo nucleophilic attack more readily than anions, it is understandable that stability in aqueous solution increases with increasing acid strength.

As for $(NSOF)_2(HNSO_2)$, the silver salt may be prepared. On alkylation, both possible isomers are observed (*165*). Alkylation takes place, however, mainly at the most nucleophilic nitrogen.

$$\tag{120}$$

3. Sulfanuryl Fluoride–Thionylimide Ions $(NSOF)_2(NSO)^-$

If mercaptans are used instead of alcohols in the reaction shown in Eq. (117), the corresponding thio salt is not obtained. Instead, the sulfur that is attacked, undergoes reduction (*162*):

$$(NSOF)_3 + 2CH_3SH + 2N(CH_3)_3 \longrightarrow$$

$(CH_3)_3NH^+ + CH_3SSCH_3 + (CH_3)_3NHF$ (121)

This salt may also be prepared by reducing $(NSOF)_3$ with phenylhydrazine (*164*).

The free acid itself is not stable; cleavage of the ring occurs in aqueous solution, iminobissulfamide being formed (*164*):

$\xrightarrow{H_2O}$ $SO_2 + O_2S$ $SO_2 + 2HF$ (122)

Mixed thiazyl–sulfanuryl halides (*86, 167*) may be obtained by reaction of the amino salt with PCl_5 or PF_5 (*164*):

$$(NSOF)_2NSO^-(CH_3)_3NH^+ \begin{cases} + PCl_5 \longrightarrow (NSOF)_2(NSCl) \\ \qquad\qquad\quad (78\%) \\ + PF_5 \longrightarrow (NSOF)_2(NSF) \\ \qquad\qquad\quad (5\%) \end{cases}$$ (123)

B. CYCLIC CATIONS

Until a short time ago the only known sulfur–nitrogen cation was $S_4N_3^+$ (*58*), but recently further species of this sort have been prepared; for example, as $S_5N_5^+$ (*4, 6, 7, 125*), $S_3N_2^+$ or $S_6N_4^{2+}$ (*5*), and $S_3N_2^{2+}$ (*120*). The range of sulfur–nitrogen–halogen cations described in the literature is even more limited. The ionic structure of $S_3N_2Cl_2$ (Fig. 7) has been established by X-ray investigations (*169*). Examination of the bond distances shows the structure to be better described by II (derived from the threefold tetracoordinated ring system IIa) than by I (derived from a ring with sulfur in 3, 4, and 5, coordination). The different nature of the two chlorine atoms in $S_3N_2Cl_2$ is also apparent in its reactions, the ionically bound chlorine being exchanged under milder conditions than is covalent chlorine (*120*):

$$S_3N_2Cl_2 \begin{cases} + XSO_3H \xrightarrow[RT]{CH_2Cl_2} S_3N_2Cl^+SO_3X^- + HCl \quad (X = Cl, F, CF_3) & (124) \\[4pt] + 2XSO_3H \xrightarrow[reflux]{CH_2Cl_2} S_3N_2(SO_3X)_2 + 2HCl & (124a) \\[4pt] + MCl_3 \xrightarrow[RT]{CH_2Cl_2} S_3N_2Cl^+MCl_4^- \quad (M = Fe, Al) & (124b) \\[4pt] + SbCl_5 \xrightarrow[RT]{CH_2Cl_2} S_3N_2Cl^+SbCl_6^- & (124c) \\[4pt] + 2SbCl_5 \xrightarrow[RT]{CH_2Cl_2} S_3N_2^{2+} (SbCl_6^-)_2 & (124d) \end{cases}$$

A fluorine compound with an analogous composition, $S_3N_2F_2$ (65, 68), has been reported in the literature. It has, however, been assigned the acyclic structure FS—N=S=N—SF; a closer study has yet to be made.

There are two possibilities for a systematic approach to the preparation of cyclic cations: (a) The corresponding chlorides may be reacted with $Ag^+MF_6^-$ (M = P, As, Sb), AgCl being split off to form the cation; or (b) as for the acyclic compounds, fluoride ions may be split off with Lewis acids.

As was discussed at the beginning of this chapter, sulfur is exclusively tetrahedrally coordinated in the halogenocyclothiazenes. Sulfur(VI) is chemically very similar to that in fluorosulfonylamides, and the sulfur-(IV) to that in fluorosulfinylamides. We may, therefore, expect that only the thiazyl group $\left(\begin{array}{c} \overset{(\cdot)}{\underset{|}{-N=S-}} \\ F \end{array} \right)$ will function as a fluoride ion donor.

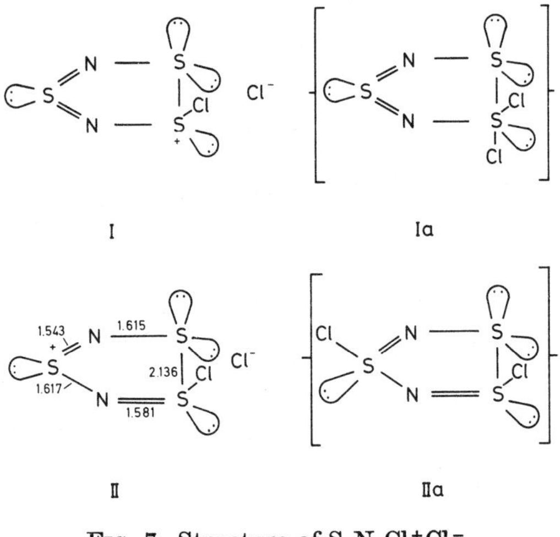

FIG. 7. Structure of $S_3N_2Cl^+Cl^-$.

The oxythiazene group $\left(\begin{array}{c} F \\ | \\ -N{=}S{-} \\ \| \\ O \end{array} \right)$ should either not react or co-ordinate through oxygen.

In the reaction of sulfanuryl fluoride with Lewis acids no stable adducts are observed, but with trithiazyl trifluoride salts are formed (99):

$$N_3S_3F_3 \begin{cases} + \ MF_5 \longrightarrow \ N_3S_3F_2{}^+ \ MF_6{}^- \ (99) \quad (M = As, \ Sb) \qquad (125) \\ + \ BF_3 \longrightarrow \ N_3S_3F_2{}^+ \ BF_4{}^- \ (56, \ 99) \qquad\qquad\qquad (125a) \end{cases}$$

The corresponding chloro cation [in $N_3S_3Cl_2{}^+SbCl_6{}^-$] may also be prepared. Both $N\bar{S}{}^+AsF_6{}^-$ and $N_3S_3F_2{}^+AsF_6{}^-$ are isolated by reaction of $(NSF)_4$ with excess AsF_5 (99):

$$N_4S_4F_4 + AsF_5 \longrightarrow \{N_4S_4F_3{}^+AsF_6{}^-\} \longrightarrow N\bar{S}{}^+AsF_6{}^- + N_3S_3F_3$$

$$\Big\downarrow AsF_5 \qquad (126)$$

$$N_3S_3F_2{}^+AsF_6{}^-$$

The $N_4S_4F_3{}^+$ cation is unstable and the eight-membered ring loses one NS unit. The remainder recombines to form the trimer rather than decompose (99).

These reactions show the $(NS)_3$ structure to be more stable than the tetramer, although $S_4N_4F_4$ is much less sensitive to hydrolysis. In the tetramer, nucleophilic attack on a ring sulfur atom is hardly possible because of the compact structure. With Lewis acids, however, reaction can occur on the ligands outside of the ring. Synthetic possibilities with these cations have not so far been examined in detail, but it is known that the reaction of $N_3S_3F_2{}^+AsF_6{}^-$ with NOCl gives only $(NSF)_3$ and $(NSCl)_3$ as isolable products (94).

It should be possible to prepare cations from the mixed thiazyl–sulfanuryl rings, $(NSOF)(NSF)_2$ (167) and $(NSOF)_2NSF$ (86, 167). Only the four-valent sulfur would, however, be capable of acting as a fluoride ion donor.

VI. Relationship between Isoelectronic Sulfur and Phosphorus Compounds

Structural studies on $CSF_3{}^+$ (47) and $OSF_3{}^+$ cations (81) show that the molecular shape does not change in going from the cation to the isoelectronic neutral compound. This similarity in molecular shape of isoelectronic species has been fully discussed (47) and is attributed to

FIG. 8. Isoelectronic sulfur and phosphorus fluorides and oxyfluorides.

orbital hybridization of the central atom, hybridization being characteristic of the periods of the periodic system to which the elements in question belong.

Since the shape of SF_3^+ and PF_3 or OSF_3^+ and OPF_3 (Fig. 8) is determined by the mutual influence of bonding and nonbonding electron pairs, it seems that the change of the effective nuclear charge of the central atom must have a similar effect on all electrons of the valence shell, whether they be bonding or nonbonding. For this reason the molecular shape of isoelectronic species will remain constant (47). If this is generally true, sulfur cations may be considered as model substances for the corresponding phosphorus compounds. Since they exist as solids at room temperature, structural studies on them are at present considerably simpler. Effects that occur at substantially lower temperatures with the phosphorus compounds are observable with the sulfur compounds in a readily accessible temperature range. Thus, for example, the NMR spectrum of $(CH_3)_2N\overline{S}F_2^+$ shows the molecule at $-15°C$ to have a semicoplanar structure (the atoms SNC_2 being in one plane that bisects the FSF angle) which is found for $(CH_3)_2N\overline{P}F_2$ in the solid state (104). Hindered rotation does not occur in the phosphorus compound down to $-90°C$. The high reactivity of low coordinated phosphorus compounds (107) and the restricted possibilities for their preparation render structural studies on these systems exceedingly difficult. The corresponding sulfur cations, R_2N—$\overset{+}{\underline{S}}$=O or R_2N—$\overset{+}{\underline{S}}$=N—R, on the other hand, are very stable and are also readily accessible.

How far these structural similarities extend to chemical behavior has not yet been investigated, but it should be possible to carry over some of the reaction principles of phosphorus chemistry to the study of sulfur cations. One outstanding question is whether sulfur(IV) cations could function as ligands to transition metals in an analogous manner to phosphorus(III) compounds. Dialkylaminodifluoro cations would be especially suitable for such a study because of their high stability.

ACKNOWLEDGEMENT

I wish to thank Professor Dr. O. Glemser for his generous support and for many helpful discussions.

REFERENCES

1. Abe, T., and Shreeve, J. M., *J. Fluor. Chem.* **3**, 17 (1973/1974).
2. Appel, R., and Lassmann, E., *Chem. Ber.* **104**, 2246 (1971).
3. Azeem, M., Brownstein, M., and Gillespie, R. J., *Can. J. Chem.* **47**, 4159 (1969).
4. Banister, A. J., and Clarke, H. G., *J. Chem. Soc., Dalton Trans.* p. 2661 (1972).
5. Banister, A. J., Clarke, H. G., Rayment, I., and Shearer, H. M. M., *Inorg. Nucl. Chem. Lett.* **10**, 647 (1974).
8. Barr, M. R., and Dunell, B. A., *Can. J. Chem.* **48**, 895 (1970).
9. Bartlett, N., and Robinson, P. L., *Chem. Ind. (London)* 1351 (1956).
10. Bartlett, N., and Robinson, P. L., *J. Chem. Soc., London* p. 3417 (1961).
11. Baum, K., *J. Amer. Chem. Soc.* **91**, 4594 (1969).
12. Becke-Goehring, M., *Advan. Inorg. Chem. Radiochem.* **2**, 159 (1962).
13. Becke-Goehring, M., and Fluck E., *Develop. Inorg. Nitrogen Chem.* **1**, 150 (1966).
14. Behrend, E., and Haas, A., *J. Fluor. Chem.* **4**, 99 (1974).
15. Breitinger, D., Brodersen, K., and Limmer, J., *Chem. Ber.* **103**, 2388 (1970).
16. Brownstein, M., Dean, P. A. W., and Gillespie, R. J., *Chem. Commun.* p. 9 (1970).
17. Buckendahl, W., and Glemser, O., to be published (Diss. W. Buckendahl, Göttingen, 1975).
18. Buckendahl, W., Glemser, O., and Saran, H., *Z. Naturforsch. B* **28**, 222 (1973).
19. Buss, B., Altena, D., Höfer, R., and Glemser, O., to be published.
20. Buss, B., Altena, D., Mews, R., and Glemser, O., to be published.
21. Chatt, J., and Dilworth, J. R., *J. Chem. Soc., Chem. Commun.* p. 508 (1974).
22. Christe, K. O., Curtis, E. C., Schack, C. J., and Pilipovich, D., *Inorg. Chem.* **11**, 1679 (1972).
23. Christe, K. O., Schack, C. J., Pilipovich, D., Curtis, E. C., and Sawodny, W., *Inorg. Chem.* **12**, 620 (1973).
24. Clauss, K., and Jensen, H., *Tetrahedron Lett.* p. 119 (1970).
25. Clifford, A. F., and Duncan, L. C., *Inorg. Chem.* **5**, 692 (1966).
26. Clifford, A. F., and Kobayashi, C. S., *Inorg. Chem.* **4**, 571 (1965).
27. Clifford, A. F., and Thompson, J. W., *Inorg. Chem.* **5**, 1424 (1966).
28. Clifford, A. F., and Zeilenga, G. R., *Inorg. Chem.* **8**, 979 (1969).
29. Connelly, N. G., *Inorg. Chim. Acta Rev.* **6**, 47 (1972).
30. Connelly, N. G., and Dahl, L. F., *Chem. Commun.* p. 880 (1970).
31. Cowan, D. O., Gleiter, R., Glemser, O., and Heilbronner, E., *Helv. Chim. Acta* **55**, 2418 (1972).
32. Cramer, R., *J. Org. Chem.* **26**, 3476 (1961).
33. Darragh, J. I., Hossain, S. F., and Sharp, D. W. A., *J. Chem. Soc., Dalton Trans.* p. 218 (1975).
34. Darragh, J. I., and Sharp, D. W. A., *Chem. Commun.* p. 864 (1969).
35. De Marco, R. A., and Shreeve, J. M., *Chem. Commun.* 788 (1971).
36. De Marco, R. A., and Shreeve, J. M., *J. Fluor. Chem.* **1**, 269 (1971/72).
37. Demitras, G. C., and MacDiarmid, A. G., *Inorg, Chem.* **6**, 1903 (1967).
38. Drullinger, L. F., and Griffiths, J. E., *Spectrochim. Acta, Part A* **27**, 1793 (1971).

39. Dudley, F. B., J. Chem. Soc. (London) p. 3407 (1963).
40. Dudley, F. B., Cady, G. H., and Eggers, D. F., J. Amer. Chem. Soc. 78, 1553 (1956).
41. Dunphy, R. F., Lau, C., Lynton, H., and Passmore, J., J. Chem. Soc., Dalton Trans. p. 2533 (1973).
42. Evans, J. A., and Long, D. A., J. Chem. Soc., A p. 1688 (1968).
43. Feser, M., Höfer, R., and Glemser, O., Z. Naturforsch. B 29, 716 (1974).
44. Feser, M., Höfer, R., and Glemser, O., Z. Naturforsch. B 30, 327 (1975).
45. Feser, M., Mews, R., and Glemser, O., to be published.
46. Froböse, R., Mews, R., and Glemser, O., to be published.
47. Gibler, D. D., Adams, C. J., Fischer, M., Zalkin, A., and Bartlett, N., Inorg. Chem. 11, 2325 (1972).
48. Gillespie, R. J., "Molecular Geometry." Van Nostrand-Reinhold, New York, 1972.
49. Glemser, O., Feser, M., von Halasz, S. P., and Saran, H., Inorg. Nucl. Chem. Lett. 8, 321 (1972).
50. Glemser, O., and Gruhl, S., unpublished results (Diss. S. Gruhl, Göttingen, 1966).
51. Glemser, O., and von Halasz, S. P., Inorg. Nucl. Chem. Lett. 4, 191 (1968).
52. Glemser, O., and von Halasz, S. P., Inorg. Nucl. Chem. Lett. 5, 393 (1969).
53. Glemser, O., von Halasz, S. P., and Biermann, U., Z. Naturforsch. B 23, 1381 (1968).
54. Glemser, O., and Höfer, R., Z. Naturforsch B 29, 121 (1974).
55. Glemser, O., and Koch, W., Angew. Chem. 83, 145 (1971); Angew. Chem., Int. Ed. Eng. 10, 127 (1971).
56. Glemser, O., and Koch, W., An. Asoc. Quim. Argen. 59, 143 (1971).
57. Glemser, O., Krebs, B., Wegener, J., and Kindler, E., Angew. Chem. 81, 568 (1969); Angew. Chem., Int. Ed. Eng. 8, 598 (1969).
58. Glemser, O., and Mews, R., Advan. Inorg. Chem. Radiochem. 14, 333 (1972).
59. Glemser, O., Mews, R., and von Halasz, S. P., Inorg. Nucl. Chem. Lett. 5, 321 (1969).
60. Glemser, O., Mews, R., and Roesky, H. W., Chem. Ber. 102, 1523 (1969).
61. Glemser, O., Mews, R., and Roesky, H. W., Chem. Commun. p. 914 (1969).
62. Glemser, O., and Richert, H., Z. Anorg. Allg. Chem. 307, 313 (1961).
63. Glemser, O., Richert, H., and Haeseler, H., Angew. Chem. 71, 524 (1959).
64. Glemser, O., Saran, H., and Mews, R., Chem. Ber. 104, 696 (1971).
65. Glemser, O., Schröder, H., and Wyszomirski, E., Z. Anorg. Allg. Chem. 298, 72 (1959).
66. Glemser, O., and Wegener, J., Angew. Chem. 82, 324 (1970); Angew. Chem., Int. Ed. Eng. 9, 309 (1970).
67. Glemser, O., and Wegener, J., Inorg. Nucl. Chem. Lett. 7, 623 (1971).
68. Glemser, O., and Wyszomirski, E., Angew. Chem. 69, 534 (1957).
69. Goehring, M., and Voigt, G., Chem. Ber. 89, 1050 (1956).
70. Hasek, W. R., Smith, W. C., and Engelhardt, V. A., J. Amer. Chem. Soc. 82, 543 (1960).
71. Hazell, A. C., Acta Chem. Scand. 26, 2542 (1972).
72. Höfer, R., and Glemser, O., Z. Naturforsch. B 27, 1106 (1972).
73. Höfer, R., and Glemser, O., Z. Naturforsch. B 30, 458 (1975).
74. Höfer, R., and Glemser, O., Z. Naturforsch. B 30, 460 (1975).
75. Jache, A. W., Advan. Inorg. Chem. Radiochem. 16, 177 (1974).

76. Jäckh, C., and Sundermeyer, W., *Angew. Chem.* **86**, 442 (1974); *Angew. Chem., Int. Ed. Eng.* **13**, 401 (1974).
77. Jolly, W. L., Lazarus, M. S., and Glemser, O., *Z. Anorg. Allg. Chem.* **406**, 209 (1974).
78. Klüver, H., and Glemser, O., to be published (Diss. H. Klüver, Göttingen, 1975).
79. Kolditz, L., and Schäfer, W., *Z. Anorg. Allg. Chem.* **315**, 35 (1962).
79a. Kramar, M., and Duncan, L. C., *Inorg. Chem.* **10**, 647 (1971).
80. Krebs, B., Meyer-Hussein, E., Glemser, O., and Mews, R., *Chem. Commun.* p. 1578 (1968).
81. Lau, C., Lynton, H., Passmore, J., and Siew, P.-Y., *J. Chem. Soc., Dalton Trans.* p. 2535 (1973).
82. Lau, C., and Passmore, J., *Chem. Commun.* p. 950 (1971).
83. Lau, C., and Passmore, J., *J. Chem. Soc., Dalton Trans.* p. 2528 (1973).
84. LeBlond, R. D., and DesMarteau, D. D., *J. Chem. Soc., Chem. Commun.* p. 555 (1974).
85. Lin, T. P., and Glemser, O., unpublished results.
86. Lin, T. P., and Glemser, O., to be published (Diss. T. P. Lin, Göttingen, 1973).
87. Lindner, E., *Angew. Chem.* **82**, 143 (1970); *Angew. Chem., Int. Ed. Eng.* **10**, 114 (1970).
88. Lindner, E., and Weber, H., *Chem. Ber.* **101**, 2832 (1968).
89. Liu, C. S., Mews, R., and Glemser, O., to be published.
90. Lustig, M., and Ruff, J. K., *Inorg. Chem.* **6**, 2115 (1967).
91. Mews, R., *Z. Naturforsch. B* **28**, 99 (1973).
92. Mews, R., *J. Fluor. Chem.* **4**, 445 (1974).
93. Mews, R., *Angew. Chem.* **87**, 669 (1976); *Angew. Chem., Int. Ed. Eng.* **14**, 640 (1975).
94. Mews, R., to be published.
95. Mews, R., and Glemser, O., *Inorg. Nucl. Chem. Lett.* **7**, 821 (1971).
96. Mews, R., and Glemser, O. *Z. Naturforsch. B* **28**, 362 (1973).
97. Mews, R., and Glemser, O., *J. Chem. Soc., Chem. Commun.* p. 823 (1973).
98. Mews, R., and Glemser, O., *Angew. Chem.* **87**, 208 (1975); *Angew, Chem., Int. Ed. Eng.* **14**, 186 (1975).
99. Mews, R., Wagner, D. L., and Glemser, O., *Z. Anorg. Allg. Chem.* **412**, 148 (1975).
100. Middleton, W. J., *J. Org. Chem.* **40**, 574 (1975).
101. Moeller, T., and Dieck, R. L., *Prep. Inorg. React.* **6**, 63 (1971).
102. Morino, Y., Kuchitsu, K., and Moritani, T., *Inorg. Chem.* **8**, 867 (1969).
103. Moritani, T., Kuchitsu, K., and Morino, Y., *Inorg. Chem.* **10**, 344 (1971).
104. Morris, E. D., and Nordman, C. E., *Inorg. Chem.* **8**, 1673 (1969).
105. Niecke, E., and Flick, W., *Angew. Chem.* **85**, 586 (1973); *Angew. Chem., Int. Ed. Eng.* **12**, 585 (1973).
106. Niecke, E., and Flick, W., *Angew. Chem.* **86**, 128 (1974); *Angew. Chem., Int. Ed. Eng.* **13**, 134 (1974).
107. Niecke, E., and Scherer, O. J., *Nachr. Chem. Tech.* **23**, 395 (1975).
108. Parshall, G. W., Gramer, R., and Foster, R. E., *Inorg. Chem.* **1**, 677 (1962).
109. Ratcliffe, C. T., and Shreeve, J. M., *J. Amer. Chem. Soc.* **90**, 5403 (1968).
110. Roderiguez, J. A., and Noftle, R. E., *Inorg. Chem.* **10**, 1874 (1971).

111. Roesky, H. W., *Angew. Chem.* **80**, 43 (1968); *Angew. Chem., Int. Ed. Eng.* **7**, 63 (1968).
112. Roesky, H. W., *Angew. Chem.* **80**, 44 (1968); *Angew. Chem., Int. Ed. Eng.* **7**, 63 (1968).
113. Roesky, H. W., *Angew. Chem.* **80**, 236 (1968); *Angew. Chem., Int. Ed. Eng.* **7**, 218 (1968).
114. Roesky, H. W., *Angew. Chem.* **80**, 626 (1968); *Angew. Chem., Int. Ed. Eng.* **7**, 630 (1968).
115. Roesky, H. W., *Inorg. Nucl. Chem. Lett.* **4**, 147 (1968).
116. Roesky, H. W., *Angew. Chem.* **81**, 493 (1969); *Angew. Chem., Int. Ed Eng.* **8**, 510 (1969).
117. Roesky, H. W., *Angew. Chem.* **83**, 252 (1971); *Angew. Chem., Int. Ed. Eng.* **10**, 265 (1971).
118. Roesky, H. W., and Babb, D. P., *Angew. Chem.* **81**, 705 (1969); *Angew. Chem., Int. Ed. Eng.* **8**, 674 (1969).
119. Roesky, H. W., and Babb, D. P., *Inorg. Chem.* **8**, 1733 (1969).
120. Roesky, H. W., and Dietl, M., *Chem. Ber.* **106**, 3101 (1973).
121. Roesky, H. W., and Giere, H. H., *Chem. Ber.* **102**, 3707 (1969).
122. Roesky, H. W., and Giere, H. H., *Inorg. Nucl. Chem. Lett.* **7**, 171 (1971).
123. Roesky, H. W., Giere, H. H., and Babb, D. P., *Inorg. Chem.* **9**, 1076 (1970).
124. Roesky, H. W., Glemser, O., Hoff, A., and Koch, W., *Inorg. Nucl. Chem. Lett.* **3**, 39 (1967).
125. Roesky, H. W., Grosse-Böwing, W., Rayment, I., and Shearer, H. M. M., *J. Chem. Soc., Chem. Commun.* p. 735 (1975).
126. Roesky, H. W., and Hoff, A., *Chem. Ber.* **101**, 162 (1968).
127. Roesky, H. W., and Holtschneider, G., *Z. Anorg. Allg. Chem.* **378**, 168 (1970).
128. Roesky, H. W., and Mews, R., *Angew. Chem.* **80**, 235 (1968); *Angew. Chem., Int. Ed. Eng.* **7**, 217 (1968).
129. Roland, A., and Sundermeyer, W., *Z. Naturforsch. B* **27**, 1102 (1972).
130. Rosenberg, R. M., and Muetterties, E. L., *Inorg. Chem.* **1**, 756 (1962).
131. Ross, D. S., and Sharp, D. W. A., *J. Chem. Soc., Dalton Trans.* p. 34 (1972).
132. Ruff, J. K., *Inorg. Chem.* **4**, 1446 (1965).
133. Ruff, J. K., *Inorg. Chem.* **5**, 732 (1966).
134. Ruff, J. K., *Inorg. Chem.* **5**, 1787 (1966).
135. Ruff, J. K., and Lustig, M., *Inorg. Chem.* **3**, 1422 (1964).
136. Ruff, J. K., and Lustig, M., *Inorg. Syn.* **11**, 138 (1968).
137. Sauer, D. T., and Shreeve, J. M., *Chem. Commun.* p. 1679 (1970).
138. Sauer, D. T., and Shreeve, J. M., *J. Fluor. Chem.* **1**, 1 (1971/1972).
139. Sauer, D. T., and Shreeve, J. M., *Z. Anorg. Allg. Chem.* **385**, 113 (1971).
140. Schack, C. J., Wilson, R. D., Muirhead, J. S., and Cohz, S. N., *J. Amer. Chem. Soc.* **91**, 2907 (1969).
141. Seel, F., *Advan. Inorg. Chem. Radiochem.* **16**, 297 (1974).
142. Seel, F., and Detmer, O., *Angew. Chem.* **70**, 163 (1958).
143. Seel, F., and Detmer, O., *Z. Anorg. Allg. Chem.* **301**, 113 (1959).
144. Seel, F., Jonas, H., Riehl, L., and Langer, J., *Angew. Chem.* **67**, 32 (1955).
145. Seel, F., and Riehl, L., *Z. Anorg. Allg. Chem.* **282**, 293 (1955).
146. Seel, F., and Simon, G., *Angew. Chem.* **72**, 709 (1960).
147. Seel, F., and Simon, G., *Z. Naturforsch. B* **19**, 354 (1964).
148. Seppelt, K., and Sundermeyer, W., *Angew. Chem.* **82**, 931 (1970); *Angew. Chem., Int. Ed. Eng.* **9**, 905 (1970).

149. Seppelt, K., and Sundermeyer, W., *Z. Naturforsch. B* **26**, 65 (1971).
150. Sheppard, W. A., *J. Amer. Chem. Soc.* **84**, 3058 (1962).
151. Stahl, I., Mews, R., and Glemser, O., *J. Fluor. Chem.* **7**, 65 (1976).
152. Stahl, I., Mews, R., and Glemser, O., to be published.
153. Sundermeyer, W., Roland, A., and Seppelt, K., *Angew. Chem.* **83**, 443 (1971); *Angew. Chem., Int. Ed. Eng.* **10**, 419 (1971).
154. Tullock, C. W., Coffman, D. D., and Muetterties, E. L., *J. Amer. Chem. Soc.* **86**, 357 (1964).
155. Tunder, R., and Siegel, B., *J. Inorg. Nucl. Chem.* **25**, 1097 (1963).
156. Varwig, J., Mews, R., and Glemser, O., *Chem. Ber.* **107**, 2468 (1974).
157. Varwig, J., Mews, R., and Glemser, O., unpublished results.
158. Varwig, J., Steinbeißer, H., Mews, R., and Glemser, O., *Z. Naturforsch. B* **29**, 813 (1974).
159. von Halasz, S. P., and Glemser, O., *Chem. Ber.* **103**, 594 (1970).
160. von Halasz, S. P., and Glemser, O., *Chem. Ber.* **104**, 1256 (1971).
161. von Halasz, S. P., Glemser, O., and Feser, M., *Chem. Ber.* **104**, 1242 (1971).
162. Wagner, D. L., Wagner, H., and Glemser, O., *Chem. Ber.* **108**, 2469 (1975).
163. Wagner, D. L., Wagner, H., and Glemser, O., *Z. Naturforsch. B* **30**, 279 (1975).
164. Wagner, D. L., Wagner, H., and Glemser, O., to be published.
165. Wagner, D. L., Wagner, H., and Glemser, O., *Chem. Ber.* **109**, 1424 (1976).
166. Wagner, H., Mews, R., Lin, T. P., and Glemser, O., *Chem. Ber.* **107**, 584 (1974).
167. Weiß, J., Mews, R., and Glemser, O., *J. Inorg. Nucl. Chem. Supplement* p. 213 (1976).
168. Yu, S. L., and Shreeve, J. M., *J. Fluor. Chem.* **6**, 259 (1975).
169. Zalkin, A., Hopkins, T. E., and Templeton, D. H., *Inorg. Chem.* **5**, 1767 (1966).

ISOPOLYMOLYBDATES AND ISOPOLYTUNGSTATES

KARL-HEINZ TYTKO and OSKAR GLEMSER

Anorganisch-Chemisches Institut der Universität Göttingen, Göttingen, West Germany

I. Introduction

The chemistry of isopolymolybdates and isopolytungstates has been an area of intense activity for several decades. Yet, even today we cannot claim that all the important processes occuring on acidification of a molybdate or tungstate solution have been elucidated. Numerous contradictory reports appearing in the literature right up to the present have repeatedly done more to confuse than to clarify the subject. Thus it is hardly surprising that not a single textbook correctly describes the situation in acidified molybdate and tungstate solutions.

The number of proposals for polyions occurring in such solutions is legion. With little exaggeration it may be said that there is no conceivable species that has not been proposed at some time or other.

Much of the effort involved in this field consists in scrutinizing the experimental conditions, the evaluation of measurements, and the interpretations given in past and present publications. The large number of methods already used in these studies presents a further difficulty. Many of the contradictions abounding in the literature arise from attempts to establish the existence of a given species from very small, often rather doubtful changes in the difference quotient of a physical quantity and the amount of H^+ added with the initial ratio H^+/MO_4^{2-} (M = Mo, W). Yet even the interpretation of obvious breaks or jumps in experimental plots can often cause considerable difficulty, as will be shown later.

This review aims to provide a critical account of the present state of research in the field of isopolymolybdates and isopolytungstates. Such an aim requires that we also take a detailed look at the methods of investigation. Several contradictions in the literature have arisen because the state of the systems was inaccurately defined. Earlier studies will be considered only in so far as the results are still valid or, even if incorrect, have played a significant role in the development of this field. Access to the earlier literature is facilitated by previous reviews or articles with lengthy introductory surveys, particularly those by Lindqvist (1), Kepert (2–4), Aveston et al. (5, 6), Glemser et al. (7), Jahr and Fuchs (8), and Sasaki and Sillén (9), as well as the monographs by Souchay (10, 11). Compounds prepared by thermal routes (e.g., from melts) that do not contain discrete anions and, therefore, do not strictly belong to the isopolyanions will not be considered here.

II. Characterization of the Species

A. QUANTITIES P, Z, AND Z^+

The overall equation for formation of isopolymolybdate and isopolytungstate ions in aqueous solution reads

$$p H^+ + q MO_4^{2-} \rightleftharpoons [H_{p-2r}M_qO_{4q-r}]^{(2q-p)-} + r H_2O \qquad (1)$$

In dealing with H^+/MO_4^{2-} systems it has proved expedient to introduce the following three quantities.

a. The degree of acidification P of a solution. This is the molar ratio of H^+ ions initially present to MO_4^{2-} ions initially present:

$$P \equiv C_{H^+}/C_{MO_4^{2-}} \qquad (2)$$

where C is the initial concentration of the species denoted by the subscript. The utility of quantity P in practical work is obvious: its value is obtainable directly from the quantities used (molar numbers, masses, volumes) and can thus also be directly adjusted.

b. The acidity Z of *a solution*. This is defined as the molar ratio of *reacted* H^+ ions to *initially* present MO_4^{2-} ions, so that

$$Z = \frac{C_{H^+} - c_{H^+} + K_W/c_{H^+}}{C_{MO_4^{2-}}} \tag{3a}$$

In this expression c is the equilibrium concentration of the species denoted by the subscript, and K_W is the ionic product of water. The degree of acidification P and the acidity Z are related by the expression

$$Z = P - \frac{c_{H^+} - K_W/c_{H^+}}{C_{MO_4^{2-}}} \tag{3b}$$

It should be noted that Eqs. (3a and b) contain the equilibrium *concentration* of H^+. This necessitates corresponding calibration of the potentiometric measuring technique (glass or quinhydrone electrode).

Quantity Z is required if the experimental results are evaluated with the aid of the law of mass action, in which case measurements have to be performed in a medium of constant high ionic strength, or if

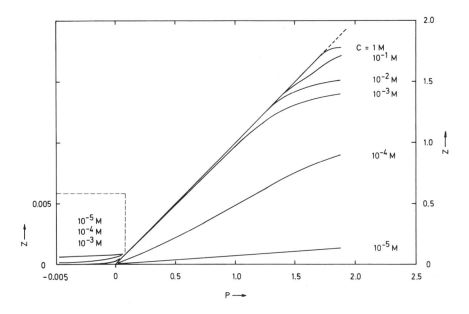

FIG. 1. Relation between Z and P illustrated for the H^+/MoO_4^{2-} system [3 M Na(ClO$_4$)] as ionic medium, 25°C].

the equilibrium constant for a reaction indicates that $P \approx Z$ will not apply for the initial ratio $P = p_i/q_i$ (subscript i denotes the polymetalate species under consideration or its formation reaction). The relation between Z and P is shown for the example of the molybdate system in Fig. 1.

c. *The Z^+ value of a polymetalate species or its overall formation reaction.* We define this quantity as the ratio of the stoichiometric coefficients of H^+ and MO_4^{2-} in the overall formation reaction (1):

$$Z_i^+ \equiv p_i/q_i \qquad (4)$$

To a first approximation, this value, written as a decimal, provides information about the conditions (Z value) for the occurrence of the species i. Expressed as an uncanceled rational fraction, it serves rather to characterize a given species in the absence of a better system of nomenclature (cf. Section II, C).

Regrettably, many authors fail to distinguish between the quantities P and Z, on the one hand, and Z and Z^+, on the other, regarding both symbolism (e.g., using the same symbol for Z and Z^+) and meaning. This is underlined by statements such as "the acidification x denotes the ratio H^+/MO_4^{2-}." The reason for equating these three quantities is that P, Z, and Z^+ do indeed assume identical numerical values in specific cases. Let us assume that we are investigating the formation of the so-called paratungstate A ion,

$$7H^+ + 6WO_4^{2-} \longrightarrow (H^+)_7(WO_4^{2-})_6 \qquad (5)$$

(cf. Sections II, C; III, B, 1; and especially V, B, 3, a concerning this formulation of this polyion). Then, assuming complete reaction [large formation constant(s)] of both H^+ and WO_4^{2-} ions, we conclude from the titration result, $P \approx 1.17$ (initial ratio H^+/WO_4^{2-} at "end point") that $Z \approx 1.17$ and, hence, assuming the occurrence of a single reaction, that $Z^+ = 1.17$. Equating P and Z, and Z and Z^+, in this way is permissible only when the system under investigation has a simple structure, i.e., sufficiently large differences in the Z^+ values of the species involved or equilibrium constants favoring the species studied but not favoring species having close Z^+ values. This problem is considered in detail further on (cf. Sections III, A, 3, a; III, A, 4, a; and III, A, 7).

B. Concentration Data for Polymetalate Ions

Concentration data for mono- and polymetalate species usually refer to 1 M atom. For example, a solution of a species $Mo_7O_{24}^{6-}$ whose concentration is given as 1 molar contains 1/7 mole of $Mo_7O_{24}^{6-}$/liter.

Only three reasons will be given for justification of this procedure: (a) the value of q is still unknown for some polymetalate species so that there is no other choice; (b) for several species of various degrees of aggregation coexisting in equilibrium, there is likewise no alternative choice; and (c) quantities such as extinction or intensity coefficients are similar in magnitude for all mono- and polymetalate anions when referred to 1 M atom. If they were not based on the concentration as defined above, then the spectrum of, e.g., the 36-molybdate ion would appear about 36 times more intense than that of the monomolybdate ion plotted on the same scale. (Logarithmic representation of the intensity is undesirable since the spectra are mainly evaluated with respect to their intensity.)

In order to distinguish the concentrations referring to 1 M atom or quantities involving concentration (e.g., extinction or intensity coefficients) from the others, special labels should be used, e.g., c', ε'. In cases complying with condition b, the symbol for the initial (total) concentration (C) is always used.

If the law of mass action is applied, then the equilibrium concentration referring to q M atoms must of course be used.

Assessment of the relative significance acquired by the various species under a set of given conditions [degree of acidification, Z value or pH value; (initial) concentration; time; temperature] makes use of the species fraction F_i, i.e., that fraction of M present as the species under discussion.

C. NOMENCLATURE

A further source of frequent misunderstanding is the nomenclature employed. For instance, we take the term tetrametalate (4-metalate) to mean a salt whose discrete (oxo-)anion contains 4 M atoms. However, the designation is also used for salts having an analytically determined base/acid ratio $A_2^I O/MO_3$ of 1:4 [the correct and unequivocal designation is A (1:4)-metalate or (1:4)-metalate] whether or not a discrete anion is present. Consequently, three different "tetrametalates" have been reported in the literature: a tetratungstate ion $[W_4O_{12}(OH)_4]^{4-}$ (discrete ion) (12, 13); a "tetramolybdate" $(NH_4)_2O \cdot 4MoO_3 \cdot 2.5H_2O$ (14, 15), that contains the discrete anion $Mo_8O_{26}^{4-}$ (15, 16) and is correctly denoted as octamolybdate; and a "tetramolybdate" $K_2Mo_4O_{13}$ (17, 18) that does not contain a discrete polyanion and, according to its structure (17, 18), can be regarded as polymeric octamolybdate containing the ion $[Mo_8O_{26}^{4-}]_n$. Salts in which discrete polyanions of hitherto unknown size occur and salts without discrete polyanions, of yet unknown structure, should only be represented in the notation $a\mathrm{A_2O} \cdot b\mathrm{MO_3} \cdot c\mathrm{H_2O}$ to avoid confusion. For nondiscrete polyanions of known structure, the formula should always give an indication that no discrete polyanions are present, e.g., in the case of potassium (1:10)-molybdate ("decamolybdate") by bracketing and addition of subscript ∞, i.e., $[\mathrm{KHMo_5O_{16}(H_2O)} \cdot \mathrm{H_2O}]_\infty$.

Frequently, the only fact known about a polymetalate species whose existence has been recognized is its Z^+ value. If known as an uncanceled rational fraction this quantity, as already mentioned, provides a good method of nomenclature or abbreviation, e.g., p,q-metalate ion (if only one number is given, it is the value of q), species p,q or (p,q), metalate species ($Z^+ = p/q$). If it is only known as a decimal (e.g., for the salts $a\mathrm{A_2O} \cdot b\mathrm{MO_3} \cdot c\mathrm{H_2O}$), then only the last approach leads to a useful designation. (Quantities Z^+, a, and b are related according to $Z^+ = 2 - 2a/b$.) Thus the 6-tungstate ion (paratungstate A ion) formed according to Eq. (5) is a 7,6-tungstate ion.

Quantity q in the formula of a polymetalate ion is termed the degree of *aggregation* of the pertinent species. We avoid the designation degree of condensation because studies on the mechanisms of formation (*13, 19–24*) have revealed the operation of various kinds of aggregation mechanisms, including pure addition mechanisms (*13, 21*). The degree of aggregation q (number of M atoms in the species) and the "degree of condensation" r (number of condensed $\mathrm{H_2O}$ molecules) do not, therefore, agree; nor do they exhibit any obvious mutual relationship (e.g., $r = q - 1$, as is normal for linear condensation reactions). We, therefore, generally speak of aggregation to give a polymetalate ion.

The degree of condensation in the sense of r is frequently unknown, and it has therefore become common practice to give the formula of a polymetalate with as few hydrogen atoms as possible, i.e., $p - 2r = 0$ or 1. For example, the paratungstate A ion is usually formulated as $\mathrm{HW_6O_{21}^{5-}}$ or $[\mathrm{W_6O_{20}(OH)}]^{5-}$. Formulas $\mathrm{H_3W_6O_{22}^{5-}}$ or $[\mathrm{W_6O_{19}(OH)_3}]^{5-}$ ($= \mathrm{HW_6O_{21}^{5-}} \cdot \mathrm{H_2O}$), $\mathrm{H_5W_6O_{23}^{5-}}$ or $[\mathrm{W_6O_{18}(OH)_5}]^{5-}$ ($= \mathrm{H\ W_6O_{21}^{5-}} \cdot 2\mathrm{H_2O}$), and $\mathrm{H_7W_6O_{24}^{5-}}$ or $[\mathrm{W_6O_{17}(OH)_7}]^{5-}$ ($= \mathrm{HW_6O_{21}^{5-}} \cdot 3\mathrm{H_2O}$) could, however, also apply. This procedure will be criticized later (Section III, B, 1). In the interest of unambiguous characterization, the formulation $(\mathrm{H^+})_p(\mathrm{MO_4^{2-}})_q$ should be adopted for polymetalate species of unknown r, as has been done for the paratungstate A ion in Eq. (5).

It is frequently the case that the existence of a (new) species has been definitely recognized [nowadays generally as a fingerprint, preferably a characteristic spectral band (cf. Section III, A, 1)], but neither Z^+ nor even p and q, nor any other data can initially be determined. In such cases it has become customary during the past few decades to append an upper-case letter to the word polymetalate or, if a species formerly regarded as homogeneous proves to be a mixture of two species, to an established trivial name (polytungstate Y, X; paratungstate A, B). No objection can be raised to this practice. However, it seems expedient to avoid using the same letter for several levels of Z^+ in order to rule out any danger of confusion.

III. Experimental Methods for Characterizing Species

The number of methods employed in elucidating the chemistry of polymolybdates and -tungstates is very large. However, this variety of methods is not manifested in a correspondingly extensive knowledge of polymolybdate and polytungstate chemistry, but rather reflects the relatively modest amount of information a single method generally affords and the complexity of the systems.

For our purposes it is appropriate to classify the methods of study according to the information they yield about the formation reaction of polymetalate ions and about the polymetalate ions themselves. Critical literature studies repeatedly show that statements about reactions and species are derived from investigations that are simply incapable of yielding such information with the methods used. It often also happens that a new experimental or evaluation technique has to be used in conjunction with results obtained by other methods, and these results (e.g., taken from the literature) were inadequate or even incorrect, or were merely incorrectly interpreted.

If separate studies show that a solution and the solid crystallizing therefrom contain the same polymetalate species (which is by no means self-evident), then the experimental results obtained for the solid can be applied to the species in solution, and vice versa. By applying the above classification, we can consider the experimental methods for solutions and solids together.

The experimental methods generally require that the species being identified or characterized should predominate (species fraction $\geqslant 0.9$–0.95) in a given range of experimental conditions, or that the species being formed (or undergoing reaction) is accompanied only by the starting species (or the reaction product). Methods suitable for detecting or even identifying several coexisting species are comparatively rare (cf. Sections III, A, 1; III, A, 3, b; and III, A, 5, b).

We shall first consider the most important static methods (Sections III, A and B) and then go on to discuss techniques for fast reactions (Section III, C).

A. Methods Providing Information about the Existence and Range of Existence of Polymetalate Species

The existence (i.e., occurrence in significant concentration) of a polymetalate species in aqueous medium at a given acidification and a given initial concentration of $MO_4{}^{2-}$ at a given time after acidification is decided at a given temperature by the p,q values, overall formation constants, and kinetic parameters of all polymetalate species occurring in significant concentration at any acidification and any initial concentration of $MO_4{}^{2-}$ at any time. With the aid of these quantities the contribution of each species to the aggregation products (species fraction) can be calculated, at least in principle, for any concentration, at any acidification, and at any time. By contrast, solids may also contain species not occurring in significant concentration in aqueous media because their formation constants are unfavorable in comparison with those of the other species. On the other hand, some polymetalate ions occurring as principal components in aqueous media have not yet proved isolable as solid salts.

1. Methods Yielding a Fingerprint of a Species

A fingerprint is a directly measurable complex quantity (e.g., a spectrum) that can be used without further processing to recognize a particular species in different circumstances. Apart from the qualitative component (e.g., the band *frequency* of a spectrum), permitting the actual identification, in solutions the quantitative component, i.e., the component proportional to the concentration (e.g., the band *intensity* of a spectrum) may also be employed, for instance, in establishing a uniform reaction course or determining Z^+ (cf. Sections III, A, 3, a and b, and III, A, 4, a). Further information can also be deduced from the qualitative component (cf. Sections III, B, 2, c; III, B, 3, a; and III, A, 1, e). Assignment of a fingerprint to a species is sometimes problematical.

Closely related to the task of obtaining a spectrum as a fingerprint of a species, or of assigning one to a species, is the question of the homogeneity of a species in a solution or solid. Solution of this problem requires observation of the spectra recorded with a certain degree of variation of the conditions of preparing the solid or, in the case of a solution, with a certain degree of variation of the Z value, the time, or, where applicable, the temperature, as well as also varying the concentration in each case. It is, therefore, expedient to combine work on this problem with questions concerning the uniformity of reaction (especially by the method of extinction difference and intensity difference diagrams) and the identity of a species in the solid and in the corresponding solution (spectroscopic methods).

a. Ultraviolet Spectra of Solutions. Frequency and shape serve for identification (*7, 25*) and the intensity for determing the concentration of a species.

b. Raman Spectra of Solutions and Solids. Here too the band frequency (and the intensity ratio) serve for identification (*5, 6, 26, 27*), and the intensity (with solutions, suitable calibration of the method being assumed) for determining the concentration (*5, 26, 27*) of a species. Furthermore, the spectra also provide information about structural questions (cf. Section III, B, 2, c).

c. Infrared Spectra of Solids. As for Raman spectra, the frequency and relative intensity can be used for identification of species (*7*) and, possibly, for checking product purity. In addition, structural information can also be gleaned from the spectra (cf. Sections III, B, 2, c and III, B, 3, a).

d. Paper Chromatography of Solutions. Polymetalate species are identified by their R_f values (*28*). The amount of substance (content of M) in the spot of the chromatogram affords the concentration; it can be readily determined with the aid of a counter tube using isotopically labeled metalate (*28*).

It should be noted that methods based on separation of a polymetalate mixture—this applies both to paper chromatography and electromigration analysis—can only be applied to systems in which re-equilibration after perturbation of the original equilibrium requires considerably longer than migration.

e. Isomatrix Electromigration Analysis of Solutions. With this technique, species identification is based on the path traveled under standard conditions. The amount of substance at the point concerned affords the concentration. Once again it can easily be determined with a counter, using labeled metalate (*29, 30*). This technique can also only be applied to systems in which re-equilibration after disturbance of the original equilibrium takes considerably longer than the migration.

The paths traveled by the polymetalate ions depend on their charge, mass, and size (or compactness), with the result that the paths, in turn, provide some estimate of *relative* charge, mass, and size provided that two of the three quantities remain constant or are known (*29*).

Paper chromatography and electromigration analysis are the only methods providing a direct indication of the number of species co-existing in a solution. However, their use should be preceded by a critical assessment of whether interference can arise due to fast equilibrations during the separation.

f. Polarography of Solutions. The half-wave potential is used for identification, and the diffusion current for quantitative determination of the polymetalate species (*31*).

g. Powder Diagrams of Solids. The positions and relative intensities of the lines are used for identification or purity control of a substance (*1, 8, 9*).

h. Chemical Reactions in Solution. This method is based on the following principle (*31*): A compound is added to the polymetalate solution to be studied (the published investigations deal with the tungstate system), which is able to form a complex—apparently with the monometalate ions present in equilibrium—whose concentration is determined by a physicochemical method [polarography (*31*), UV spectrophotometry (*32*), etc.]. It then depends on the thermodynamic and kinetic stability of the polymetalate ions whether (and if so to what extent) and how fast the complex is formed at the expense of the polymetalate ions.

It has proved possible, by skillful combination of variously acidified and aged metalate solutions and by use of various complexing agents capable of reacting with the monometalate ions, to identify various polymetalate species and to determine their contribution. Particular use has been made of 11-tungstatosilicic acid (formation of the 12-tungstatosilicate ion) (*31*), the vanadatophosphate ion (formation of the vanadatotungstatophosphate ion) (*33*), and pyrocatechol (formation of the 1:1 complex) (*32*) as complexing agents.

2. Methods for Establishing the Identity of a Species in a Solid and in Solution

a. Spectroscopic Methods. Comparison of solution with solid-state spectra (*34, 5, 6, 26, 27*) is presently the most reliable method for establishing the identity of a species in solution and in the solid state since two fingerprints are being compared. Only Raman measurements have so far been used in the molybdate and tungstate system. Good agreement of the spectra [e.g., in the case of the hepta- and 36-molyb-date ion (27)] constitutes proof of identity. If they show reasonable overall agreement [as in the case of the paratungstate B ion (27)] and a plausible explanation is available for poorer agreement at certain positions (e.g., unresolved bands due to experimental line broadening in the dissolved state, and splitting of bands that are degenerate in solution when spectra are recorded for a crystal lattice), then identity can be assumed in this case too (6, 27).

A variant of this method is employed when identity in solution and the solid state is proved for one of a series of similar compounds and the similarity of all the species of the series in solution has been demonstrated by the similarity of their spectra [in such cases UV studies have also been performed (7)]. For example, the metatungstate ion could be shown to have the same structure in solution as the dodecatungstatosilicate ion in the solid state (cf. Sections V, B, 3, b and III, A, 2, b).

b. Direct Comparison of the X-Ray Structure Analysis of a Solid with That of a Solution. Recently, the first comparison of this kind for an isopolymetalate ion became available (35). Hitherto, only examples from heteropolyanion chemistry were known (36, 37); however, the structural identity of the above-mentioned dodecatungstatosilicate ion (36) in solution and in the solid state has just been demonstrated in this way.

c. Comparison of Stoichiometric Coefficients. A further technique that can provide in certain cases strong evidence for identity, if not conclusive proof, is available. If the accurately known (!) stoichiometric coefficients p and q of the overall formation reaction for a solid and a species in solution are equal, then there is some justification for assuming the same species to be present. Although the probability of the same p,q pair occurring coincidentally for solids and dissolved species is statistically relatively low, mechanistic studies, nevertheless, show that for geometrical reasons several structures can be given for certain pairs of p,q values (e.g., 12,8) (23) and are apparently realized (27), whereas for other pairs [e.g., 7,6 (19) and 8,7 (23)] only one or possibly two very closely related structures can indeed be given.

The Z^+ values of solids and species in solution can also be compared in place of p,q values. However, the uncertainty is then greater because the ranges of all integral multiples of q_{min} (determined from the Z^+ value as the denominator of the rational fraction with the lowest divisor) must be taken into consideration.

Both variants are frequently employed, especially in cases where q and p or Z^+ are accurately known for a solid from structural analysis

(1). In general, however, the experimenter's main interest (often unexpressed) was to work backward in order to decide between several possible values of p,q pairs or Z^+ values all of which lie within the limits of experimental accuracy (cf. Sections IV, B, 3, a and V, B, 3, a).

d. Criterion for Determining Time Required for Crystallization (27). An additional criterion for establishing identity is provided by the time required for crystallization of the solid. If a species occurs as the main component in solution, then its salt should crystallize rapidly (provided that the solubility product permits crystallization). Very protracted crystallization usually indicates prior formation of a new species. For example, this is assumed to be the case for the octamolybdate $(NH_4)_4Mo_8O_{26} \cdot 5H_2O$, having the structure determined by Lindqvist (cf. Sections IV, A, 1 and IV, B, 3, b), and has been definitely established for the polymetalates $(NH_4)_{4n}[Mo_4O_{14}]_n$, $(NH_4)_{6n}[Mo_8O_{27}]_n \cdot 4nH_2O$, $A_2O \cdot 3MoO_3 \cdot yH_2O$, $(NH_4)_2O \cdot 4MoO_3$, $Na_2O \cdot 4MoO_3 \cdot 6H_2O$, $(NR_4)_2Mo_6O_{19}$, and $[AHMo_5O_{16}(H_2O) \cdot H_2O]_\infty$ (27).

3. Methods for Proving Uniform Course of Formation and Further Reaction of a Polymetalate Species

There are several methods for demonstrating the uniformity of a reaction for a given acidification range. The most reliable results are probably available nowadays from extinction and intensity, or extinction difference and intensity difference diagrams (*27*, *26*, *38*). They have the added advantage of being applicable also to time-dependent reactions (*39–41*). In general, use of these methods simultaneously yields additional quantities such as Z^+ (cf. Section III, A, 4, a) or p and q and the overall formation constants β (cf. Sections III, A, 5, b and III, A, 6).

a. Concentration-Proportional Titrations. This method is based on measurements of a physical quantity X, which is proportional to the concentration of one or more reactants j, at constant initial concentration of the reactant MO_4^{2-} as a function of the "acidification" x:

$$X(x) = \sum_j \chi_j c_j \qquad (C_{MO_4^{2-}} = \text{const.}) \qquad (6)$$

[χ_j = proportionality constant valid for the species j (polymetalate and other species)]. A straight line is obtained for a uniform reaction course. If the first uniform reaction is followed by a second one, then another straight line results. Since the second reaction usually starts before the first one has gone to completion, the two straight lines are joined together by a curved line. If several parallel reactions occur or if the consecutive reactions show pronounced overlapping then curves are obtained. After completion of all reactions, a final straight line results.

This method affords a curved line in spite of a uniform reaction course if the formation constant of the polymetalate species is small and the quantity chosen as variable x is proportional to the *initial* concentration of H^+, e.g., P. A plot versus a quantity proportional to the *reacted* concentration, e.g., Z, still gives a straight line for unfavorable formation constants (cf. also the discussion in Section II, A).

Apart from the methods suitable for investigating solutions listed in Section III, A, 1, in which the quantitative component of a complex quantity provides a measure for the concentration, it is of course possible to employ methods in which noncomplex quantities (lacking the qualitative component) are measured. Conductivity titrations (*42–47*), as illustrated in Fig. 2, represent the commonest variant of this

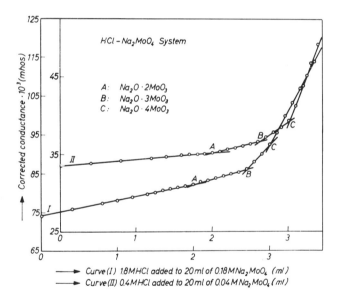

Curve (I) 1.8 M HCl added to 20 ml of 0.18 M Na₂MoO₄ (ml)
Curve (II) 0.4 M HCl added to 20 ml of 0.04 M Na₂MoO₄ (ml)

FIG. 2. Example of a conductivity titration (*47*).

method, which is generally used for determining the Z^+ value of a polymetalate species. Grave errors are often committed in practical application of this method (see Section III, A, 4, a, where further quantities that have been followed during titration are also given).

b. Extinction Difference and Intensity Difference Diagrams as well as Extinction and Intensity Diagrams. Extinction difference (ED) and intensity difference (ID) diagrams are obtained by calculating the differences in extinction and intensity, respectively, of a series of spectra with increasing conversion on addition of a reactant (*27, 26, 38, 48*) or for a time-dependent reaction (*39–41*) relative to a reference spectrum (appropriately the spectrum of the initial state) at two different wavelengths or wave numbers, and plotting the differences against each other (Fig. 3). Extinction (E) and intensity (I) diagrams result when

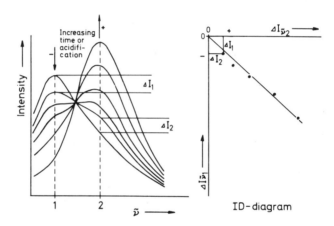

Fig. 3. Construction of an intensity difference (ID) diagram.

the extinctions and intensities, respectively, themselves are plotted against each other for two different wavelengths. The validity of Beer's law is assumed for the absorption spectra; suitable measures have to be adopted to ensure linearity between the concentration of a species and the intensity of its spectrum in the case of intensity spectra (e.g., Raman spectra). If a reaction proceeds in a uniform manner,* then a

* The concept of a (spectroscopically) uniform reaction should not be taken as meaning that only a single reaction takes place. Several intermediates can, of course, occur (mechanism) if the requirements of a quasi-steady state are fulfilled. Moreover, the term also embraces the case of two (or even more) parallel reactions provided that the reaction products are always formed together in the same ratio regardless of the concentration. For time-dependent reactions this will require the same order of reaction; and equal stoichiometric coefficients will be required for p and q in the reaction equation in the case of equilibrium reactions (titrations). In practice these requirements mean that only reactions having the same kind of mechanism can occur as parallel reactions, such as the formation of planar and compact hexatungstate (para A) ion (*19*) (cf. Fig. 16).

straight line is obtained in the ED (E) or ID (I) diagram, whereas curved lines appear if several parallel reactions take place. Depending on the p,q values and the pertinent equilibrium constants of reactions occurring on successive acidification or the magnitude of the rate constants of reactions occurring with increasing time, one or several regions in which a single reaction predominates can be recognized in a system (*38, 41*).

The occurrence of two (and only two) linear independent reactions parallel to each other can be ascertained with the aid of extinction (intensity) difference quotient diagrams (EDQ and IDQ diagrams, respectively) which then yield a straight line (*41*). These diagrams are obtained by calculating the extinction (intensity) differences at three different wavelengths and plotting $\Delta E_{\lambda_2}/\Delta E_{\lambda_1}$ versus $\Delta E_{\lambda_3}/\Delta E_{\lambda_1}$ (or the quotients of intensity differences, respectively).

Diagrams ED, ID, E, I, EDQ, and IDQ have two particular advantages over all other methods (*38*).

1. For each reaction taking place and each species appearing in a system, suitable choice of pairs of wavelengths permits construction of a diagram having optimal information content for the pertinent reaction or species.

2. When merely employed to check uniformity of reaction(s), the diagrams are completely insensitive to errors in the Z value.

c. *The* (Z, log c_{H^+}, $C_{MO_4^{2-}}$) *Equilibrium Curves* (*49*).   An indication of a possibly uniform reaction course is obtained from (Z, log c_{H^+},

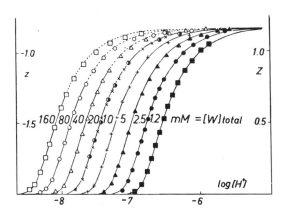

FIG. 4. The (Z, log c_{H^+}, $C_{WO_4^{2-}}$) equilibrium curves for the H^+/WO_4^{2-} system in 3 M Na(ClO$_4$) at 25°C within a short time of acidification. Occurrence of just a single polytungstate ion, 7,6-(paratungstate A) ion, and its formation constant were deduced from the curves (*50*).

$C_{MO_4^{2-}}$) equilibrium curves if the $(Z, \log c_{H^+})$ curves for the various MO_4^{2-} (initial) concentrations lie parallel to one another within a given range of Z. If the curves are parallel, then only so-called core + links complexes can be present. In order to establish whether a single core + links complex is formed, all the curves are superposed by an appropriate coordinate transformation and the new curve analyzed either by comparison with calculated theoretical curves or directly. As an example, the $(Z, \log c_{H^+}, C_{MO_4^{2-}})$ diagram for the tungstate system in the range $0 < Z \leqslant 1.17$ shortly after acidification is depicted in Fig. 4 (50).

4. Methods for Determining the Value of Z^+ (As a Decimal)

a. *Concentration-Proportional Titrations* (*"Break" Titrations*).   The Z value of the intersection of the extrapolated straight lines obtained by the method described in Section III, A, 3, a (and Section III, A, 3, b) affords the Z^+ value of the species formed (see Fig. 2). The P value may only be used if it is numerically equal to the Z value.

Apart from electrical conductivity, the following quantities have also been measured in the course of titrations: UV absorption (51–53); intensity of Raman bands (27); Rayleigh scattering (54); the refractive index (55, 56); the volume dilation of the solution (57–59); the increase in temperature (43, 60, 61); the cryoscopic depression (42, 31, 8); the radioactivity of labeled tungsten in individual fractions after separation of the reaction mixture by paper chromatography (28); and the diffusion current on polarographic reduction of the polymetalate species (62).

This method is one of the most commonly employed techniques for investigations in isopolymetalate chemistry, although its accuracy is not particularly high. The most frequent errors committed in its application are plots versus P (or analogous parameters) instead of Z; errors in extrapolation on determining the intersection of straight lines, especially when only short sections of a line are available (which in most cases are not really and simply cannot be straight lines at all) (cf. Section III, A, 7); overestimation of the accuracy of the method with regard to determination of p_{min} and q_{min} (see Section III, A, 5, a).

b. *Potentiometric Titrations*.   These differ from break titrations in that (a) only one species is responsible for the effect measured, and (b) the measured effect is not proportional to the concentration itself but to its logarithm. For experimental reasons, measurements have so far been limited to the potential due to the H^+ concentration (or H^+

activity) (glass or quinhydrone electrode) (7, 47, 52, 63, 64; see Refs. 2 and 9 for additional references).

The occurrence of one of the reactions describable by Eq. (1) is apparent from the specific shape of the (log c_{H^+}, P) curve from which the (log c_{H^+}, Z) curve may be calculated. The most striking feature of these curves is the potential jump observed at $Z \approx Z^+$. Basically, however, the entire curve is characteristic of the reaction concerned. Thus valuable additional information is lost if the curves are evaluated only with respect to the Z value (or even merely the P value) at which the potential jump occurs. If several reactions take place, then several potential jumps are found, provided that the reactions are not extensively superposed. This requires that the Z^+ values of the species occurring should be sufficiently different and that the overall formation constants should have favorable values. In principle, however, even in those cases where no further (resolved) potential jumps appear the maximum possible information is still present in the overall curve. Mere determination of Z^+ values from the potential jumps cannot, therefore, be regarded as a fitting application of the method.

With additional titration curves for other initial concentrations of MO_4^{2-}, mathematical analysis of the curves (see Sections III, A, 5, b and III, A, 6) affords, in principle, the p,q values and overall formation constants of all species occurring in significant concentration in the range under study.

c. *Analysis of Solids.* If analysis of a substance indicates a formula $aA_2O \cdot bMO_3 \cdot cH_2O$, its Z^+ value* follows from

$$Z^+ = 2 - (2a/b) \qquad (7)$$

Here, too, the accuracy with regard to determination of p_{min} and q_{min} is grossly overestimated. For the maximum error in Z^+ not to exceed about 0.5% (cf. Section III, A, 5, a), the sum of the relative errors in the analytical values for A_2O and MO_3 should not exceed 0.5%. Salts containing much water of crystallization pose the additional sampling problem of using samples with the same water of crystallization content in the individual analyses because the water is easily given up. This problem led to protracted discussions whether the paramolybdates and paratungstates are (3:7)- or (5:12)-metalates (65, 66, 31, 67, 68).

* This formula requires that only A^+ may be present as cation in the substance $aA_2O \cdot bMO_3 \cdot cH_2O$. If H^+ (in hydrated form) also occurs as cation (which cannot be established by chemical analysis), then spurious (too high) Z^+ values will be obtained. For example, the substance $A_2O \cdot 10MO_3 \cdot 5H_2O$ might be the polymetalate $A_2M_{10}O_{31} \cdot 5H_2O$ ($Z^+ = 1.80$), $A_2M_{10}O_{30}(OH)_2 \cdot 4H_2O$ ($Z^+ = 1.80$) or $A_2(H_3O)_2M_{10}O_{32} \cdot 2H_2O$ ($Z^+ = 1.60$).

d. Ion Exchange Processes. The amount of metalate removed from a polymetalate solution on contact with an ion exchanger and the amount of anion released by the latter, which is equivalent to the amount of charge of the metalate ions, can be determined by chemical analysis of the solution. Hence, assuming a single polymetalate species to be present, the ratio of charge to degree of aggregation of the polymetalate ion $1/R \equiv (2q - p)/q = 2 - p/q = 2 - Z^+$ can be established. If several species are present and the exchange process is nonselective, then an average value for all the species in the solution is obtained. In cases where selection does occur the average value applies to the adsorbed species. In order to decide whether just one or several species are taken up by the ion exchanger, the method would have to be combined with another one (presence of a single species could be deduced from the occurrence of a plateau in R curves [e.g., $R(pH)$] in appropriate cases). However, even when this additional method has been found, the problems have by no means been solved. Concentration changes resulting from adsorption (general drop in concentrations, selection, etc.) drastically modify the equilibria so that fast re-equilibration (during the exchange process) further confuses the situation *(69, 70)*. Moreover, it is also possible that the pore size and charge distribution in the matrix of the exchanger may favor formation and adsorption of species occurring only in minimal concentrations, if at all, in solution *(69, 70)*. Thus the results obtained for the molybdate system by the ion exchange process *(71, 72)* fail to agree with those determined by other methods.

5. *Methods for Determining Degree of Aggregation*

a. Determination of Molar Mass of Dissolved Ions with the Ultra-centrifuge. Three different methods are available: the sedimentation equilibrium technique *(5, 6, 26, 73–75)*; the sedimentation velocity technique *(7, 52, 75)*; and the Archibald technique (measurement in transition state) *(7)*. All three methods have actually been used. Since the effect of ionic charge cannot be established with certainty and since numerous possibilities exist for formulating the M—O skeletons of the polyions without introducing further assumptions (for instance, the formulation of the 36-molybdate ion can range between $Mo_{36}O_{112}$ and $Mo_{36}O_{144}$ if no restrictive assumptions are made), and a broad range must consequently be considered for the "theoretical" or "calculated" molar mass, errors of 5 to 10% generally have to be expected in the resulting degree of aggregation.

Ultracentrifuge studies can be used to great advantage in combination with X-ray studies on crystals when a polymetalate occurs both in solution and as a crystalline solid. The molar mass of possible formula units can be calculated from the volume of the unit cell, the density, and the various possible values of z (number of formula units in the unit cell) for the space group and compared with that obtained from analysis and studies with the ultracentrifuge. Thus, ultracentrifuge studies indicate the order of magnitude, and X-ray analysis the precise value (26). In a similar manner, although with less accuracy, the Z^+ value can also be combined with ultracentrifuge studies. Conversion of Z^+ into the rational fraction with lowest divisor yields p_{min} and q_{min}. p and q are then equal to p_{min} and q_{min} or integral multiples thereof. The order of magnitude of q is again obtained from ultracentrifuge studies and the precise value from Z^+ $(7, 52, 76)$. This combination is employed frequently. It is limited to low degrees of aggregation: even 7,6- $(Z^+ = 1.167)$ and 8,7-metalate ions $(Z^+ = 1.143)$ $(\Delta Z^+ = 0.024 \triangleq 2.1\%)$ can, if at all (cf. Section III, A, 7), only be distinguished by very careful measurements (maximum error in Z^+ about 0.5%). Diffusion $(43, 77-79)$ and dialysis methods (80) fail to give accurate values for the molar mass of dissolved ions $(81, 82)$. Salt cryoscopy $(42, 31)$ has a similar reputation (83).

b. *Mathematical Analysis of* $(Z,$ log $c_{H^+},$ $C_{MO_4^{2-}})$ *Equilibrium Curves* $(84–86)$.   In principle, this is a very elegant method for identification of species present in solution by means of their p and q values and provides the opportunity of observing several species coexisting in equilibrium and of calculating their formation constants too. However, unequivocal results are obtained only on very accurate measurement of the triple set of values (log $c_{H^+},$ $Z,$ $C_{MO_4^{2-}}$). Evaluation rests solely on application of the law of mass action and observation of mass and charge balance, and requires that measurements are conducted at constant, high ionic strength. The calculations are extremely involved. The solution of the problem is the set of p,q values and the corresponding overall formation constants that give the best fit between experimental and calculated Z values. The standard deviation σ_Z is, therefore, the criterion for assessment. As an example, Fig. 5 shows the $(Z,$ log $c_{H^+},$ $C_{MO_4^{2-}})$ diagram for the molybdate system from which the occurrence of the species $(1,1),$ $(2,1),$ $(8,7),$ $(9,7),$ $(10,7),$ $(11,7),$ $(34,19),$ and $(5,2)$ [alongside $(0,1)$] was deduced. The points indicate experimental values and the solid lines were calculated assuming occurrence of the above species. The limitations of the technique are reached when several sets of species with their corresponding

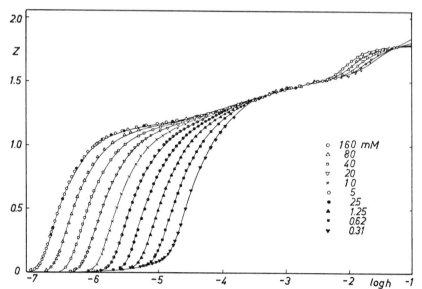

Fig. 5. The $(Z, \log c_{H^+}, C_{MoO_4^{2-}})$ equilibrium curves for the H^+/MoO_4^{2-} system in $3\ M$ Na(ClO$_4$) at 25°C. Occurrence of the species (1,1), (2,1), (8,7), (9,7), (10,7), (11,7), (34,19), and (5,2) was deduced and the overall formation constants of the species determined from the curves $(9, 87)$. Abscissa: $\log h = \log[H^+]$.

formation constants lead to similarly small standard deviations (cf. the discussion in Sections IV, B, 3, a–c and V, B, 3, a and b).

An additional method, developed especially for evaluation of measurements at small Z values, permits recognition and elimination of certain systematic errors that play an increasingly significant role with decreasing Z value $(88, 89)$.

c. X-Ray Structure Analysis of Salts and of Species in Solution.

The most reliable results are obtained by X-ray structure analysis of crystalline polymetalates $(90–95, 15, 16, 96–106)$. The findings are transferable to the situation in solution provided that separate investigations have established the presence of the same species in solution and in the solid state. In certain cases, X-ray structure analysis can also be performed on solutions.

6. Determination of Overall Formation Constants of Polymetalate Ions

Mathematical analysis of $(Z, \log c_{H^+}, C_{MO_4^{2-}})$ equilibrium curves $(84–86, 88, 89)$ furnishes the stoichiometric coefficients p and q and, at the same time, the overall formation constants of polymetalate ions. The relevant data are listed in Tables I and II (see Sections IV, B, 1 and V, B, 1, respectively).

7. Computational Methods for Assessing Possible Existence of Isopolymetalate Ions

We have found a very useful aid in assessing the possible existence of species proposed in the literature and in studying the structure of H^+/MO_4^{2-} systems to be a computer program (107) permitting calculation of the Z value and the distribution of the species (species fraction) for a given set of species and their overall formation constants as a function of the initial concentration $C_{MO_4^{2-}}$ and of log c_{H^+}. The most important conclusion (108) drawn from such calculations is that use of the method in Sections III, A, 3, and A, 4, a does not permit detection of any number of successive species with increasing Z value, as would appear to be the case on considering the species and reactions proposed in the literature. For a section of a plot to be identifiable as a straight line, it is essential that the Z^+ values of successive species differ by about 0.2 units. Apart from the errors already mentioned, failure to observe this fact is one of the most frequent sources of error in the method of Section III, A, 4, a.

8. Determination of Kinetic Data

The methods given in Sections III, A, 1; III, A, 3, a and b; and III, A, 4, a, which are based on measurement of quantities or, in the case of spectra, functions proportional to concentration, may be employed not only for determining the Z or pH dependence but also, as mentioned on various occasions, for investigating the time dependence of reactions. Standard methods of evaluating kinetic data then permit the derivation of rate laws and rate constants, and, where appropriate, activation energies. Special methods are available for fast reactions (see Section III, C).

B. METHODS FOR DETAILED CHARACTERIZATION OF POLYMETALATE SPECIES

1. Methods for Determining the (Empirical) Formula of a Polymetalate Ion

The methods described so far furnish the degree of aggregation (q) and the charge number ($2q - p$) of a polyion. Formulation of the M—O skeleton (number of O atoms) also requires knowledge of r, the stoichiometric coefficient of the condensed water. At the same time the number of H atoms in the ion (which are bonded to oxygen) is established. Experimental determination of r can be performed only by examining a solid.

For species known only in solution and also for solids when r could not be determined, it has become common practice, as already mentioned, to give the formula of a polymetalate species with as few H atoms as possible, i.e., $p - 2r = 0$ or 1. Regrettably, the same approach is often adopted when r is known. Experimental findings provide hardly any basis for rules governing the number of H atoms (OH groups) present in a polymetalate ion (e.g., as a function of the degree of aggregation or Z^+ value). Attention has already been drawn to the numerous possible formulations of the paratungstate A ion, and to the resulting difficulties encountered in the interpretation of molar mass determinations (cf. Section III, A, 5 a). Theoretical approaches to the formation mechanisms and structures of polymetalate ions merely indicate that the number of OH groups in stable polymetalate ions can never be large (23). In order to distinguish whether a formula is the true (empirical) formula or that having the lowest possible number of H atoms, the latter should always be given. It is most expedient to adopt the proposed notation $(H^+)_p(MO_4^{2-})_q$ when r is unknown and a formula is desired.

　　a. X-Ray Structure Analysis of Solids.　If we know the structure of a polymetalate ion, then we also know the formula of its M—O skeleton (90–95, 15, 16, 96–106), from which r and, knowing p, the number of H atoms in the polyion can be determined.

　　b. Analysis of the Solid.　Analysis indicates the formula as $aA_2O \cdot bMO_3 \cdot cH_2O$, where a, b, and c are only *proportional* numbers. In the cases $c = 0$ and $b = 2q$ for $c = 1$, the formula of the M—O skeleton is unequivocally known: The M/O ratio in the M—O skeleton is $b/(3b + a + c)$.

Attempts have also been made to prepare crystalline polymetalates by controlled hydrolysis of tungstic and molybdic tetraesters in organic solvents, expecting the products to contain no water of crystallization and all the hydrogen present to be in the form of OH groups (8, 109–112). In this way the formula $W_5O_{16}^{2-}$, or a multiple thereof, has been established for the polytungstate Y ion (112), and the number of possibilities (cf. Section II, C) for the paratungstate A ion reduced to two, namely $HW_6O_{21}^{5-}$ {or $[W_6O_{20}(OH)]^{5-}$} and $H_3W_6O_{22}^{5-}$ {or $[W_6O_{19}(OH)_3]^{5-}$} (109) (see Section V, B, 3, a).

c. *Thermal Degradation of Polymetalates.* Assuming that the water of crystallization of polymetalates is released more readily than the so-called water of constitution participating as OH groups in the structure of the M—O skeleton, and that no fundamental changes in the polymetalate ion skeleton occur, various polymetalates have been subjected to stepwise thermal degradation. The individual degradation steps were examined for intactness of the M—O skeletons by means of differential thermal analysis, X-ray analysis, IR spectra, etc. The formula $[H_2W_{12}O_{40}]^{6-}$ for the metatungstate ion was confirmed (*113*, *114*) in this way, and strong evidence obtained in favor of the formula $[H_2W_{12}O_{42}]^{10-}$ for the paratungstate B ion (*8*, *115–117*). However, it should always be borne in mind that there are OH groups that are eliminated as H_2O more readily than the water of crystallization is released. Location of the OH group within the W—O framework rules out this complication in the case of metatungstate and paratungstate B ions (*2*, *118*, *119*, *98*, *99*, *21*).

2. Methods for Structural Elucidation of Polymetalate Ions

a. *X-Ray Structure Analysis of Solids.* Nowadays X-ray structural analysis of crystalline solids permits a comprehensive structure determination [geometry of the M—O skeleton; positions of cations and water of crystallization; structure of the hydrogen bond system (*94*, *95*, *98–100*, *103*, *104*, *106*)]. Within limits, the positions of H atoms (present as OH groups) can also be established by indirect ways (*98*, *99*, *106*).

The structure of the metatungstate ion could also be determined from its isomorphism with the dodecatungstatophosphate ion (*120*), whose structure had been elucidated by X-ray analysis (*121*) (cf. Section V, A, 1).

Figures 6 and 9 (see Sections IV, A, 1 and V, A, 1, respectively) show the structures found so far for isopolymolybdate and isopolytungstate ions (*90–95*, *15*, *16*, *96–106*).

b. *X-Ray Structure Analysis of Solutions.* In certain cases, X-ray structure analysis can also be performed on solutions. Up to now, the positions of the M atoms in the M—O skeletons of one isopolymetalate ion, i.e., the heptamolybdate ion (*35*), and of two heteropolymetalate ions, i.e., the dodecatungstatosilicate (*36*) and pentamolybdatodiphosphate ion (*37*, *35*), have been determined. However, the dodecatungstatosilicate ion is of significance in the structural elucidation of the metatungstate ion (cf. Sections III, A, 2, a and b).

c. Vibrational Spectroscopy. Assignment of the vibrational spectra of polymetalate ions is rather difficult owing to the large number of atoms participating in the structure and the low symmetries usually encountered in these species. Unequivocal assignment of all the bands of a polymetalate structural type has only been accomplished very recently by normal coordinate analysis of the highly symmetric M_6O_{19} ions (*122*). However, for the present it appears possible to prove the presence of certain groups of atoms, e.g., $M-O_b-M$ bridges and terminal $M-O_t$ and $M\begin{smallmatrix} \diagup O_t \\ \diagdown O_t \end{smallmatrix}$ groups, (*123–126*), where subscript b signifies a bridging O atom, and t terminal O atoms.

d. Exchange Reactions with $*MO_4^{2-}$ *and* $H_2{}^{18}O$. These exchange reactions relate to the $M-O$ skeleton of the polymetalates. The data obtainable are primarily mechanistic in nature and can, therefore, provide structural information.

Exchange between $*WO_4^{2-}$ and the paratungstate B ion (relatively slow exchange) (*127, 128*) and between (acidified) $*WO_4^{2-}$ and the metatungstate ion (no exchange) (*129, 2*) have been studied.

Measurements on the paratungstate B ion performed with $H_2{}^{18}O$ (*127, 130*) are too limited to permit any conclusion.

3. Methods of Determining Kind of Bonding and Positions of H Atoms

a. Infrared Spectra. Attempts were made to determine the kind of bonding of H atoms from IR spectra (*116, 117, 131, 132*). Overlapping with the bands due to water of crystallization precluded detection by this method of OH groups in various polytungstates (*7*). It could, however, be shown (*7*) that H_3O^+ ions are not present, as proposed (*131, 132*), in paratungstates B.

Experimentally, it should be mentioned that care must be exercised when pressing with KBr samples of the salts having high water of crystallization contents. Even grinding with an embedding agent is sometimes not permissible.

b. ¹H Nuclear Magnetic Resonance. It could be established with the aid of ¹H NMR spectra that the H atoms of various polymetalate ions are present in the form of OH groups (*7*). In individual cases, e.g., for the paratungstate B (*119*) and metatungstate ion (*118*), the positions of the H atoms could even be determined.

c. Exchange Reactions with D_2O. Hydrogen atoms lying in the interior of a polyion skeleton should undergo only relatively slow exchange. However, studies performed with paratungstate B and metatungstate ions, which were based on this reasoning, failed to yield conclusive results (*127, 133*).

4. Determination of Thermodynamic Data

a. Determination of Heats and Entropies of Reaction for Formation of Polymetalate Ions. Mathematical analysis of enthalpy titration curves has afforded ΔH_r^0 and ΔS_r^0 values for the overall formation reactions of those polymolybdate and polytungstate ions that were identified by mathematical analysis of (Z, log c_{H^+}, $C_{MO_4^{2-}}$) equilibrium curves (*134, 135*). The data are listed in Tables I and II.

Methods of determining overall formation constants have already been considered (see Section III, A, 6).

5. Reducibility of Polymetalate Ions

According to Pope (*136*) the ease of reduction of hetero- and isopoly-anions to "molybdenum blue" and "tungsten blue" depends on the types of octahedra present in the structure: Only those polyanions that possess octahedra with monooxo terminal groups (octahedra with one free corner) are reducible. Thus the presence of octahedra with one free corner can be deduced for polyanions of unknown structure if they are reducible.

Numerous other publications on reduced polymetalates also exist (e.g., Refs. *137–141*; see also footnote on p. 292).

C. METHODS FOR STUDYING FAST REACTIONS

The methods discussed so far belong to the so-called static methods. Those applicable to solutions can be used for reactions having half-reaction times down to 1 min, depending on the nature of the method. For faster reactions, flow (rapid-mixing) and relaxation methods are available. Application of these methods has hitherto only yielded quantities indicating the existence and range of existence. However, the continuous flow method should also permit studies leading to detailed characterization of polymetalate species.

1. Flow (Rapid-Mixing) Methods

Flow methods permit (at least in principle) the techniques described so far to be used within 10^{-3} sec to 1 min after acidification, wherever this is compatible with their nature. With the continuous-flow method, a physical quantity X (a single quantity, e.g., pH value, or a function, e.g., a spectrum) can be measured at certain selected times after mixing of the reactants. In this way X can be investigated as a function of the acidification P and the time, and the information given in Sections III, A and B can be obtained for all applicable methods. By contrast, the stopped-flow method permits continuous measurement of a single quantity as a function of time for the mixture of reactants having a certain P value. The information obtained corresponds to that of methods in Section III, A, 8, i.e., kinetic data. A serious difficulty of this technique often lies in the assignment of the information obtained to a particular reaction or species (see, e.g., Section IV, B, 3, b and Ref. *142*).

With the continuous-flow method, it has become possible to study the simple protonation reactions (duration $\approx 10^{-8}$ sec) of the mono-metalate ions (formation of the species HMO_4^- and H_2MO_4) in isolation from the subsequent aggregation reactions. However, it was also necessary to delay the aggregation reactions by use of very low initial concentrations of MO_4^{2-} (*143, 144*). This method has so far afforded the following information: the stoichiometric coefficients p and q of special species [(1,1) and (2,1) for both the molybdate and tungstate system] and their formation constants by mathematical analysis of rapid-titration curves; the Z^+ value ($= 1.25$) of a short lived (primary) aggregation product in the tungstate system that was viewed as 5,4-tungstate ion; the order of magnitude of the protonation constants of various polytungstate ions [paratungstate A and B ions (cf. Section V, B, 3, a), metatungstate ion]; and the order of magnitude of the rate constants for primary aggregation steps (*143, 144*). Use of the stopped-flow method furnished rate laws and rate constants for hydrolytic disaggregation of various polymetalate ions (cf. Section IV, B, 4).

2. Relaxation Methods

The process of relaxation to a new equilibrium state following a very brief perturbation of a chemical system at equilibirum can be studied as a function of chemical parameters, e.g., of the initial concentration $C_{MO_4^{2-}}$ and the Z value, and subjected to mathematical analysis. In the studies performed so far (*145–149*), perturbation was effected by a temperature jump, and the relaxation process was followed by spectrophotometry of a coupled pH indicator equilibrium. Apart from rate constants, the stoichiometric coefficients p and q of several, generally primary aggregation reactions were obtained (cf. Sections IV, B, 2 and V, B, 2).

Very fast reactions (with rates approaching the limits set by diffusion processes) can be studied by ultrasonic methods based on continuous perturbations of equilibria. Protonation reactions of mono- and polymetalate ions have been investigated in this way (150) (cf. Sections IV, B, 1 and IV, B, 3, b).

IV. Isopolymolybdates

A. SOLID ISOPOLYMOLYBDATES

The classic method of preparing isopolymolybdates and isopoly-tungstates is crystallization of alkali (including ammonium), and sometimes also alkaline earth, salts from acidified aqueous metalate solutions. A variant of the preparation with regard to acidification consists in the use of "molybdic acid" or "tungstic acid" ($MO_3 \cdot H_2O$, $MO_3 \cdot 2H_2O$) (7) and, with regard to deposition, in the precipitation of the ions from acidified aqueous alkali salt solutions as sparingly soluble salts (K, Ba, tri- and tetraalkylammonium salts, etc.) (7, 31, 74, 151). Commercially available "ammonium molybdate," $(NH_4)_6Mo_7O_{24} \cdot 4H_2O$, is obtained by crystallization after dissolution of molybdenum trioxide in aqueous ammonia solution. During the past decade some other methods have also become known: controlled hydrolysis of the tetraesters of the metallic acids in organic solvents and in the presence of organic bases or salts (cf. Section III, B, 1, b) (8, 109–112, 152); reaction of the metal oxides or oxide hydrates with organic bases, either as liquids or dissolved in organic solvents (occasionally also in water) (153, 103, 124); and reaction of polymetalate salts with organic cations to give those with inorganic cations (74, 124).

Salts having inorganic cations are hardly soluble in organic solvents, whereas those with organic cations dissolve well. The latter can, therefore, be recrystallized from organic solvents with comparative ease (74, 111). A further advantage is that equilibrium reactions between polymetalate ions (reactions involving H_2O) are suppressed in the absence of water.

1. Crystallization or Precipitation from Aqueous Solution (27)

Whereas normal molybdates ($A_2MoO_4 \cdot yH_2O$) crystallizing from "neutral" [$P(!) = 0$] aqueous molybdate solutions contain the tetrahedral ion $MoO_4{}^{2-}$ (1, 154), all salts of known structure hitherto obtained from acidified molybdate solutions comprise polyanions made up of MoO_6 octahedra. Ammonium (1:2)-molybdate obtained from solutions having $P \approx 0$ consists of MoO_6 octahedra and MoO_4 tetrahedra.

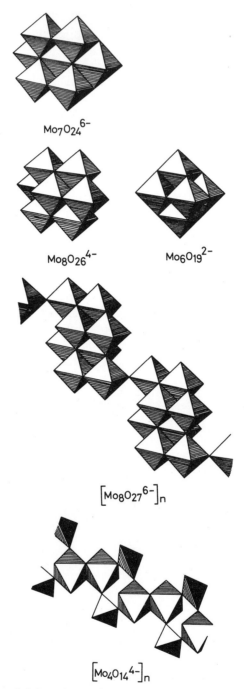

$Mo_7O_{24}^{6-}$

$Mo_8O_{26}^{4-}$ $Mo_6O_{19}^{2-}$

$\left[Mo_8O_{27}^{6-}\right]_n$

$\left[Mo_4O_{14}^{4-}\right]_n$

FIG. 6. Polymolybdate ions of known structure. (The structure of the "decamolybdate" ion has been omitted. It is too complicated to be depicted in the same manner.)

(3:7)-Molybdates (65, 66), the so-called paramolybdates, crystallize from solutions having $Z \approx 1.1$. They are heptamolybdates containing the discrete ion $Mo_7O_{24}^{6-}$ ($Z^+ = 8/7 = 1.14$), whose structure has been determined and repeatedly confirmed (90–95) (Fig. 6). The known salts include $Na_6Mo_7O_{24} \cdot 22H_2O$, $Na_6Mo_7O_{24} \cdot 14H_2O$, $(NH_4)_6Mo_7O_{24} \cdot 4H_2O$, $K_6Mo_7O_{24} \cdot 4H_2O$ (155, 9, 1, 27).

At $Z \approx 1.3$, the so-called "trimolybdates," (1:3)-molybdates of formula $A_2O \cdot 3MoO_3 \cdot yH_2O$, are deposited as a very voluminous, fibrous mass of crystals (155, 1, 9, 156). Only the Z^+ value (1.33, from analysis) and fingerprints [X-ray powder diagram (156), IR (156) and Raman spectra (27)] are known. The fingerprints show that the same polymetalate species is present in all cases. The known salts include $Na_2O \cdot 3MoO_3 \cdot yH_2O$, $(NH_4)_2O \cdot 3MoO_3 \cdot yH_2O$, $K_2O \cdot 3MoO_3 \cdot yH_2O$. Some uncertainty attaches to the value of y owing to the voluminous texture of the products. For the sodium salt having $y = 3$, a chain-type structure was postulated (156).

Solutions of $Z \approx 1.5$ afford the crystalline metamolybdates or "tetramolybdates," (1:4)-molybdates of formula $A_2O \cdot 4MoO_3 \cdot yH_2O$ ($Z^+ = 1.50$). The known salts are $Na_2O \cdot 4MoO_3 \cdot 6H_2O$ (yellowish), $(NH_4)_2O \cdot 4MoO_3 \cdot 2H_2O$ or $\cdot 2.5H_2O$, and $(NH_4)_2O \cdot 4MoO_3$ (without H_2O) (155, 1, 9, 27). A potassium salt cannot be prepared; under the appropriate conditions the "trimolybdate" if always formed instead (14, 27). According to their Raman and IR spectra the three salts contain different polymetalate ions (27). The structure of salt $(NH_4)_2O \cdot 4MoO_3 \cdot 2.5H_2O$ or $\cdot 2H_2O$ is known: It contains discrete octamolybdate ions having the formula $Mo_8O_{26}^{4-}$ (15, 16) (Fig. 6). However, mechanistic studies also allow the presence of (structurally different) $Mo_8O_{26}^{4-}$ (cf. Section III, A, 2, c) in the other two salts (23).

At $Z \approx 1.8$, crystallization occurs of (1:9)-molybdates (26, 27) containing a discrete highly aggregated molybdate ion of probable formula $Mo_{36}O_{112}^{8-}$ ($Z^+ = 64/36 = 1.78$) (26). The known sodium, ammonium, potassium, and barium salts crystallize with $\approx 80H_2O$ and display identical Raman spectra (27). The structure of the ion has yet to be determined. A structural proposal (26) compatible with all available data received no support from mechanistic studies because no explanation can be given for the fast formation of this structure. A new proposal, which appears to satisfy this aspect too (23), is put forward in Fig. 7.

Solutions having $P = 2.0$ ($Z \approx 1.8$) to $P = 4$ (the acidification of such solutions can no longer be satisfactorily described by a Z value; cf. Fig. 1) precipitate the so-called decamolybdates, (1:10)-molybdates of formula $A_2O \cdot 10MoO_3 \cdot yH_2O$ ($Z^+ = 1.80$) as small hexagonal

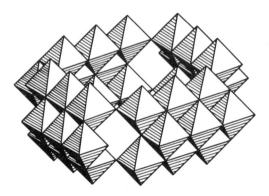

FIG. 7. New proposed structure for the $Mo_{36}O_{112}^{8-}$ ion.

prisms (*157, 43, 27, 106*). The known sodium, ammonium, and potassium salts are shown by their Raman and IR spectra (*27, 106*) to represent defined substances based on the same polymolybdate ion. No discrete polyanion is present; the basic structural units are double chains of MoO_6 octahedra linked together in three dimensions via shared octahedral corners (*106*). Structural studies indicate $[AHMo_5O_{16}$ $(H_2O)\cdot H_2O]_n$ to be the most appropriate formulation.

The polymolybdates described so far crystallize readily from the acidified alkali (including ammonium) molybdate solutions; although the "trimolybdates," metamolybdates, and "decamolybdates" require a little time to do so (several days). Several further polymolybdates can be obtained on observation of certain conditions (from aqueous solution).

After several hours of heating of an ammonium molybdate solution of $-0.5 \leqslant P \leqslant 0.5$ at $> 60°C$ (sealed vessel or replenishment of evaporating ammonia), a crystalline (1:2)-molybdate of formula $(NH_4)_2O\cdot 2MoO_3$ (*158, 1, 27, 159, 125*) appears. This salt does not contain discrete polyanions but a chain-type polytetramolybdate ion $[Mo_4O_{14}^{4-}]_n$ ($Z^+ = 4n/4n = 1.00$) (*105, 24*) (Fig. 6). The corresponding sodium and potassium salts are unknown.

If an ammonium heptamolybdate solution is allowed to stand for a very long period (*160, 161*) or acidified to $Z \approx 1.20$ and treated with ammonium chloride (*162*), it deposits crystals of a (3:8)-molybdate having the formula $3(NH_4)_2O\cdot 8MoO_3\cdot 4H_2O$. This is likewise devoid of discrete polyanions, containing a chain-type polyoctamolybdate ion $[Mo_8O_{27}^{6-}]_n$ ($Z^+ = 10n/8n = 1.25$) (Fig. 6) (*104, 22*). The corresponding sodium and potassium salts are again unknown.

From solutions of $Z \approx 1.7$, addition of tetrabutylammonium salts precipitates a yellow (1:6)-molybdate; complete agreement of its IR spectrum with that of (1:6)-tungstate indicates the presence of the $Mo_6O_{19}^{2-}$ ion ($Z^+ = 10/6 = 1.67$) (*112*) (Fig. 6). The salt crystallizes without water.

Apart from the alkali salts of the normal monomolybdate, the heptamolybdate, and the 36-molybdate ion, all these substances are sparingly soluble in water (*27*).

According to older literature (*163*), many more polymolybdates should be obtainable from aqueous solution. However, since several decades ago only inadequate methods of identification were available —products were characterized practically only by elemental analysis and appearance—the same products were probably often described as being different, owing to analytical errors (cf. Section III, A, 4, c), and mixtures of different products or mixed crystals as homogeneous substances*.

2. Preparation by Hydrolysis of Esters

Since a pure molybdic ester soluble in organic solvents is not yet available, the ester–ammonia adduct $3MoO_2(OC_2H_5)_2 \cdot 2NH_3$ has been used. In this way, salts $(NH_4)_6Mo_7O_{24}$ (without H_2O) ($Z^+ = 1.14$), $(NH_4)_2O \cdot 4MoO_3$ (without H_2O) ($Z^+ = 1.50$), and $2(NH_4)_2O \cdot 6MoO_3 \cdot H_2O$ ($Z^+ = 1.33$) could be prepared (*152*).

3. Preparation by Reaction of Molybdenum Trioxide with Organic Bases

Reaction of molybdenum trioxide with hexakis(dimethylamino)-cyclotriphosphazene in aqueous suspension affords the 10,6-molybdate $[HN_3P_3(N(CH_3)_2)_6]_2Mo_6O_{19}$ whose anion has the structure shown in Fig. 6 (*103*). (Remarkably, the proton of the base is located on a ring nitrogen atom.)

B. Isopolymolybdate Ions in (Aqueous) Solution

Studies on aqueous alkali metalate solutions have been performed in some instances without adopting any special measures regarding the ionic medium and in others at greater or lesser ionic strengths. Measurements in which the law of mass action is applied (e.g., methods of Sections III, A, 3, c and A, 5, b) require constant ionic strength for the whole set

---

* The polymolybdate types listed above are those that we (*27*) too have obtained from aqueous solution on systematic variation of the conditions of preparation (acidification, crystallization time, temperature, initial concentration of molybdate) and by using cations Na^+, NH_4^+, and K^+.

of experimental data (49). High ionic strengths favor aggregation reactions, equilibrium constants become larger (5, 164, 165), and the measured effects more pronounced (potential jumps, breaks in titrations, etc.) (31, 52, 63). Rate constants also increase (166); however, in disaggregations not only positive (166) but also negative effects (167, 168) have been observed. These effects apparently arise by complex formation between the polyions and the alkali metal ions (2, 5, 6, 169, 166, 144). Nevertheless, since obviously all aggregation reactions are affected in a similar manner, the same species always appear regardless of the ionic strength (cf. Section IV, B, 3, b). Differences reported in the literature are no doubt largely due to the authors' interpretation of their measurements. The sole experimentally confirmed case of the ionic medium having an influence on a reaction seems to be that of the "acidic paratungstate ion" (170).

Equilibration of the reactions occuring in molybdate solutions is complete within several minutes at the latest (5, 9, 27, 143, 171). In this respect, the molybdate system differs considerably from the tungstate system. Subsequent reactions are related to the gradual precipitation of sparingly soluble solids (27, 172) (cf. Sections III, A, 2, d and IV, B, 3).

1. Protonated Monomolybdate Species

On acidification of a monomolybdate solution, the MoO_4^{2-} ion is protonated to give $HMoO_4^-$ and subsequently H_2MoO_4. Even though the occurrence of these reactions must be regarded as self-evident, their experimental proof is very difficult since the aggregation reactions following simple mono- and diprotonation are very fast (requiring use of flow methods) and quantitative (requiring action concerning concentrations) (143).

Studies using a flow technique and plotting potentiometric titration curves as function of the concentration at a time of 10^{-2} sec after acidification showed that no changes in the curves occur only for (initial) concentrations of MoO_4^{2-} of 2×10^{-4} M and below, that is, separation of the pure protonation reactions (of the monomolybdate ion) from the aggregation reactions could be accomplished (143). (The concentration limit would lie higher for shorter times.) Equilibrium constants for the mono- and diprotonation reaction could be estimated from the shape of the curve (see Table I). Ultrasonic relaxation methods (150) afforded a rate constant of 5×10^9 M^{-1} sec^{-1} for monoprotonation of the MoO_4^{2-} ion. This value is about one power of ten lower than that of diffusion-controlled reactions and is explained in terms of structural changes of the monomolybdate ion on protonation (see below).

TABLE I

OVERALL FORMATION CONSTANTS β, OVERALL STANDARD REACTION ENTHALPIES ΔH_r°, AND OVERALL STANDARD REACTION ENTROPIES ΔS_r° FOR FORMATION OF MOLYBDATE SPECIES (p,q)

Parameter	(1,1)	(2,1)	(8,7)	(9,7)	(10,7)	(11,7)	(12,8)	(34,19)	(64,36)	(5,2)	Medium	Temp. (°C)	Ref.
Log β	≈3.88	7.75	—	—	—	—	—	—	—	—	0.1 M (KCl)	20	143
Log β	4.00	8.21	—	—	—	—	—	—	—	—	μ = 0.0023	20	175
Log β	4.68	8.68	—	—	—	—	—	—	—	—	μ = 10.5 (?)	—	176
Log β	3.89	7.50	57.74	62.14	65.68	68.21	—	196.30 (373.69[a])	—	≈19	3 M Na(ClO$_4$)	25	87
Log β	3.55	7.20	52.81	57.39	61.02	63.40	—	—	—	—	1 M Na(Cl)	25	5
Log β[b]	3.53	7.26	52.80	57.42	60.84	—	71.56	—	—	—	} 1 M Na(Cl)	25	
ΔH_r° (kcal)	14 ± 7	—	−56.0	−53.4	−52.6	−53.2	—	—	—	—	}		
ΔS_r° (cal deg^{-1})	65 ± 23	—	76	105	124	134	—	—	—	—	3 M Na(ClO$_4$)	25	134

[a] This value was calculated assuming that a 64,36 ion (87) and not a 34,19 (26) is formed.

[b] Alternative set of species with a 12,8 ion instead of the 11,7 ion.

Detection of the two protonated species with static methods, that is, at equilibrium since in the H^+/MoO_4^{2-} system after a few minutes changes no longer occur, has all the more chance of success the greater the contribution made by the pertinent species to the sum of all molybdate species (with the exception of the MoO_4^{2-} ion). As a consequence of the law of mass action, this is the case for minimum possible initial concentrations of MoO_4^{2-} and minimum possible Z values for the $HMoO_4^-$ ion and maximum possible Z values for H_2MoO_4 (about 1–2). The lower limit for the MoO_4^{2-} initial concentration and the Z value are set by the sensitivity of the method of investigation. Using this approach [evaluation of potentiometric (Z, log c_{H^+}, $C_{MoO_4^{2-}}$) equilibrium curves], both $HMoO_4^-$ and H_2MoO_4 could be detected (87, 5) and their formation-constants and thermodynamic parameters ΔH_r^0 and ΔS_r^0 (134) (Table I) determined.

The two consecutive dissociation constants of molybdic acid $K_1 = \beta_{2,1}/\beta_{1,1}$ and $K_2 = \beta_{1,1}$ are practically equal while they differ by a factor of about 10^{-5} for other polybasic acids (173, 174). This is generally rationalized in terms of a structural change accompanying protonation, specifically a transition from tetrahedral to octahedral coordination of M according to (143, 175, 176, 150)

$$MoO_4^{2-} \xrightarrow{\ +H^+,\ +2H_2O\ } [Mo(OH)_5]^- \xrightarrow{\ +H^+\ } Mo(OH)_6 \qquad (8)$$

Our own studies (177) of this problem have shown that the occurrence of $[Mo(OH)_5]^-$ and $Mo(OH)_6$ is even less likely than that of $[MoO_3(OH)]^-$ and $MoO_2(OH)_2$ but that the problem can be solved quite well by assuming the species $[MoO_3(OH)(H_2O)_2]^-$ and $MoO_2(OH)_2(H_2O)_2$.

2. Primary Aggregation Products

We use the term *primary aggregation products* to mean (a) transient species with $q > 1$ requiring investigation techniques suitable for fast reactions, and (b) species detectable also in equilibrium by static methods having $1 < q < q_H$ and $Z^+ \lesssim Z^+_H$ and, additionally, those with $q = q_H$ and $Z^+ < Z^+_H$, where H is the initial major product occurring on successive acidification of the solution—in this case the $Mo_7O_{24}^{6-}$ ion. Since such species must be fundamentally capable of existence for mechanistic reasons, there has been no lack of attempts to detect them. The detection of di-, tri-, tetra-, and hexamolybdate ions is reported in the literature (178, 164, 169, 58, 61, 179, 59, 145, 146, 148; additional data in Refs. 1, 5, 52), these species being described, however, in some cases as major products and not as primary products.

On the basis of model calculations (108) with the computer program mentioned in Section III, A, 7, it will now be shown that primary aggregation products cannot be recognized by static methods, especially those given in Sections III, A, 3, a and A, 4, a, if their Z^+ value differs by less than 0.2–0.3 units from that of the main product, as is the case for the 6,6 ion (*58, 179, 59*). Those methods permitting recognition of several species coexisting in equilibrium and their characterization with respect to p and q represent exceptions; the only methods of this kind to be used so far are those based on mathematical analysis of $(Z, \log c_{H^+}, C_{MO_4^{2-}})$ equilibrium curves given in Section III, A, 5, b. For example, assuming that $Mo_7O_{24}^{6-}$ is the initial major product of acidification, as is generally accepted nowadays, we introduce this ion and its overall formation constant, and a second species p,q whose Z^+ value lies between 0.9 and 1.2 (e.g., species 6,6) into the computer program and vary the formation constant for this second ion. We then find that for none of the species p,q does there exist an equilibrium constant that would lead to c_i values giving straight-line sections subject to unequivocal characterization in an $X(Z) = \Sigma_j \chi_j c_j$ plot for $C_{MoO_4^{2-}} = $ const. in the range $0 < Z \lesssim 1.2$.

No primary aggregation products were found on mathematical analysis of $(Z, \log c_{H^+}, C_{MoO_4^{2-}})$ equilibrium curves. Whenever a species with $q < 7$ has been postulated (*178, 164, 169*) in such studies, it has actually been a main product occurring at higher Z values. Very detailed studies of the range $0 < Z \lesssim 1.1$ by means of intensity difference diagrams of Raman spectra failed also to indicate primary aggregation products (*27, 180*). Thus even the very recently reported detection of a 1,2- (*58, 61, 59*) or a 6,6-molybdate ion (*58, 179, 59*) in a break titration cannot be correct. Such species are clearly at a thermodynamic disadvantage compared to the heptamolybdate ion, as is also apparent from mechanistic studies (*23*).

Relaxation studies by the temperature jump method (*145, 146*) on 7×10^{-3}–2.5×10^{-2} M molybdate solutions at acidifications $P = 0.01$–0.11 [pH $= 5.3$–5.6; medium, 0.3 M $Na(NO_3)$] indicated the existence of a very rapidly formed tetramolybdate ion ($Z^+ = 6?/4$) as primary aggregation product. At higher ionic strength [3 M $Na(ClO_4)$] but otherwise similar conditions, however, a hexamolybdate ion ($p = 6$–7, $q = 6$) was found (*148*).

3. The Main Reaction Products

The term *main product* will be used to describe all polymetalate species that can be detected by static techniques, with the exception of those defined as primary aggregation products. Under certain conditions (initial concentration of MoO_4^{2-}, acidification, time, temperature), they practically all occur at some time in significant concentrations, although this does not follow from the above definition. Apart from the main products, the solutions will certainly contain numerous other species in such low concentrations that they are undetectable. The existence of such species is also indicated by mechanistic reasoning and investigations (13, 19–21, 27, 23). In particular, the discrete polyions evading detection in aqueous solution but crystallizing out of solution as salts (e.g., the ammonium octamolybdate having the structure determined by Lindqvist and tetrabutylammonium 10,6-molybdate) must already be present in solution (cf. also Section III, A, 2, d). The same applies to polymetalates whose crystal lattice is not composed of discrete polyions if the macropolyion itself, however, is clearly formed from discrete polyions [e.g., ammonium polyoctamolybdate (22) and ammonium polytetramolybdate (24)]. Extensive formation of these salts can be explained in terms of their sparing solubility (which is, however, only rarely demonstrable directly) and removal of their constitutent species from an equilibrium that is unfavorable for them (22, 24, 27).

  a. *Range* $0 < Z \leqslant 1.14$. Very precise investigation of the range $0 < Z \leqslant 1.1$ by means of intensity difference diagrams of Raman spectra (see preceding Section) reveal that a spectroscopically uniform reaction occurs, within the limits of experimental accuracy (27, 180).

  Formation of 7,6- ($Z^+ = 1.167$) or 8,7-molybdate ions ($Z^+ = 1.143$) as initial aggregation product in solution was deduced from breaks in pH, conductivity, and other titration curves (references cited in Refs. 1, 5, 9, 52). It is very difficult to distinguish between these two ions because their Z^+ values differ only slightly, thus necessitating an accuracy superior to $\approx 0.5\%$ in the Z value of the titration break. Moreover, these studies provide no information about the degree of aggregation: 14,12- or 16,14-Molybdate ions or even higher aggregates could just as easily be present. However, molar mass determinations by ultracentrifugation of solutions with $Z \approx 1.15$ have indicated the degree of aggregation to be 7 or 8, although 6 could not be entirely ruled out (5, 52).

  Mathematical analysis of potentiometric (Z, log c_{H^+}, $C_{MoO_4^{2-}}$) equilibrium curves led Sasaki and Sillén (87, 9) to postulate a hydrolysis scheme involving the 8,7-molybdate ion as initial main product. However, Aveston et al. (5) were able to show that a scheme involving the 9,8-molybdate ion as first main product leads only to an insignificantly greater standard deviation in Z.

Preference for the 8,7-ion over the 7,6- and the 9,8-ion apparently arises because the paramolybdates, whose structure has been known since 1950 (*90*), contain $Mo_7O_{24}^{6-}$ ions ($Z^+ = 8/7$) and crystallize from solutions with $Z = 1.14$.

The good agreement found on comparison of the Raman spectrum of a solution acidified to $Z = 1.14$ with those of crystalline heptamolybdates (*5*, *27*) was first utilized by Aveston *et al.* to demonstrate that the $Mo_7O_{24}^{6-}$ ion (see Fig. 6) is, indeed, the first main product to appear in solution. We can thus formulate the overall formation equation for the heptamolybdate ion as

$$8H^+ + 7MoO_4^{2-} \longrightarrow Mo_7O_{24}^{6-} + 4H_2O \tag{9}$$

This is the only equation we shall give for the molybdate system. Experimental results only permit formulation of overall formation equations or equations in which the species of the initial and final states appear. Such equations are merely of interest so far as techniques of conducting or evaluating the reaction are concerned. In extreme cases the polymetalate species of the initial state can be the direct precursor of that of the final state; however, it may also undergo extensive prior degradation. Statement of a reaction equation for the first main product appearing in solution can be justified inasmuch as the overall formation equation and the equation relating the initial and final state of the system are then identical, i.e., the degradation reactions are redundant.

Apart from the overall formation constant (*87*, *9*, *5*) for the heptamolybdate ion, ΔH_r^0 and ΔS_r^0 (*134*) for the overall formation reaction could be determined by mathematical analysis of enthalpy titration curves (see Table I).

b. Range $1.14 \leqslant Z \leqslant 1.6$. Investigation of the range $1.14 \leqslant Z \leqslant 1.6$ with the aid of intensity difference diagrams of Raman spectra showed that on successive acidification there is no uniform reaction but instead at least three extensively superimposed consecutive reactions take place (*27*).

Before this direct evidence for the occurrence of several reactions became available, Sasaki and Sillén (*87*, *9*) had already drawn up a hydrolysis scheme with several species for this region based on the mathematical analysis of potentiometric (Z, log c_{H^+}, $C_{MoO_4^{2-}}$) equilibrium curves. They assumed formation of 8,7-, 9,7-, 10,7-, and 11,7-molybdate ions. However, Schwing *et al.* (*178*, *164*) postulated 8,7-, 8,6-, and 9,6-molybdate ions again on the basis of mathematical analysis of potentiometric equilibrium curves, and Aveston *et al.*

using the same method found that, although the hydrolysis scheme of Sillén and Sasaki with only heptamolybdate ions gives the best fit with their own experimental data, schemes with a 12,8 instead of the 11,7 ion or only with octamolybdate ions (9,8, 10,8, 11,8, and 12,8 ions) also fitted nearly as well (5). Sillén subsequently found that Schwing's data were most readily compatible with his hydrolysis scheme too (181). Thus, it is seen that, in spite of the precise and, apart from the effects of varying ionic media [Sasaki and Sillén, 3 M Na(ClO$_4$); Schwing, 3 M Na(Cl); Aveston 1 M Na(Cl)], mutually consistent experimental data of the three research groups no unequivocal conclusion can be drawn about the species occurring in the range under discussion. In addition to protonated heptamolybdate ions and unprotonated and protonated octamolybdate ions, individual hexa- and 9-molybdate ions (in conjunction with the other ions) are certainly also compatible with the experimental data.

The hydrolysis scheme for 1 M Mg(ClO$_4$)$_2$ as ionic medium with 6,6-, 9,8-, and 8,6-molybdate ions (169), which the authors believe to arise because of more pronounced complex formation of Mg^{2+} with the polyion than occurs with Na$^+$, is probably also identical with that having sodium salts as ionic medium, as shown by the standard deviations of other sets of species. Meanwhile, we have been able to prove this for the initial main product by means of Raman spectroscopy: The 8,7-molybdate ion is, indeed, formed also in this case (180).

Hydrolysis schemes proposed on the basis of breaks in titration curves and including more than three species (46, 58, 59, 62) certainly lack a secure experimental foundation. As a consequence of the superposition of several reactions, deduced from both intensity difference diagrams and model calculations ($\Delta Z^+ < 0.2$; see Section III, A, 7), there are no sufficiently defined straight-line sections for determination of intersections (breaks). Schemes containing only two species (52, 61) must, of course, also be incorrect.

An indication that protonated heptamolybdate ions occur might be seen in the minimal changes of the UV spectrum of the solution reported by Pungor and Halasz (53) for the range $1.14 \leqslant Z \leqslant 1.43$. Support for a 12,8-molybdate ion was deduced from a certain degree of resemblance between the Raman spectrum of a molybdate solution having $Z = 1.50$ with the spectrum of crystalline (NH$_4$)$_4$Mo$_8$O$_{26} \cdot$4H$_2$O ($Z^+ = 1.50$), as reported by Aveston and co-workers (5). In fact, the presence of 12,8-molybdate ions in solution at $Z = 1.50$ had already been deduced from the very existence of this ammonium 12,8-molybdate (1, 51). Although Aveston himself expresses considerable caution concerning the existence of a 12,8 ion having the structure found by

Lindqvist on the basis of the above-mentioned spectral comparison, it is accepted as well-founded (*182, 183, 53, 147, 58, 61*) by most research groups since publication of these experiments.

Our own studies of the solid-state and solution Raman spectra of several polymolybdates and polytungstates revealed much better agreement between the spectra when the same species is clearly present in the solid state and in solution than in the case in question (*27*). Thus the presence of a 12,8 ion, with the structure found by Lindqvist, in solution (at room temperature) seems to be at least "unproven." Recently, using a stopped-flow spectrophotometer, Mellström and Ingri (*142*) found two signals on disaggregation of the polyions present in molybdate solutions at $0 < Z < 1.5$ with excess OH^- ions, one of which was assignable to degradation of the $Mo_7O_{24}^{6-}$ ion and, if occurring, its protonated forms (which are not distinguishable by this method). The signal appearing at higher Z values indicates the presence of a further unidentified (and probably difficult to identify) species. This could be an octamolybdate ion; however, the possible presence of the molybdatosilicate ion formed from silicate contaminants could not be excluded. Acidified molybdate solutions being studied by the temperature-jump method (*147*) exhibited two relaxation signals, assigned to formation of $Mo_7O_{24}^{6-}$ and $Mo_8O_{26}^{4-}$. Since these investigations rest on the foundation of previous work, namely equilibrium studies [by Aveston and co-workers (*5*)], the finding that a further species appears alongside $Mo_7O_{24}^{6-}$ (and possibly its protonated forms) warrants special attention, more so than does the claim of its identification as a $Mo_8O_{26}^{4-}$ ion.

For mechanistic reasons (*23*) we also consider the occurrence of 12,8 ions to be highly probable. However, the ion having the structure found by Lindqvist cannot play a very important role (at room temperature). We would definitely rule out the presence of 6,6-, 8,6-, 9,6-, 9,8-, and 11,8-molybdate ions in significant concentration.

A rate constant of $7 \times 10^{10} M^{-1} sec^{-1}$ was deduced for protonation of the 8,7-molybdate ion from ultrasonic relaxation measurements (*150*). This value corresponds to that of diffusion-controlled reactions.

c. Range $1.6 \leqslant Z < 2$. By using intensity difference diagrams of Raman spectra, we found that yet another uniform polymolybdate ion surprisingly appears in the range $Z > 1.6$ at $Z \approx 1.8$ (*26*). This range had formerly been little investigated, no doubt because of difficulties besetting precise determination of Z. The equilibrium concentration of H^+ is already rather high so that errors in its potentiometric determination are consequently large; a further serious difficulty lies in the fact that the equilibrium concentration of H^+ steadily approaches the order of magnitude of the initial concentration, with the result that the reacted H^+ concentration, as the small difference between two large numbers, cannot be determined very accurately. However, we have already pointed out that intensity difference diagrams are completely insensitive to errors in Z values if they are employed merely to identify a uniform species.

The degree of aggregation of the species could be specified as 34–40, a surprisingly high value, by molar mass determination in an ultracentrifuge (sedimentation equilibrium technique) (26).

Additional information about the species was gained by studying the salts crystallizing out of the solutions, since complete agreement of the Raman spectra indicates the presence of the same polymolybdate species in solution and in the salts (26). Thus, combining the results of molar mass determination by ultracentrifugation with X-ray studies (see Section III, A, 5, a) yielded the degree of aggregation of $q = 36$, and the stoichiometric coefficient for the H^+ ion could be determined as $p = 64$ (even number) from the Z^+ value (1.77 $\pm$ 0.03) resulting from analysis of the solid (26). Structural models developed on the basis of mechanistic consideration led to the formula $Mo_{36}O_{112}^{8-}$ (26, 23) (cf. Section IV, A, 1 and Fig. 7).

Earlier attempts to determine the molar mass [salt cyroscopy (8), ultracentrifugation (5), dialysis techniques (184)] had merely shown that the degree of aggregation and the polydispersity increase steeply beyond $Z = 1.5$. Mathematical analysis of potentiometric (Z, log c_{H^+}, $C_{MoO_4^{2-}}$) equilibrium curves led to a best fit between calculated and experimental data assuming participation of a 34,19-molybdate ion (87). Our own calculations (26) in which the 34,19 ion was replaced by the 64,36 ion gave only a slightly greater standard deviation in Z for the set of species with the 64,36 ion.

At still higher acidification, cationic species appear in the solution (9, 11, 53, 185–187).

4. Disaggregation with OH^-

All polymolybdate ions are rapidly transformed by OH^- ions into polyions corresponding to the Z value of the solution. Accordingly, an excess of OH^- ions effects complete disaggregation to MoO_4^{2-}. Disaggregation of the $Mo_7O_{24}^{6-}$ ion is a first-order reaction with regard to both $Mo_7O_{24}^{6-}$ and OH^- ions (167, 168, 188, 142). The reaction has also been studied with various organic bases (168, 188, 189) (cf. also Section IV, B, 3, b).

5. Isopolymolybdate Ions Occurring in Solution and in the Solid State

Some polymolybdate ions occur only in solution, and others only in solids. However, several ions exist both in solution and in the solid state. The polyions of the last-mentioned group are the most important

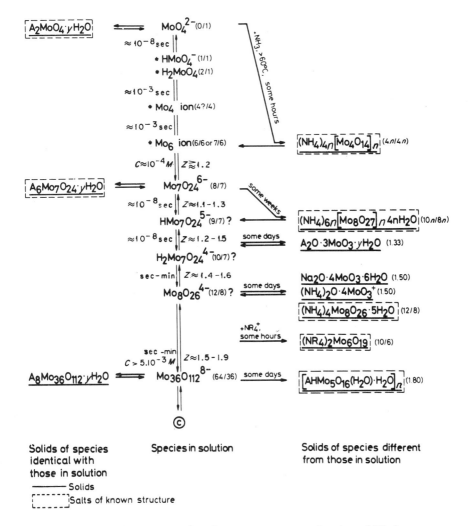

Fig. 8. Reaction scheme for the processes occurring in acidified aqueous molybdate solutions. In field c, probably only cationic species occur. In parentheses: p/q resp. Z^+. *Short-lived species that occur at equilibrium in small amounts only. †For this species reverse reaction does not take place.

for deriving reaction mechanisms: On the one hand these ions, as main products, are the thermodynamically preferred species since they are subject to the law of mass action in solution, and on the other hand, more experimental data are available for them, their structural data being of particular interest. Apart from the tetrahedral (*190*) MoO_4^{2-} ion, only the $Mo_7O_{24}^{6-}$ ion (see Fig. 6) and the $Mo_{36}O_{112}^{8-}$ ion occur both in (aqueous) solution and in the solid state.

The experimental results of Section IV are summarized as a reaction scheme in Fig. 8.

V. Isopolytungstates

A. Solid Isopolytungstates

1. Crystallization or Precipitation from Aqueous Solution (27)

The normal tungstates $A_2WO_4 \cdot yH_2O$ crystallizing from "neutral" ($P = 0$) aqueous tungstate solutions contain the tetrahedral WO_4^{2-} ion (*154, 2*),* whereas polyanions made up of WO_6 octahedra occur in the salts obtained from acidified tungstate solutions, insofar as their structures have been determined. In this respect there is no difference between molybdates and tungstates. Differences do occur, however, in the course of the aggregation reactions, with the result that there is hardly a single polyionic species in either of the systems that has its counterpart in the other.

Solutions having $Z \approx 1.17$ deposit crystalline (5:12)-tungstates (*31, 67, 68*), the so-called paratungstates B,† crystallization requiring several days. Paratungstates B are dodecatungstates containing the discrete

---

* Salt $7Li_2WO_4 \cdot 4H_2O$ is an exception. It contains at the center of the primitive cubic unit cell a $W_4O_{16}^{8-}$ group made up of WO_6 octahedra (Fig. 9) and WO_4^{2-} tetrahedra at the mid-points of the edges (*191, 8, 192*).

† Up to the mid-1940s only paratungstates are mentioned in the literature. In 1943, Souchay recognized that paratungstate solutions ($Z = 1.17$) are not homogeneous but contain two species in equilibrium. They were distinguished as paratungstate A and B ions. Solutions freshly acidified to $Z = 1.17$ were reported to contain only the para A ion, and the freshly prepared solutions of crystalline paratungstate ($Z^+ = 1.17$) only the para B ion. Subsequently, a further species was postulated for solutions having $Z = 1.17$ and designated as para B ion, Souchay's original para B ion being renamed as para Z ion. In the present article the designation introduced by Souchay is retained and other ions on the paratungstate level are designated A or B with subscripts, depending on whether the ion is regarded as a hexameric or a dodecameric particle. This is also expedient in view of the fact that we can hardly reserve the entire alphabet for the paratungstate level. Meanwhile, other species that are not precursors of the para Z ion have been designated Y and X.

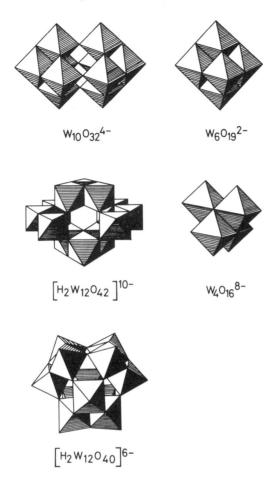

$W_{10}O_{32}^{4-}$ $W_6O_{19}^{2-}$

$[H_2W_{12}O_{42}]^{10-}$ $W_4O_{16}^{8-}$

$[H_2W_{12}O_{40}]^{6-}$

FIG. 9. Polytungstate ions of known structure.

ion $[H_2W_{12}O_{42}]^{10-}$ $(Z^+ = 14/12 = 1.17)$, whose structure has been determined by X-ray methods (Fig. 9) (*96, 193, 97–100*). The two H atoms are located in the interior of the W—O framework (*98, 99, 119*)* (cf. Section III, B, 1, c). The known salts include $Li_{10}[H_2W_{12}O_{42}]\cdot$

* Although they are, in principle, acidic according to our theory of formation mechanisms and structures of polymetalate ions, the two H atoms are practically unable to exert an acid function owing to their location in the interior of the W—O skeleton. As a result of the structures of the W—O skeletons the shielding of the H atoms is less effective in the paratungstate B ion than in the metatungstate ion, which is also apparent from rapid alkalization experiments (*144*) (cf. Sections V, B, 3, a and b).

$32H_2O$, $Na_{10}[H_2W_{12}O_{42}]\cdot 27H_2O$, $(NH_4)_{10}[H_2W_{12}O_{42}]\cdot 10H_2O$, $(NH_4)_{10}$ $[H_2W_{12}O_{42}]\cdot 4H_2O$, $K_{10}[H_2W_{12}O_{42}]\cdot 11H_2O$ (2, 7, 194, 195).

From solutions having $Z \approx 1.5$, the X-ray amorphous pseudo-(ψ-) metatungstates can be precipitated (31, 7, 151, 196). They are (1:4)-tungstates for which a probable degree of aggregation of about 24 has been deduced from the identity of the Raman spectra of solids and of solutions (27) and from molar mass determinations conducted with solutions (cf. V, B, 3, b). Salts $3K_2O\cdot 12WO_3\cdot 15H_2O$ and $3BaO\cdot 12WO_3\cdot 24H_2O$ ($Z^+ = 1.5$) are known (31, 7). Recent investigations have shown the potassium salt to deviate somewhat from the ratio $a{:}b = 1{:}4 = 3{:}12$ [$2.58(K,Na)_2O\cdot 12WO_3\cdot 17H_2O$ ($Z^+ = 1.57$) (151), $3K_2O\cdot 11.2WO_3\cdot 12.9H_2O$ ($Z^+ = 1.46$) (196)].

After standing for several months or prolonged heating of a solution with $Z \approx 1.5$, again (1:4)-tungstates crystallize; this time the so-called true metatungstates are formed. They are dodecatungstates, shown by their isomorphism (120) with the 12-tungstatophosphate ion (121) of known structure to contain the discrete $[H_2W_{12}O_{40}]^{6-}$ ion ($Z^+ = 18/12 = 1.50$) (Fig. 9). Further proof of structure comes from the close resemblance among UV spectra of metatungstate, dodecatungstatophosphate, and dodecatungstatosilicate ions in solution (7). By X-ray structural analysis on aqueous solutions of the last-mentioned species, it could be shown that the structure of the $[SiW_{12}O_{40}]^{4-}$ anion is retained on dissolution in water (36). The two H atoms are situated in the interior of the W—O skeleton (118, 119) (cf. Section III, B, 1, c; see also footnote on p. 281). The known salts include $Na_6[H_2W_{12}O_{40}]\cdot 21H_2O$ and $K_6[H_2W_{12}O_{40}]\cdot 11H_2O$ (7). (5:24)-Tungstates can be precipitated from the metatungstate solution by addition of trialkylammonium acetates (alkyl, e.g., pentyl and isopentyl) (74). They are acidic metatungstates having the formula $(HNR_3)_5[H_3W_{12}O_{40}]$ (without water of crystallization) ($Z^+ = 19/12 = 1.58$). Their appearance is probably due to the pronounced bulk of the trialkylammonium ion (74). One of the H atoms of the anion must be "peripherically" bonded.

If a tungstate solution is rapidly acidified to $Z \gtrsim 1.6$, then the yellowish (1:5)-tungstates, the so-called polytungstates Y can be precipitated (74, 151). These salts contain the discrete decatungstate ion $W_{10}O_{32}^{4-}$ ($Z^+ = 16/10 = 1.60$) whose structure was first deduced from theoretical considerations (197) and has meanwhile been experimentally verified (101) (see Fig. 9). Among these salts, we know $K_4W_{10}O_{32}\cdot 4H_2O$, $(HNR_3)_4W_{10}O_{32}$, and $(NR_4)_4W_{10}O_{32}$ (without water of crystallization) (R = propyl, butyl, pentyl, isopentyl) (151, 74).

Addition of acetone to tungstate solutions freshly acidified to $Z \approx 1.17$, which contain the paratungstate A ion ($Z^+ = 7/6 = 1.17$)

(cf. Section V, B, 3, a) whose salts do not, however, crystallize, precipitates a salt, regarded as paratungstate A, that is contaminated with the salt of the mineral acid used for acidification (198).

Concerning the preparation of a salt of another polytungstate species having $Z^+ = 1.50$, see footnote on p. 292 (Section V, B, 3, b).

2. Preparation by Hydrolysis of Esters

Controlled hydrolysis of tungstic tetraesters in organic solvents usually in the presence of organic bases or salts (8, 109, 111, 112) opens a route to polytungstates, which in some cases can only be prepared impure, if at all, from aqueous solution. Which polyion actually is formed depends mainly on the solvent and the strength of the base. The salts generally crystallize without water (cf. Section III, B, 1, b).

It was possible to prepare the following salts: guanidinium monotungstate (109), ammonium, piperidinium, and pyridinium paratungstate A $\{A_5[HW_6O_{20}(OH)_2]$ or $A_5[HW_6O_{21}]$ (109)}; ammonium paratungstate A_1 (8; cf. footnote †, p. 280); quinolinium and quinaldinium ψ-metatungstate (109); trialkylammonium polytungstates Y (alkyl = ethyl, n-propyl, n-butyl, n-pentyl, isopentyl) (112); and tetraalkylammonium hexatungstates $(NR_4)_2W_6O_{19}$ (R = ethyl, n-propyl, n-butyl) (111). In the last-named salts, (1:6)-tungstates, the discrete $W_6O_{19}^{2-}$ ion (102) ($Z^+ = 10/6 = 1.67$) is present; its structure is depicted in Fig. 9. This ion cannot be detected in aqueous solution.

Salts of the polytungstate ions formed by aging (paratungstate B ion, metatungstate ion) have not yet been obtained by this method (109).

3. Preparation by Reaction of $WO_3 \cdot H_2O$ with Organic Bases

Piperidinium paratungstate A has been obtained by reaction of $WO_3 \cdot H_2O$ with piperidine as a medium-strength organic base (153).

4. Preparation of Polytungstates Having an Inorganic Cation from Those Having an Organic Cation

According to this method, the tributylammonium salt of the hydrogen metatungstate ion, $(HNR_3)_5[H_3W_{12}O_{40}]$ (cf. Section V, A, 1), dissolved in acetone was reacted with a solution of potassium iodide in acetone to form anhydrous potassium metatungstate $K_6[H_2W_{12}O_{40}]$, and, inter alia, the hydrogen metatungstate ion identified as such (74).

B. ISOPOLYTUNGSTATE IONS IN (AQUEOUS) SOLUTION

In the tungstate system too, investigations have been performed in some cases without adopting any special measures regarding the ionic medium and in others at greater or lesser ion strengths. The remarks made in the case of the molybdate system, where references pertinent to the tungstate system are also cited, also apply in the present section.

In contrast to the molybdate solutions, very fast reactions (*143, 144, 12*) and those reaching completion within a few minutes at the latest (*31, 2, 7, 166, 27*) are accompanied by very slow reactions (*31, 2, 7, 166, 27*) in tungstate solutions.

1. Protonated Monotungstate Species

On acidification of a monotungstate solution, the $WO_4{}^{2-}$ ion is protonated to $HWO_4{}^-$ and thence to H_2WO_4. Occurrence of these reactions is even more difficult to prove than in the analogous molybdate case, since the subsequent aggregation reactions start at even lower metalate concentrations and, thus, also proceed even faster (*143, 144*).

Studies employing a flow method (*143, 144*), as indicated for the molybdate system, showed no changes in the shape of curve for a time 5×10^{-3} sec after acidification only when the initial concentration of $WO_4{}^{2-}$ was $\approx 7 \times 10^{-5}$ M or less (separation of the pure protonation reactions of the monotungstate ion from the aggregation reactions). Here, too, the equilibrium constants of the mono- and diprotonation reaction could be estimated from the shape of the curve (see Table II).

With static methods [evaluation of potentiometric (Z, log c_{H^+}, $C_{WO_4{}^{2-}}$) equilibrium curves], H_2WO_4 could not be detected at all, and detection of the $HWO_4{}^-$ ion and determination of its formation constant (see Table II) could be accomplished only by the method developed especially for measurements at low Z values (*88*).

Concerning the two dissociation constants of tungstic acid and the conclusions drawn from their magnitude regarding the structure $\{[WO_3(OH)]^-$, $[WO(OH)_5]^-$, or $[WO_3(OH)(H_2O)_2]^-$ and $WO_2(OH)_2$, $W(OH)_6$, or $WO_2(OH)_2(H_2O)_2\}$, the same situation applies as in the molybdate system.

2. Primary Aggregation Products

In the tungstate system too, there has been no lack of attempts to detect primary aggregation products. In particular, the occurrence of di- and tritungstate ions (*54, 199, 200*) has been repeatedly claimed. Since paratungstate B and metatungstate ions can be pictured as being made up of tritungstate units, it was considered necessary to stipulate tritungstate ions for structural reasons (*200*). The feasibility of detecting such species is subject to the same restrictions as have been mentioned

TABLE II

OVERALL FORMATION CONSTANTS β, OVERALL STANDARD REACTION ENTHALPIES ΔH_r°, AND OVERALL STANDARD REACTION ENTROPIES ΔS_r° FOR FORMATION OF TUNGSTATE SPECIES (p,q)

Parameter	Species							Medium	Temp. (°C)	Ref.
	(1,1)	(2,1)	(1,2)	(6,6)	(7,6)	(14,12)	(18,12)			
Log β	≈4.05	8.1	—	—	—	—	—	0.1 M (KCl)	20	143
Log β	≈3.5	8.1	—	—	—	—	—	0.1 M (NaClO$_4$)	20	144
Log β	≈3.8	—	≈5.8	≈51.7	≈59.7	—	—	3 M Na(NO$_3$)	22	88
Log β	—	11.3(?)	—	52.46	60.76	123.24	—	3 M Na(ClO$_4$)	25	201
Log β	—	—	—	—	53.98	111.03	132.51	3 M Li(Cl)	50	6
Log β	—	—	—	—	59.98	—	—	3 M Na(Cl)	25	165
ΔH_r° (kcal)	—	—	—	−57.1	−62.5	−126.9	—	3 M Na(ClO$_4$)	25	135
ΔS_r° (cal deg^{-1})	—	—	—	49	68	138	—			

in our discussion of the molybdate system. Accordingly, all attempted detections except those considered in the following lack a firm experimental foundation.

Mathematical analysis of $(Z, \log c_{H^+}, C_{WO_4^{2-}})$ equilibrium curves revealed a 6,6- (201, 88) and a 1,2-tungstate ion (88). The former appears only at very low Z values and, according to our concepts of the formation mechanisms and structures of polymetalate ions, should be formulated as $[W_6O_{20}(OH)_2]^{6-}$ (19). It has no acid/base relationship with the paratungstate A ion (see Section V, B, 3, a). The 1,2-tungstate ion, which we formulate as $[W_2O_7(OH)]^{3-}$ (13), could only be detected by the method developed especially for measurements at low Z values. It occurs at relatively high tungstate concentrations but only with very low Z values. (As an exception to the rule that polymetalate ions are built up of MO_6 octahedra, this ion must also contain an MO_4 tetrahedron according to our ideas about the formation mechanism of polymetalate ions.)

Relaxation measurements by the temperature-jump method (149) on 0.5×10^{-2} to 2.2×10^{-2} M tungstate solutions at Z values up to 0.2 ($-\log c_{H^+} \approx 7$) indicated the existence of a very rapidly formed tetratungstate ion. Our mechanistic concepts (cf. Section VI, A) require that this species be formulated as $[W_4O_{12}(OH)_4]^{4-}$ ($Z^+ = 4/4 = 1.00$) (13).

Schwarzenbach and co-workers (143, 144) had likewise detected a tetratungstate ion ($Z^+ = 5/4 = 1.25$) using the flow technique by recording potentiometric titration curves at 5×10^{-3} sec after acidification. We formulate this ion as $[HW_4O_{12}(OH)_4]^{3-}$. It occurs in a different pH range (pH ≈ 3) from that found by the temperature-jump method (pH ≈ 7), thus explaining protonation beyond $Z^+ = 1$.

Apart from the overall formation constants for the 1,2- (88) and 6,6-tungstate ion (201, 88), mathematical analysis of enthalpy titration curves also afforded ΔH_r^0 and ΔS_r^0 of the overall formation reaction for the 6,6 ion (135) (see Table II).

Remarkably enough, more primary aggregation products have been detected in the tungstate system than in the molybdate system, although it might be thought an easier task in the latter case owing to the less pronounced tendency to undergo aggregation (19).

3. The Main Reaction Products

a. Range $0 < Z \leqslant 1.17$. Precise investigations of the range $0 < Z \leqslant 1.17$ using intensity difference diagrams of Raman spectra showed that a spectroscopically uniform reaction occurs, within the limits of experimental accuracy, if the time of acidification of the solution is adopted as reference point (27).

Breaks in pH, conductivity, and other titration curves, in conjunction with molar mass determinations by the diffusion technique and by salt cryoscopy led to postulation of 7,6-tungstate ions ($Z^+ = 1.167$) (method as described in Section III, A, 5, a; references cited in Refs. 2, 6, 7). 8,7-Tungstate ions ($Z^+ = 1.143$) have only been postulated occasionally, although precisely the same problem is met as in the molybdate system: The Z^+ values of the two species lie so close together that the Z value of the break has to be known to within less than $\approx 0.5\%$. Molar mass determinations by ultracentrifugation, which were performed because of the inaccuracy of the diffusion technique and salt cryoscopy, confirmed a degree of aggregation of 6 to 8 (6, 7, 75). [This polytungstate ion is one of the few species for which the degree of aggregation previously determined by the diffusion method (60, 77, 202) or by salt cryoscopy (31) was confirmed as being of the correct order.]

Mathematical analysis of potentiometric (Z, log c_{H^+}, $C_{WO_4^{2-}}$) equilibrium curves led to the postulation of hydrolysis schemes in which a 7,6-hexatungstate ion again features as the initial main product (50, 201, 6).

The polytungstate ion under discussion is usually formulated as 7,6 ion "$HW_6O_{21}^{5-}$." This formula contains the lowest possible number of H atoms. Controlled hydrolysis of tetraethyltungstate in ethanol led, in the presence of ammonia, piperidine, and pyridine as bases, to crystalline polytungstates, which displayed the same properties as the above tungstate ion on dissolution in water and may, therefore, be regarded as salts of this ion (109). According to the results of analysis, these salts can be formulated as 7,6-tungstates containing the ions $[HW_6O_{21}]^{5-}$ or $[HW_6O_{20}(OH)_2]^{5-}$ ("$HW_6O_{21}^{5-} \cdot H_2O$") (with r as 3 or 2) (109). Bearing in mind the analytical difficulties to be surmounted with the readily crystallizing heptamolybdates (65, 66) and paratungstates B (67, 68, 31) in order to establish whether they are (5:12)- or (3:7)- (i.e., 7,6- or 8,7-)metalates, there is no choice but to accept an alternative interpretation as 8,7-tungstates for these poorly crystallizing and not very stable substances. The (Z, log c_{H^+}, $C_{WO_4^{2-}}$) equilibrium curves certainly allow also a set of species including an 8,7 ion (cf. the difficulties attending a decision between the hepta- and octamolybdate species).

Thus we see that the presence of 7,6-tungstate ions in solutions acidified to $Z \approx 1.17$ is by no means experimentally established, although their occurrence is accepted as certain by nearly all authors. A theoretical approach to the mechanisms of formation and structures of polymetalates (see Sections VI, B, 1–4) has led us to propose a structure (see Fig. 16) having the formula $[HW_6O_{20}(OH)_2]^{5-}$ $(r = 2)$ (*19*). Accordingly we can write the formation equation for the ion as

$$7H^+ + 6WO_4{}^{2-} \longrightarrow [HW_6O_{20}(OH)_2]^{5-} + 2H_2O \qquad (10)$$

(cf. the comments regarding the formation equation for the hepta-molybdate ion).

In solution, the 7,6- (or possibly 8,7-)tungstate ion is subject to slow conversion into another ion having the same Z^+ value until equilibrium is reached (*31, 7, 44, 27*). Since tungstates having $Z^+ = 1.17$ are designated as paratungstates, the 7,6 ion occurring in tungstate solutions freshly acidified to $Z = 1.17$ has been called paratungstate A, and the other, paratungstate B (*31*). At high concentrations $(C_{WO_4{}^{2-}} \geqslant 0.1\ M)$ equilibrium lies on the para B side, and, at low concentrations $(C_{WO_4{}^{2-}} \leqslant 0.1\ M)$, on the side of the para A ion (*6, 27*). As a consequence of the law of mass action, this concentration dependence provides some indication that the para B ion has a considerably higher degree of aggregation than the para A ion. Mathematical analysis of potentiometric $(Z, \log c_{H^+}, C_{WO_4{}^{2-}})$ equilibrium curves identified the para B ion as a 14,12 ion $(Z^+ = 1.167)$ (*6*). However, this finding requires confirmation by another method of study, for a 13,11 or a 15,13 ion $(Z^+ = 1.182$ and $1.154)$ no doubt fits the experimental data just as well as a 14,12 ion (cf. the difficulties attending a decision between the hepta- and octamolybdate species).

Freshly prepared solutions of the paratungstates crystallizing from tungstate solutions at $Z = 1.17$ exhibit the same behavior as the paratungstate B ion (*31, 7, 27*). In these solutions equilibrium between paratungstate A and B ions is approached from the para B ion (*31, 7, 27*). Molar mass determinations on freshly prepared solutions in an ultracentrifuge again showed the presence of a 12-tungstate ion (*7, 6, 75*). However, the experimental error of these measurements could extend this result to an 11- or 13-tungstate ion too. There appear to be several reasons favoring a 12-tungstate ion: (a) As has long (*96*) been known, the paratungstates crystallizing from solution contain discrete 12-tungstate ions; (b) interconversion of hexa-para A and dodeca-para B ions could be explained by a dimerization step in the conversion mechanism; (c) an ion having a degreee of aggregation that is already so high should (since it is probably made up of several small structural units of the same kind) have a symmetrical structure, i.e., an even degree of aggregation {cf., however, the highly symmetrical ion $[Al_{13}O_4(OH)_{24}(H_2O)_{12}]^{7+}$ (*203*)!}. Comparison of the Raman spectra recorded for crystalline paratungstates with those of their (freshly prepared) solutions provided experimental evidence for the 12-tungstate ion (6, 27). Thus, the paratungstate B ion is assigned the formula $[H_2W_{12}O_{42}]^{10-}$ (see Fig. 9), and the crystalline paratungstates are paratungstates B.

The equilibrium between the paratungstate A and B ions can be formulated as

$$2[HW_6O_{20}(OH)_2]^{5-} \rightleftharpoons [H_2W_{12}O_{42}]^{10-} + 2H_2O \qquad (11)$$

Nevertheless, it should be clearly pointed out that no experimental proof has been obtained for a dimerization step in the interconversion of paratungstate A and B ions.

Fast acidification experiments in a flow apparatus (144) showed that the para A ion is hardly protonated down to a pH value of 3, affording a pK_a value of $\leqslant 2$ for $[H_2W_6O_{20}(OH)_2]^{4-}$ [ionic medium 1 M Na(Cl)]. By contrast, protonation of the para B ion starts at considerably higher pH values [with 1 M Na(Cl) as ionic medium at pH ≈ 6] and the pK_a values are 5.0 for the $[H_3W_{12}O_{42}]^{9-}$ ion and 4.3 for the $[H_4W_{12}O_{42}]^{8-}$ ion. Fast alkalization experiments (144) afforded a pK_a value of ≈ 11 [1 M Na(Cl)] for the $[H_2W_{12}O_{42}]^{10-}$ ion. The authors give no pK_a for the $[HW_6O_{20}(OH)_2]^{5-}$ ion but a value > 11 [1 M Na(Cl)] can be derived from the titration curve reported, which is readily compatible with our structural proposal for the para A ion (19). On the other hand, a pK_a value of 8.30 [3 M Na(ClO$_4$)] and 8.0 [3 M Na(NO$_3$)] would result from the overall formation constants for the 7,6 (para A) and 6,6 ion if the two ions are regarded as a conjugate acid/base pair. We, therefore, consider the experimentally detected 6,6 ion not to be the deprotonated para A ion, as does Arnek (135), but an ion of different structure for which we have put forward proposals (19).

Among the thermodynamic data, in addition to the overall formation constants for the para A (50, 6, 201) and para B ion (6, 201), ΔH_r^0 and ΔS_r^0 of the overall formation reactions have been established by mathematical analysis of enthalpy titration curves (135) (see Table II).

According to Boyer and Souchay (170, 166), at low tungstate concentrations and low ion strengths, an "acidic paratungstate" ion ($Z^+ \approx$ 1.33) appears in place of the paratungstate A ion ($Z^+ \approx 1.17$). This ion is stabilized by $N(CH_3)_4^+$ ions and destabilized by Na^+ owing to ion pair formation between Na^+ and para A ions. At higher tungstate concentrations (> 0.1 M) and on use of the sodium salt, the acidic paratungstate ion no longer appears because the Na^+ ion of the salt itself will suffice for ion pair formation with the para A ion.

b. *Range* 1.17 $\leqslant Z \leqslant$ 1.6. In the tungstate system too, aggregation reactions are not complete at $Z \approx$ 1.17. Up to $Z \approx$ 1.6, further reactions occur which, in contrast to the molybdate system, exhibit a pronounced time dependence.

The "acidic paratungstate" ion ($Z^+ \approx 1.33$), which appears in place of the para A ion under certain conditions, has already been discussed in the preceding section.

Conductivity, pH, and other titration curves show a break at $Z \approx 1.50$ (references cited in Refs. 2, 6, and 7). A precise investigation has shown that not just one, but three or even four species having a Z^+ value close to 1.50 appear consecutively (44, 25, 166).

The first to appear directly (ca. 1 min) after acidification is designated as the metatungstate A ion (44). It is distinguished unequivocally from the subsequently appearing species, but characterized only by its approximate Z^+ value. The species is assumed to be the doubly protonated paratungstate A ion (44); however, the opinion has recently been voiced that it is identical with the polytungstate Y ion described below (124).

The metatungstate A ion is transformed within several hours into the ψ-metatungstate ion (44). This transformation manifests itself in a change in the shape of pH and conductivity titration curves, the shift of the long-wave UV absorption edge toward shorter wavelengths, and the appearance of a polarographic half-wave potential (44). The degree of aggregation is still a matter of dispute. Salt cryoscopy indicated sixfold tungstate aggregation (31). Investigations with an ultracentrifuge (Archibald technique) surprisingly yielded 24-aggregation (7).This result ($q = 24 \pm 3$) was confirmed by the same research group using the sedimentation equilibrium technique (204). A degree of aggregation of 24 has for a short period also been postulated by another group (205), but later this group claimed the ψ-metatungstate ion to have $q = 12$ (sedimentation equilibrium and velocity methods) (75). The ψ-metatungstate ion can be precipitated as the potassium and the barium salt (31, 7). Comparison of Raman spectra shows that the same polytungstate species are present in the solid state and in solution (27). According to ^{1}H NMR studies the ion still possesses OH groups (7, 119). The Z^+ value can be given approximately as 1.50 (cf. Section V, A, 1).

The ψ-metatungstate ion is irreversibly transformed in the course of weeks or months—faster on heating—into the true metatungstate ion, which appears to be the thermodynamically most stable species at $Z = 1.50$ (31, 44, 7). This transformation can be readily followed by UV (7, 25) and Raman spectroscopy (27), as well as by the disappearance of the polarographic half-wave potential of the ψ-meta ion (31).

The metatungstate ion was identified by molar mass determinations in the ultracentrifuge as a dodecatungstate ion (7, 73, 75), and, thus, an 18,12-tungstate ion. Mathematical analysis of potentiometric (Z, log c_{H^+}, $C_{WO_4^{2-}}$) equilibrium curves of aged tungstate solutions (6) likewise indicated an 18,12 ion. As in other cases, the experimental accuracy of the molar mass determinations and of the potentiometric measurements is certainly not so great as to rule out an 11- or 13-tungstate ion. However, the presence of a dodecatungstate ion is verified by the good agreement between Raman spectra recorded for the solution and the solid state (34, 27), and by the agreement existing among the UV spectra of solutions of metatungstate, dodecatungstatophosphate, and dodecatungstatosilicate ions (7). As already mentioned, it was demonstrated for the last-named species that the structure of the anion $[SiW_{12}O_{40}]^{4-}$ is retained on dissolution in water (36, 206). Accordingly, the metatungstate ion should also be formulated as $[H_2W_{12}O_{40}]^{6-}$ ($Z^+ = 18/12 = 1.50$) in solution (see Fig. 9) [the two H atoms are located in the interior of the W—O skeleton (118, 119)]. Fast acidification and alkalization experiments show the metatungstate ion to be aprotic in the pH range of 3 to 10 (144). Salts of the protonated ions can, however, be obtained from metatungstate solutions by precipitation with trialkylammonium acetate solution (74) (cf. Sections V, A, 1 and V, A, 4).

Interestingly enough, the metatungstate ion is not only formed in (aged) solutions having $Z \approx 1.5$ (or in the range $Z \approx 1.17$–1.50). At low tungstate concentrations, it also appears in considerable concentrations in (aged) solutions at $Z \approx 1.17$ (and lower) alongside the paratungstate A and B ions co-existing in equilibrium (6, 7, 207). This is simply a consequence of the pronounced thermodynamic stability of the metatungstate ion and the law of mass action: In dilute metalate solutions, the pH value for the same Z value lies considerably lower than in concentrated solutions, so that the high H^+ concentration required for formation of the 18,12 ion is present. Under these conditions, the ψ-metatungstate ion does not occur at all. Since the metatungstate ion possesses a higher Z^+ value than would correspond to the Z value of the solution, a further species having $Z^+ \ll 1.17$ must also be present in such solutions. This is the unprotonated monotungstate ion WO_4^{2-} ($Z^+ = 0$).

It has recently been found that yet another species, called polytungstate X, intervenes between the ψ-metatungstate and the metatungstate ion (25). Molar mass determinations in an ultracentrifuge show that it is probably also a dodecatungstate ion (75) like metatungstate, and the Z^+ value is most likely also 1.50. The UV spectrum of the solution and the polarographic half-wave potential are available as fingerprints (25).*

If a paratungstate A (or a monotungstate) solution is *rapidly* acidified to $Z \geqslant 1.6$, a further species is formed, the so-called polytungstate Y ion (7), which is characterized by typical UV (7, *151, 74*) and Raman spectra (27). With regard to the acidification, the optimum conditions for the occurrence of this species lie at $Z \approx 1.6$ (*166*, 27). Anhydrous trialkylammonium, tetraalkylammonium (alkyl = propyl, butyl, pentyl, isopentyl), and tetraphenylphosphonium salts of composition $A_2O \cdot 5WO_3$ (*74, 124*) as well as the potassium salt (*151*) $K_2O \cdot 5WO_3 \cdot 4H_2O$ (*74*) can be precipitated from solution. Raman spectra show the same polytungstate species to be present in the salts as in their aqueous solutions (27). For mechanistic reasons (*197*), a species having $c = 0$ and a base/acid ratio of 1:5 can only be a decatungstate ion, $W_{10}O_{32}^{4-}$ ($Z^+ = 16/10 = 1.60$) whose structure is shown in Fig. 9. Molar mass determination in water and in acetone as solvent (*74*) and X-ray structure determination (*101*) confirm the presence of decatungstate ions. Thus, contrary to former assumptions (7, *151, 75, 166*), dodecameric ions are not present. In (freshly) acidified aqueous solutions, the polytungstate Y ion is always accompanied by the ψ-metatungstate ion and is not stable for any length of time (at room temperature); it is transformed into the ψ-metatungstate ion within a few hours (7, *151, 74,* 27). In fact, however, the two ions are in a temperature-dependent equilibrium that lies on the side of the ψ-meta ion at room temperature and on the side of the poly Y ion at higher temperatures (at 80°C the equilibrium in a $C = 1.2 \times 10^{-1} M$ solution lies to the extent of 60% on the side of the poly Y ion) (*205*).

Possibly, the polytungstate Y ion is identical with the metatungstate A ion. Evidence in support of this conclusion is provided not only by approximately equal rates of conversion into the ψ-meta ion and the unidirection shift of the long-wave UV absorption edge, but also by the break in the conductivity titration curve that lies at $Z = 1.56$ (*44*) instead of exactly at $Z = 1.50$.

Transformation of polytungstate ions Y into ψ-meta, ψ-meta into X,

---

* The authors were able to employ a trick to prepare pure polytungstate X (*138*): They reduced a ψ-metatungstate solution under defined conditions at a mercury cathode, allowed the reduced ψ-metatungstate ion to transform into the reduced polytungstate X ion, oxidized it to the polytungstate X ion, and precipitated it as the cesium salt after prior removal of small amounts of the decomposition products of the ψ-meta and poly X ion by precipitation as the potassium salts.

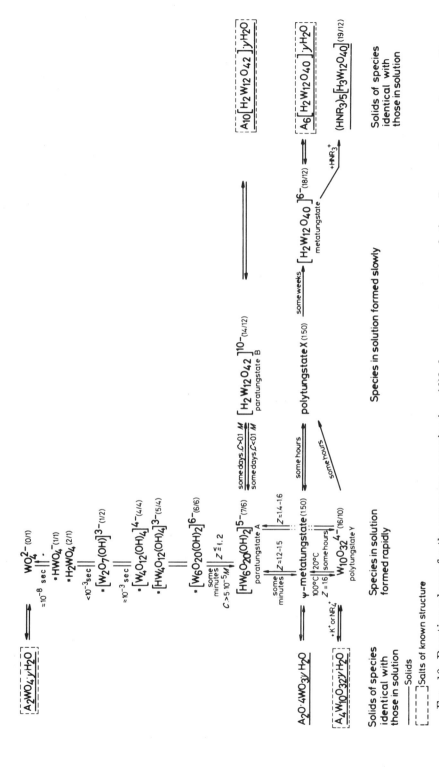

Fig. 10. Reaction scheme for the processes occurring in acidified aqueous tungstate solutions. In parentheses: p/q resp. Z^+. *Short-lived and other thermodynamically unfavorable species that occur at equilibrium in small amounts only.

Y into X, X into meta, meta into para A, X into para A, X into ψ-meta, ψ-meta into para A, and Y into para A has also been studied at defined pH values in buffered solution (25, 151, 166). It was found that all reactions are first-order in the polytungstate ion. Transformation of poly Y ion into the ψ-meta ion and into the para A ion is also first-order with respect to the OH$^-$ ion. The reacton rate is independent of OH$^-$ concentration for conversion of polytungstate X into meta, X into ψ-meta, and X into para A. (In the last case there is no OH$^-$ dependence because the ψ-meta ion occurs as intermediate and its formation is the rate-determining step.)

4. Disaggregation with OH$^-$

In unbuffered solution all polytungstate ions are converted by OH$^-$ ions into those polyions corresponding to the Z values of the solution and the time. There appears to be a rule that slowly formed anions also disaggregate slowly. With an excess of OH$^-$, ions, complete disaggregation to WO$_4^{2-}$ takes place. Disaggregation of the metatungstate ion is first-order in metatungstate and in OH$^-$ ion (208).

5. Isopolytungstate Ions Occurring in Solution and in the Solid State

The importance of possessing maximum information about precisely those species that occur both in solution and in the solid state has already been mentioned in the context of the isopolymolybdates. Apart from the tetrahedral (190) WO$_4^{2-}$ ion, such species comprise the [H$_2$W$_{12}$O$_{42}$]$^{10-}$, the ψ-metatungstate, the [H$_2$W$_{12}$O$_{40}$]$^{6-}$, and the W$_{10}$O$_{32}^{4-}$ ions. Within certain limits this also applies to the paratungstate A (27) and polytungstate X ion (138).

The experimental results of Section V are summarized in the reaction scheme in Fig. 10.

VI. Formation Mechanisms of Polymetalates

We shall now broach the questions of why usually well-defined species occur under certain conditions instead of complex polymer mixtures, why particular structures are formed, why completely different species appear in the molybdate and tungstate systems, why some species are formed very rapidly and others very slowly, etc.

The structure of heteropolyanions was first considered by Miolati (209) who applied Werner's coordination theory to this class of compound. This theory was then extended by Rosenheim and co-workers (210) and applied to isopolyanions too. Once the first structural elucidation of a (hetero)polyanion, with the discovery of octahedral coordination of the M atoms with six O atoms, had become available (121), the theory was seen to be untenable. However, even prior to that discovery, Pauling (211) had become convinced that heteropolymetalates contain MO_6 octahedra with shared edges, even though his proposal was found to be incorrect for (1:12)-heteropolymetalates. The further development of polymetalate chemistry was decisively influenced by the first structural studies on isopolymetalates by Lindqvist (90, 15, 96), from whom the first ideas about the systematics of isopolymolybdate structures also originate (1). Similar views concerning heteropolymolybdate structures were later expressed by Kepert (212), who also attempted to derive the existence of the various (iso- and hetero)polymetalate structures solely on the basis of the size and charge of the metal atoms (212).

A first attempt to elucidate the mechanism of aggregation, i.e., the geometrical course of the individual successive aggregation steps is due to Freedmann (200). Starting from the [unsound (2, 213)] experimental finding that a tritungstate ion should be the initial aggregation product on hydrolysis of a WO_4^{2-} solution (which seemed to be confirmed by the presence of W_3 in the polytungstate structures then known), the author presents a series of geometrical steps purported to lead from the protonated tetrahedral monotungstate ion, via an increase in coordination from tetrahedral to octahedral, to a tritungstate as the preliminary product according to the building principle adopted. The result of these geometrical considerations could not be correct because, apart from other objections (23), the author had incorrectly counted the O atoms in the structures (13). Another attempt to derive the existence and isotope exchange behavior of a trimetalate group from particular geometric aspects (214) is also geometrically incorrect (23). By contrast, the geometrical building principle proposed by Kepert (2), which entails an increase in tungsten coordination from tetrahedral to octahedral with simple addition of a bidentate ligand, appears highly reasonable, even though in this case too (again owing to an error in counting of the O atoms) no adequate result was obtained (13).

Yet another attempt to put forward certain structural and mechanistic (partial) conclusions about isopolymetalates is due to Chojnacki (215, 216). He proposed a "core + links" hypothesis resembling that of Sillén (49). The principal deviations from the latter are that the "core" may also be made up of several M atoms; that the core possesses the local Z^+ value of 2 (or possibly 3) and its M atoms, therefore, display octahedral coordination toward oxygen; that the number of "links" having the local Z^+ value of 1 (or possibly 0) and tetrahedral coordination with oxygen is limited to the number of peripheral O atoms in the core; and that information about the pH range of existence of a species is derived from the ratio of the M atoms in the core to those in the links. Since a polyion composed entirely of octahedra accordingly has to have a Z^+ value of at least 2 and no polyions exist for $Z^+ = 1$, all structures known so far contradict this hypothesis.*

We shall now present our own mechanistic views (13, 19–23) on the formation of isopolymetalates, which embrace not only purely geometrical aspects but also consideration of acid/base properties of polymetalate ions and of their thermodynamic and kinetic stability.

A. The Addition Mechanism (13, 23)

According to the studies we have performed so far, the addition mechanism is the simplest of all conceivable aggregation mechanisms. As such it serves especially for the formation of primary aggregation products.

1. Construction of Octahedra by the Addition Principle; Fundamental Geometrical Alternatives (13)

A prerequisite for the occurrence of aggregation is doubtless the very fast ($\approx 10^{-8}$ sec) protonation of the $MO_4{}^{2-}$ ions to $HMO_4{}^-$ and possibly H_2MO_4. In both species, M—O bonding is extended owing to protonation (2), and there is a possibility of attaining octahedral coordination for the M atom by addition of two monodentate or one bidentate ligand (Fig. 11a). A possible monodentate ligand is H_2O, which would yield species $[MO_3(OH)(H_2O)_2]^-$ and $MO_2(OH)_2(H_2O)_2$ proposed earlier (cf. Section IV, B, 1). However, reaction with biden-

---

* Assumption of a core + links principle, also with a polynuclear core, is not irrelevant. If our following theory of the formation mechanisms and structures of isopolymetalate ions is correspondingly applied to heteropolymetalate ions (23a), the heteroatom polyhedron [e.g., $Te(OH)_6$] acts as aggregation nucleus ("core") and the $HMO_4{}^-$ or H_2MO_4 function as "links" according to the addition or condensation principle (i.e., with raising of the coordination number of M). Hence an MO_6 octahedron [e.g., $MO_2(OH)_2(H_2O)_2$] can, at least in principle, also be taken as an aggregation nucleus (23). The final result on operation of the addition and condensation mechanism is the same as for the tetranuclear tetrametalate ion (formed only by the addition mechanism) as core, because the final product actually formed is decided by the acid/base properties (acidic and nonacidic H atoms) and the thermodynamic and kinetic stability of the competing species, granted that the same geometrical principles govern the constructional steps.

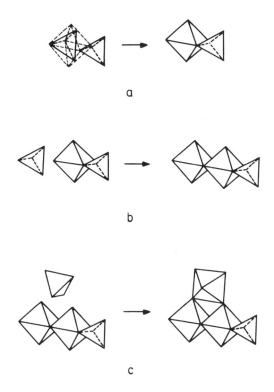

FIG. 11. Construction of an octahedron by the addition principle. (a) Addition of an MO_4 tetrahedron (HMO_4^-, H_2MO_4) to a tetrahedral edge (MO_4^{2-}, HMO_4^-, H_2MO_4), (b) to an octahedral edge (polymetalate ion already formed having MO_6 octahedra), (c) to the corners of two adjacent octahedra (larger polymetalate ion already formed).

tate ligands is no doubt more favorable. We assume the MO_4^{2-} ion, a further HMO_4^- ion or H_2MO_4 molecule (Fig. 11a), and already formed polymers containing MO_6 octahedra (Figs. 11b and c) to be capable of acting as such. Numerous polymeric species can be built up in compliance with this principle. We can classify them according to structure as follows.

A. Open (chainlike) polymetalates. Apart from MO_6 octahedra these also contain a MO_4 tetrahedron.

　　1. Unbranched (straight or angled) chains
　　　　a. Tetrahedral group in terminal position
　　　　b. Tetrahedral group within the chain

2. Branched chains
   a. Octahedron as branching point
   b. Tetrahedron as branching point
   c. Branching from the corners of two octahedra.

B. Closed (cyclic) polymetalates. These are made up exclusively of octahedra, and thus contain no tetrahedron. They arise on occurrence of intramolecular addition, i.e., when two O atoms of a polymeric ion act as dentic groups toward the tetrahedral group of the same ion (Fig. 12).

1. Unbranched (planar or puckered) rings
   a. Four-membered "rings"
   b. Six- and higher-membered rings
2. Rings bearing side chains (including branched ones)
   a. Side-chain linkage via an octahedral edge
   b. Side-chain linkage via the corners of two octahedra

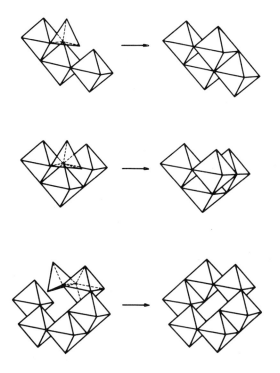

FIG. 12. Intramolecular addition with ring closure illustrated for formation of the two four-membered "rings" and a six-membered ring.

Among these numerous polymetalate structures, the cyclic structural units (without side chains) are distinguished from the others, *inter alia*, by requiring simultaneous disruption of two octahedra on removal of an MO_4 group, whereas only one octahedron is destroyed in all other cases. Moreover, of the cyclic structures, the four-membered "rings" are favored for purely statistical reasons by a factor of 10^3–10^4 over the next-highest representatives, the six-membered rings. This is simply explained: The probability that, of all the numerous possible addition sites, precisely the two ends of a long, angled, and relatively rigid chain should unite is extremely low.

If the (initial) concentration of MO_4^{2-} is very small, the aggregation reactions are disfavored by the law of mass action and H_2O plays an increasing role as ligand (cf. Sections IV, B, 1 and V, B, 1). The same applies also to the small amounts of protonated monometalate ions necessarily present in equilibrium (*23*).

2. Acid/Base Functions of Polymetalate Ions; Number and Positions of H Atoms (*19*)

The H atoms of the HMO_4^- ions forming the polymetalate ion are essential for the maintenance of octahedral coordination (Fig. 13) and,

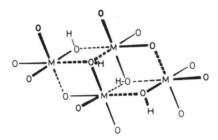

Fig. 13. "Meshing" of four HMO_4^- ions to form the $[M_4O_{12}(OH)_4]^{4-}$ ion.

therefore, cannot be acidic as a general principle. It is worthwhile to consider which positions these H atoms will occupy and whether more H atoms than dictated by the ratio $1H^+/MO_4^{2-}$ (for an ion made up exclusively of octahedra) will be required for any particular reason.

Each octahedron is built up from an HMO_4^- ion and two further O atoms (belonging to 1 or 2 other HMO_4^- ions) (cf. Fig. 13) and thus bears a negative charge, i.e., it corresponds to a basic function in the aggregate. Now there are two kinds of O atoms in each octahedron: nonterminal O atoms (O atoms in M—O—M bridges and in OH groups) and terminal O atoms. In the pH range in which aggregation occurs, only terminal O atoms participate in charge acceptance. Inspection of all possible structures reveals that the following octahedra occur:

$$M(O_6)^- \quad M(O_5)O^- \quad M(O_4)O_2^- \quad M(O_3)O_3^-$$

The O atoms shown in parentheses are those belonging to M—O—M bridges and OH groups. It is seen from this formulation that the more stable base functions, i.e., those that have many O atoms outside the parentheses, occur when the (nonacidic) H atoms occupy positions on the O bridges. [Another reason why such positions are more favorable for H atoms than the others will become apparant later (Section VI, A, 3, b)]. Thus, the positions of the nonacidic H atoms are largely determined.

We can estimate the pKa values of the acid functions corresponding to these base functions by comparison with those of the mononuclear oxo acids. Table III shows that, in the pH range where significant aggregation takes place, the base functions $M(O_6)^-$ and $M(O_5)O^-$ can only occur in the protonated form. Hence the number and positions of further H atoms are established, which are, however, acidic.* Provided that protonation of the $M(O_4)O_2^-$ and $M(O_3)O_3^-$ functions does not take place, the base and acid functions are largely independent of each other (23).

3. Energetics (19, 20, 23)

a. Reaction and Structural Components of Favorable and Unfavorable Energy. We must now examine what factors favor aggregation and what factors disfavor the process.

Coordination of the bidentate ligands to HMO_4^- or H_2MO_4, i.e., construction of an octahedron, is doubtless the most important energy-producing process; a large portion of this energy will certainly come from the chelate effect. Since the oxygen has only four bonding orbitals it will avoid a coordination number of 5 wherever possible.

Both the repulsive forces acting among the negative charges and those among the M atoms as the central atoms of the octahedra are to be regarded as unfavorable. These interactions have a pronounced detrimental effect on the construction of highly aggregated polymetalates of compact structure. In particular, these compact structures also restrict the ability of the M atoms to minimize mutual repulsion by

---

* The $M(O_6)H$ function can, however, be regarded as practically nonacidic (19). Polymetalate ions are no longer formed at the high pH values required for its deprotonation.

TABLE III

RELATION BETWEEN FORMULA OR STRUCTURE AND $p\bar{K}_a$ VALUE OF OXO ACIDS AND ACID FUNCTIONS OF POLYMETALATE IONS

Acid/base	pK_a	Examples		
		Chlorine acids [X = Cl]	Polymetalate octahedra [X = M(O$_n$)]	
$XO_3(OH)/XO_4^-$	Very small	$ClO_3(OH)/ClO_4^-$	Nonexistent	
$XO_2(OH)/XO_3^-$	$-3\cdots0$	$ClO_2(OH)/ClO_3^-$	$M(O_3)O_2(OH)/M(O_3)O_3^-$	$(3O_t)$
$XO(OH)/XO_2^-$	$1\cdots4$	$ClO(OH)/ClO_2^-$	$M(O_4)O(OH)/M(O_4)O_2^-$	$(2O_t)$
$X(OH)/XO^-$	$7\cdots10$	$Cl(OH)/ClO^-$	$M(O_5)(OH)/M(O_5)O^-$	$(1O_t)$
Nonexistent for mononuclear oxo acids	>11	—	$M(O_6)H/M(O_6)^-$	$(0O_t)$

moving away from the octahedral (or tetrahedral) centers. The repulsive forces among the negative charges are reduced by the additional protonation of the $M(O_6)^-$ and $M(O_5)O^-$ functions, when these base functions are present in a structure. The base functions should be considered relative to one another. The more stable, i.e., energetically favorable, functions are those with many terminal O atoms.

b. Free Energy of Reaction for the Various Aggregates. We have seen that there are energetically favorable and unfavorable reaction and structural components among the polymetalate ions. They appear in varying proportions for the different structures. If a structure and the reaction pathway by which it is formed is resolved to a sufficient extent, then the free energy of reaction $\Delta G^0_{r,i}$ for species i can be thought of as made up of increments for the individual reaction and structural components:

$$\Delta G^0_{r,i} = \sum_e h_{e,i} \, g_e \tag{12}$$

Term $h_{e,i}$ is the frequency of the component e for species i, and g_e is the increment of $\Delta G^0_{r,i}$ for the single occurrence of component e. (For components whose frequency cannot be counted out on the structure or reaction equation, e.g., the energy of electrostatic repulsion among negative charges, an appropriate standardization must be performed.)

However, we now have to render comparable the $\Delta G^0_{r,i}$ values for the various species. This is done by standardizing all $\Delta G^0_{r,i}$ values in such a manner that they correspond to experimental or imagined conditions. In the acidification range up to $Z \approx 1$, for instance, this means that the frequencies of the reaction and structural components have to be normalized for the initial amount of H^+ ions (h) as the reactant in which the system is deficient:

$$\Delta G^{0,h}_{r,i} = \sum_e h^h_{e,i} \, g_e \tag{13}$$

Components to be treated in this way include: construction of an octahedron (O); presence of an octahedron formed by coordination of one or two H_2O molecules (T_1, T_2); occurrence of coordination number 5 for oxygen (K); electrostatic repulsive energy among the negative charges (E); electrostatic repulsive energy among the M atoms (Z'); differing abilities of the M atoms to undergo displacement from the octahedral centers for octahedra with three adjacent, two adjacent, one, or no free corners (A_3, A_2, A_1); occurrence of an $M(O_3)O_3^-$ and $M(O_4)O_2^-$ base function (in the structure regarded as final product) (N_3, N_2); and protonation of an $M(O_5)O^-$ function appearing during the course of aggregation (H).

In order to obtain from Eq. (13) information about the thermo-dynamic stability of the numerous species possible according to the addition mechanism, we adapt it to the use of *relative* g_e values, e.g.,

$$\Gamma_i^h \equiv \Delta G_{r,i}^{0,h}/g_0 = \sum_e h_{e,i}^h (g_e/g_0) \equiv \sum_e h_{e,i}^h \gamma_e \tag{14}$$

Now we can apply the equation to the aggregation products, estimating limiting values with an adequate safety factor for each γ_e. We find that in the case of *a scarcity of* H^+ *ions* ($Z \leqslant 1$) the maximum Γ^h (i.e., minimum $\Delta G_r^{0,h}$) occurs for the cyclic aggregates ($Z^+ = q/q = 1$), except for the compact four-membered "ring" ($Z^+ = 5/4 = 1.25$), within wide limits for the γ_e values .The estimated limiting values for γ_e were ($\gamma_0 \equiv 1$):

$$\gamma_{T_1} = -0.75\cdots-0.5 \qquad \gamma_{A_3} \approx 1.5\gamma_{A_2}$$
$$\gamma_{T_2} = -0.5\cdots0 \qquad \gamma_{A_1} \approx 0.5\gamma_{A_2}$$
$$\gamma_K \approx -0.5 \qquad \gamma_{N_3} = 0\cdots0.5$$
$$\gamma_E = -0.15\cdots0 \qquad \gamma_{N_2} = 0\cdots\gamma_{N_3}$$
$$\gamma_{Z'} = 0\cdots0.7\gamma_E \qquad \gamma_H = -0.2\cdots0.2$$
$$\gamma_{A_2} = -0.4\gamma_{Z'}\cdots0$$

The reason the planar 4,4-tetrametalate ion (and the other cyclic but statistically disfavored polymetalate ions) are so favorable thermodynamically is to be seen in the minimal Z^+ value of only 1 for this species; for nearly all other species constructued entirely from octahedra, $Z^+ > 1$. However, if Z^+ is unity the maximum possible number of octahedra (high energy gain) can be formed for a given amount of H^+ ions. If $Z^+ > 1$, then H^+ ions are required for simple protonation of the base functions $M(O_6)^-$ and $M(O_5)O^-$; although this is also energetically advantageous because it reduces the electrostatic repulsion among the negative charges, the energy gain cannot compare with that resulting on construction of an octahedron.

The necessity of protonating the base functions $M(O_6)^-$ and $M(O_5)O^-$ is a further reason for the nonacidic H atoms occupying O-bridge posi-tions (cf. Section VI, A, 2) since this reduces the number of $M(O_5)O^-$ functions.

The reaction pathway leading to the planar 4,4-metalate ion is shown in Fig. 14. The principle governing the course of aggregation from the thermodynamic aspect when there is a scarcity of H^+ ions is to achieve, for a given number of protons, a maximum number of octahedra, preferably having two and three terminal O atoms, on the one hand, and a minimum accumulation of negative charge and of M atoms, on the other.

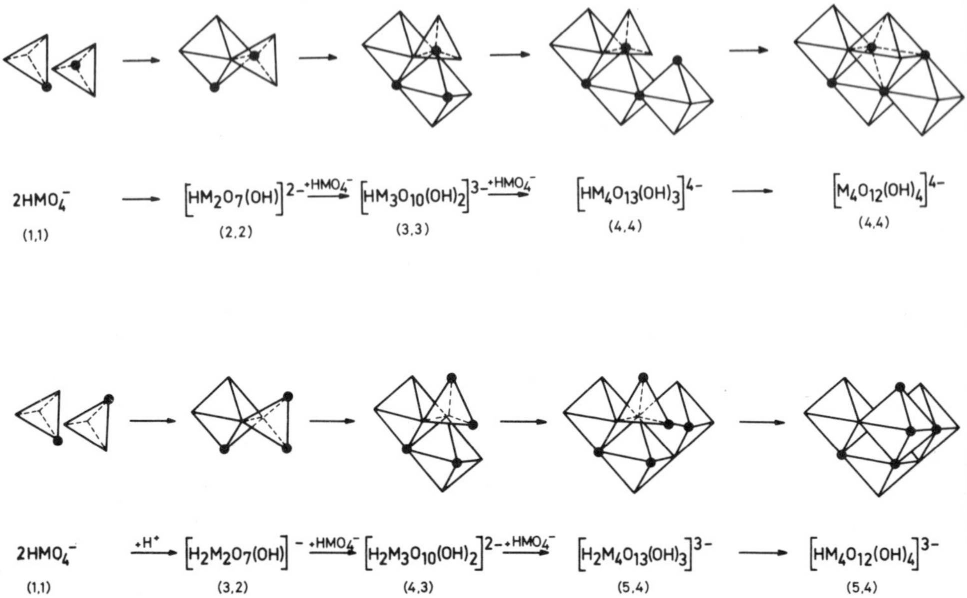

$$2HMO_4^- \longrightarrow \left[HM_2O_7(OH)\right]^{2-} \xrightarrow{+HMO_4^-} \left[HM_3O_{10}(OH)_2\right]^{3-} \xrightarrow{+HMO_4^-} \left[HM_4O_{13}(OH)_3\right]^{4-} \longrightarrow \left[M_4O_{12}(OH)_4\right]^{4-}$$

(1,1)   (2,2)   (3,3)   (4,4)   (4,4)

$$2HMO_4^- \xrightarrow{+H^+} \left[H_2M_2O_7(OH)\right]^- \xrightarrow{+HMO_4^-} \left[H_2M_3O_{10}(OH)_2\right]^{2-} \xrightarrow{+HMO_4^-} \left[H_2M_4O_{13}(OH)_3\right]^{3-} \longrightarrow \left[HM_4O_{12}(OH)_4\right]^{3-}$$

(1,1)   (3,2)   (4,3)   (5,4)   (5,4)

FIG. 14. Reaction pathway to the (planar) 4,4- and the (compact) 5,4-metalate ion according to the addition mechanism. (●) Positions of H atoms.

If there is an *adequate supply of* H^+ *ions*, the higher Z^+ value of the compact 5,4-metalate ion no longer opposes its formation. For geometrical reasons (cf. Section VI, A, 4), even the compact tetrametalate ion is now favored. The reaction pathway leading to the compact 5,4-metalate ion is also depicted in Fig. 14.

4. Kinetics (13, 23)

Geometrically, construction of octahedra according to the addition principle is a very simple process. This explains the high rate of formation of polymetalate ions formed by addition: The rate constants of the primary bimolecular aggregation steps lie in the region of 10^6 M^{-1} sec^{-1} and higher.

The assumption that aggregation is favored when the (protonated) monomeric metalate species already possess octahedral coordination is sometimes found, accompanied by references to the construction of polymetalate ions from MO_6 octahedra, in the literature (147, 176) (cf. Section IV, B, 1). However, the above geometrical considerations demonstrate that this is by no means necessary. On the contrary, aggregation even becomes more difficult for the postulated monomers

$[MO(OH)_5]^-$ and $M(OH)_6$ since simple addition is no longer possible. If our proposed monomer structures $[MO_3(OH)(H_2O)_2]^-$ and $MO_2(OH)_2$ $(H_2O)_2$ were to apply, on the other hand, they could readily be incorporated into the aggregation scheme of the addition mechanism (13).

Comparison of the reaction steps leading from the (same) trimetalate structure to the planer and the compact tetrametalate structures reveals that from a geometrical standpoint the step leading to the compact structure is the more favorable of the two [see Fig. 4 in Tytko and Glemser (13)]. With an adequate supply of H^+ ions, when the more favorable Z^+ value is not the crucial factor, formation of the compact 5,4-tetrametalate ion is thus favored.

As previously mentioned, removal of an MO_4 group from a cyclic structural unit leads to simultaneous destruction of at least two octahedra, but of only one in all other cases. This means that elimination of an HMO_4^- ion from a ring is more difficult, i.e., the cyclic structural units are kinetically more stable than all other structural groups. When viewed with regard to the number of sites where elimination of an HMO_4^- ion can occur with disruption of the minimum number of octahedra, the compact tetrametalate structure is again favored (13).

5. Comparison with Experimental Results

Comparison of these results obtained on the basis of purely theoretical concepts with the experimental results for tungstate solutions indicates excellent agreement. The addition mechanism predicts not only the fast formation of primary aggregation products (and their structures) but also their p,q values, i.e., 4,4 for scarcity of, and 5,4 for a sufficient supply of H^+ ions. For the molybdate system (see Section IV, B, 2), agreement with the (6,4) species for a scarcity of H^+ ions exists for q but not for p. Presumably the experimental value of $p = 6$ is incorrect, since no other species having a Z^+ value far in excess of unity is formed in either the molybdate or the tungstate system when there is a scarcity of H^+ ions.

B. THE CONDENSATION MECHANISM (19, 20, 23)

1. Octahedron Formation by the Condensation Principle

Further aggregation to give higher polymetalates requires a new constructional principle. Stepwise buildup of new octahedra is possible only via condensation of H_2O. However, here too, addition

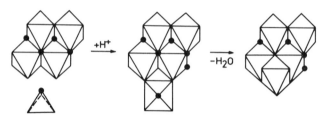

FIG. 15. Construction of an octahedron according to the condensation principle. (●) Positions of H atoms.

is the preliminary step and only then does the actual condensation occur (Fig. 15), which requires an OH group in a suitable position of the growing polymetalate ion. We can regard the growing polymetalate ion as a tridentate ligand: Two coordination sites on HMO_4^- are occupied by addition, thereby expanding the coordination sphere of the M atom; occupation of the third coordination site requires prior expulsion of the "monodentate ligand" OH from the $(HO)MO_3^-$ ion. This contructional principle can also lead to a multitude of polymetalate ions.

2. Acid/Base Functions of Condensed Polymetalate Ions; Significance of OH Groups

The function of the nonacidic H atoms in octahedra formed by the addition route, i.e., extension of the M—O bond as a prerequisite for octahedral coordination of the M atoms, is now exerted either partly or entirely by M atoms, according to the number of condensed H_2O molecules. This is why condensation products possess fewer nonacidic H atoms.

The number and positions of the acidic H atoms depend (in the pH range ≥ 3) on the $M(O_6)^-$ and $M(O_5)O^-$ base functions. Whereas the occurrence of $M(O_6)^-$ and $M(O_5)O^-$ functions in the addition mechanism is unfavorable for a given species if there is a scarcity of H^+ owing to the consumption of H^+ beyond $Z^+ = 1$ (cf. Section VI, A, 3, b), such a generalization is invalid for the condensation mechanism. Once the four nonacidic H atoms present in the tetrametalate ion have been consumed, the H atoms of the $M(O_6)H$ and $M(O_5)OH$ functions satisfy the requirements for continued condensation.

3. Energetics

Adoption of the same approach as applied to the addition mechanism requires inclusion of an additional reaction component, namely condensation of an H_2O molecule (W), which should make a considerable contribution to the stability of the polymetalate ions. The limiting values adopted were

$$\gamma_W = 0 \cdots 2$$

a. Condensation Reactions without Participation of H_2MO_4 and with a Scarcity of H^+ Ions ($Z \lesssim 1.1$) (Formation of the Paratungstate A Ion). We first examined which species are possible for reasons of geometry and acid/base properties, assuming that only HMO_4^-, and not H_2MO_4, participates in condensation. We then applied Eq. (14) to these species. The calculation [Figs. 7 and 8 in Tytko and Glemser (*19*)] shows that, in the range of limiting values for γ_e given for the addition mechanism and the foregoing limits for γ_W, there exist regions in which $\Gamma_i^h = \Gamma_{max}^h$ only for the species (4,4) (i.e., no condensation at all), (7,6), and (14,10). Formation of species other than (7,6) or (14,10) is not to be expected according to the condensation mechanism for a scarcity of H^+ ions ($Z \lesssim 1.1$), provided that only HMO_4^-, and not H_2MO_4, participates in condensation. However, the paratungstate A ion is a 7,6 ion. Apparently, the correct γ_e values lie in the range determined for a 7,6 ion. Above all, the size of γ_W and γ_E is important. For the tungstate system, γ_W is ≈ 0.7 and $\gamma_E \lesssim 0.04$.

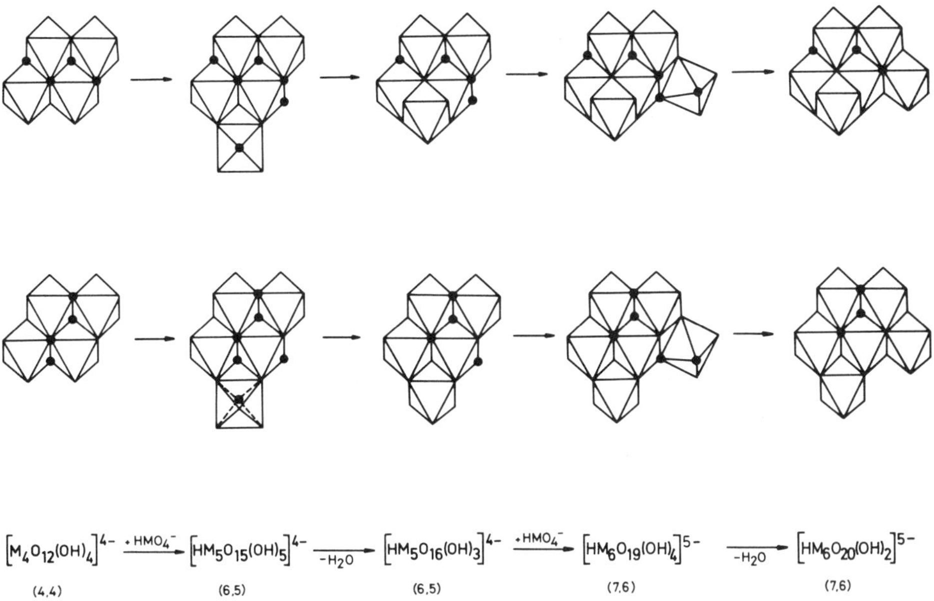

$$\left[M_4O_{12}(OH)_4\right]^{4-} \xrightarrow{+HMO_4^-} \left[HM_5O_{15}(OH)_5\right]^{4-} \xrightarrow[-H_2O]{} \left[HM_5O_{16}(OH)_3\right]^{4-} \xrightarrow{+HMO_4^-} \left[HM_6O_{19}(OH)_4\right]^{5-} \xrightarrow[-H_2O]{} \left[HM_6O_{20}(OH)_2\right]^{5-}$$

(4,4)　　　　　　(6,5)　　　　　　(6,5)　　　　　　(7,6)　　　　　　(7,6)

FIG. 16. Two of the reaction pathways leading to 7,6-metalate ions. ($\bullet$) Positions of H atoms.

The principle governing the thermodynamics of the course of aggregation for a scarcity of H^+ ions is, with a given number of protons, the formation of a maximum number of octahedra (preferably with three and two terminal O atoms) and the condensation of a maximum number of H_2O molecules, on the one hand, and accumulation of a minimum of negative charge and of M atoms, on the other.

Two of the reaction pathways leading to 7,6 ions are shown in Fig. 16, which reveals that the formula $[HW_6O_{20}(OH)_2]^{5-}$ applies to the para A ion.

b. Condensation Reactions with Participation of H_2MO_4 and Scarcity of H^+ Ions ($Z \leqslant 1.1$) (Formation of the Heptamolybdate Ion) (20, 23). If not only HMO_4^- but also H_2MO_4 participate in the condensation mechanism, it is possible that two H_2O molecules can be lost in one condensation step. Consideration of all the structures possible owing to participation of one (or more) such condensation steps and their comparison with those resulting on exclusive participation of HMO_4^- leads, for a scarcity of H^+ ions, to an 8,7-metalate ion having the formula $M_7O_{24}^{6-}$ as the species of most favorable energy. Its precursor is a 7,6 ion. The reaction pathway leading to the 8,7-metalate ion is depicted in Fig. 17.

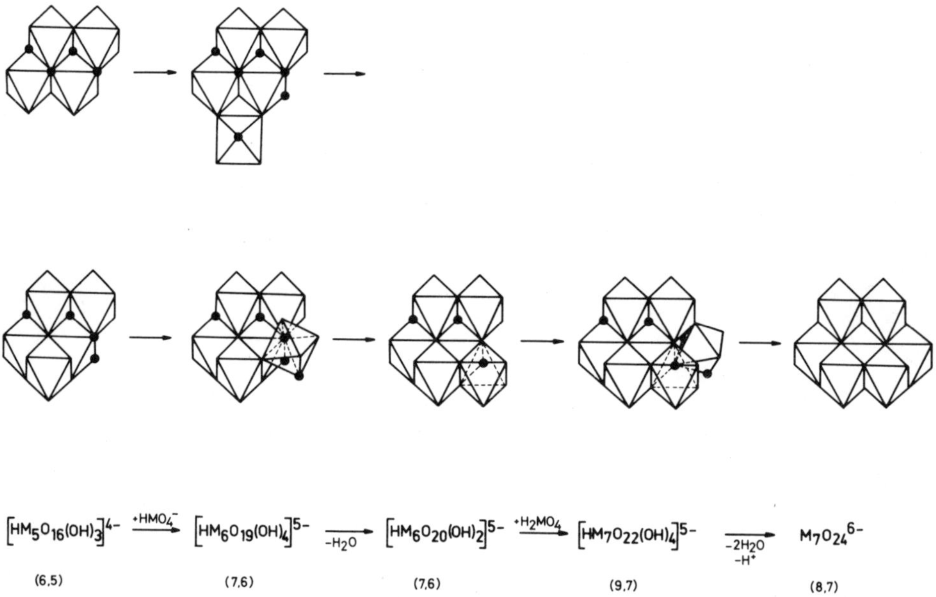

$$[HM_5O_{16}(OH)_3]^{4-} \xrightarrow{+HMO_4^-} [HM_6O_{19}(OH)_4]^{5-} \xrightarrow[-H_2O]{} [HM_6O_{20}(OH)_2]^{5-} \xrightarrow{+H_2MO_4} [HM_7O_{22}(OH)_4]^{5-} \xrightarrow[-H^+]{-2H_2O} M_7O_{24}^{6-}$$

(6,5)   (7,6)   (7,6)   (9,7)   (8,7)

FIG. 17. Reaction pathway leading to the 8,7-metalate ion. (●) Positions of H atoms.

4. Kinetics

The enhanced geometrical complexity of the condensation steps explains the lower rate of formation of the condensation products.

5. Comparison with Experimental Results

Comparison of the results obtained from the condensation mechanism on the basis of purely theoretical concepts with experimental findings indicates agreement for both the tungstate and the molybdate system: In the pH range in which the paratungstate A ion is formed (pH $\geqslant$ 6), less than 1% of the protonated monomer is present as H_2WO_4; in the pH range in which the heptamolybdate ion is formed (pH $\geqslant$ 4.5), up to 25% of the protonated monomer is present as H_2MoO_4. Hence a rapidly formed 7,6 ion is to be expected for the tungstate system, and a likewise rapidly formed 8,7 ion for the molybdate system. The most impressive result is the correct prediction of the structure of the heptamolybdate ion. Formation of the analogous planar heptametalate structure is impossible owing to the absence of H atoms in positions suitable for the condensation step leading to loss of two H_2O molecules (23).

C. OTHER FORMATION MECHANISMS

A systematic search for further geometrical principles for building up (iso)polymetalate ions has revealed the following possibilities (21): union (essentially dimerization) of larger aggregates by addition (21, 23); union of larger aggregates by condensation of water (22, 23); rearrangements without condensation of water (23); and rearrangements with condensation of water (21, 23). For instance, the paratungstate B ion is formed by simple dimerization (without condensation of water) of two enantiomeric 7,6-tungstate ions, with the more compact of the structures derived as above, and subsequent rearrangement of a WO_6 octahedron in each half-structure with condensation of one H_2O molecule each (21) .The polyoctamolybdate ion arises by polycondensation of octamolybdate ions of formula $[H_2Mo_8O_{28}]^{6-}$, formed by the condensation mechanism, which are themselves energetically very unfavorable (22).

REFERENCES

1. Lindqvist, I., Nova Acta Regiae Soc. Sci. Upsal. 15, No. 1 (1950).
2. Kepert, D. L., Progr. Inorg. Chem. 4, 199 (1962).

3. Kepert, D. L., "The Early Transition Metals." Academic Press, New York, 1972.
4. Kepert, D. L., in "Comprehensive Inorganic Chemistry" (A. F. Trotman-Dickenson et al., eds.), Vol. 4, p. 607. Pergamon, Oxford, 1973.
5. Aveston, J., Anacker, E. W., and Johnson, J. S., Inorg. Chem. 3, 735 (1964).
6. J. Aveston, Inorg. Chem. 3, 981 (1964).
7. Glemser, O., Holznagel, W., Höltje, W., and Schwarzmann, E., Z. Naturforsch. B 20, 725 (1965).
8. Jahr, K. F., and Fuchs, J., Angew. Chem. 78, 725 (1966); Angew. Chem., Int. Ed. Engl. 5, 689 (1966).
9. Sasaki, Y., and Sillén, L. G., Ark. Kemi 29, 253 (1968).
10. Souchay, P., "Polyanions et Polycations." Gauthier-Villars, Paris, 1963.
11. Souchay, P., "Ions minéraux condensés," Masson, Paris, 1969.
12. Glemser, O., and Tytko, K. H., Z. Naturforsch. B 24, 648 (1969).
13. Tytko, K. H., and Glemser, O., Chimia 23, 494 (1969).
14. Wempe, G., Z. Anorg. Chem. 78, 298 (1912).
15. Lindqvist, I., Ark. Kemi 2, 349 (1950).
16. Atovmyan, L. O., and Krasochka, O. N., Z. Strukt. Chim. 13, 342 (1972).
17. Gatehouse, B. M., and Leverett, P., Chem. Commun. p. 740 (1970).
18. Gatehouse, B. M., and Leverett, P., J. Chem. Soc. A p. 2107 (1971).
19. Tytko, K. H., and Glemser, O., Z. Naturforsch. B 26, 659 (1971).
20. Tytko, K. H., Angew. Chem. 83, 935 (1971); Angew. Chem., Int. Ed. Engl. 10, 860 (1971).
21. Tytko, K. H., Abstr. Meeting Int. Soc. Study Solute-Solute-Solvent Interactions, 1st, 1972.
22. Tytko, K. H., Z. Naturforsch. B 28, 272 (1973).
23. Tytko, K. H., unpublished investigations on the mechanisms of formation and structures of isopolymetalate ions; presented at the Universities of Gothenborg and Umeå, KTH Stockholm, and at the Gordon Research Conference on Inorganic Chemistry, New Hampton, New Hampshire, 1974.
23a. Tytko, K. H., unpublished investigations on the mechanisms of formation and structures of heteropolymetalate ions; presented at the University of Umeå (Sweden) and at the 16th International Conference on Coordination Chemistry, Dublin 1974.
24. Tytko, K. H., Z. Naturforsch. B 31 (1976) (in press).
25. Souchay, P., Chauveau, F., and Le Meur, B., C. R. Acad. Sci., Ser. C 270, 1401 (1970).
26. Tytko, K. H., Schönfeld, B., Buss, B., and Glemser, O., Angew. Chem. 85, 305 (1973); Angew. Chem., Int. Ed. Engl. 12, 330 (1973).
27. Tytko, K. H., Schönfeld, B., Cordis, V., and Glemser, O., presented at the Int. Conf. Coord. Chem., 15th, 1973, partly published in Z. Naturforsch. B 30, 471, 834 (1975).
28. Duncan, J. F., and Kepert, D. L., J. Chem. Soc. (London) p. 5317 (1961); 205 (1962).
29. Chojnacka, J., J. Inorg. Nucl. Chem. 33, 1345 (1971).
30. Wolff, C. M., and Schwing, J. P., C. R. Acad. Sci., Ser. C 268, 571 (1969).
31. Souchay, P., Ann. Chim. [11] 18, 61, 73, 169 (1943).
32. Le Meur, B., and Chauveau, F., Bull. Soc. Chim. Fr. p. 3834 (1970).
33. Charreton, B., Chauveau, F., Bertho, G., and Courtin, P., Chim. Anal. 47 17 (1965).

34. Cordier, M., Murgier, M., and Theodoresco, M., *C. R. Acad. Sci.*, **211**, 28 (1940); M. Theodoresco, *ibid.* **210**, 297 (1940).

35. Johansson, G., Pettersson, L., and Ingri, N., *Acta Chem. Scand.*, *Ser. A* **28**, 1119 (1974).

36. Levy, H. A., Agron, P. A., and Danford, M. D., *J. Chem. Phys.* **30**, 1486 (1959).

37. Pettersson, L., Lyhamn, L., Strandberg, R., Hedman, B., and Ingri, N., *Proc. Int. Conf. Coord. Chem.*, *15th*, *1973*, p. 524.

38. Tytko, K. H., unpublished results.

39. Tytko, K. H., Thesis, Aachen, 1967.

40. Tytko, K. H., unpublished.

41. Mauser, H., *Z. Naturforsch. B* **23**, 1021, 1025 (1968).

42. Dumanski, A. V., Buntin, A. P., Dijatschkovski, S. Y., and Kniga, A. G., *Kolloid-Z.* **38**, 208 (1926).

43. Jander, G., Jahr, K. F., and Heukeshoven, W., *Z. Anorg. Allg. Chem.* **194**, 383 (1930).

44. Jander, G., and Krüerke, U., *Z. Anorg. Allg. Chem.* **265**, 244 (1951).

45. Cannon, P., *J. Inorg. Nucl. Chem.* **13**, 269 (1960).

46. Beltrán, J., and Puerta, F., *An. R. Soc. Esp. Fis. Quim.*, *Ser. B* **59**, 271 (1963).

47. Saxena, R. S., and Saxena, G. P., *Z. Phys. Chem.* [N.F.] **29**, 181 (1961).

48. Blume, R., Lachmann, H., Mauser, H., and Schneider, F., *Z. Naturforsch. B* **29**, 500 (1974).

49. Sillén, L. G., *Acta Chem. Scand.* **8**, 299, 318 (1954).

50. Sasaki, Y., *Acta Chem. Scand.* **15**, 175 (1961).

51. Lindqvist, I., *Acta Chem. Scand.* **5**, 568 (1951).

52. Glemser, O., Holznagel, W., and Ali, S. I., *Z. Naturforsch. B* **20**, 192 (1965).

53. Pungor, E., and Halász, A., *J. Inorg. Nucl. Chem.* **32**, 1187 (1970).

54. Bettinger, D. J., and Tyree, S. Y., *J. Amer. Chem. Soc.* **79**, 3355 (1957).

55. Liska, M., and Plsko, E., *Chem. Zvesti* **11**, 390 (1957).

56. Ripan, R., and Szekely, Z., *Acad. Rep. Pop. Rom.*, *Fil. Cluj, Stud. Cercet. Chim.* **8**, 187 (1957).

57. Liska, M., *Chem. Zvesti* **10**, 549 (1956).

58. Wiese, G., and Böse, D., *Z. Naturforsch. B* **27**, 897 (1972).

59. Wiese, G., and Böse, D., *Z. Naturforsch. B* **29**, 630 (1974).

60. Jander, G., and Heukeshoven, W., *Z. Anorg. Allg. Chem.* **187**, 60 (1930).

61. Jespersen, N. D., *J. Inorg. Nucl. Chem.* **35**, 3873 (1973).

62. Lagrange, P., and Schwing, J. P., *Bull. Soc. Chim. Fr.* p. 718 (1967).

63. Britton, H. T. S., *J. Chem. Soc. (London)* p. 147 (1927).

64. Szarvas, P., and Kukri, E. C., *Z. Anorg. Allg. Chem.* **305**, 55 (1960).

65. Lindqvist, I., *Acta Chem. Scand.* **2**, 88 (1948).

66. Sturdivant, H. J., *J. Amer. Chem. Soc.* **59**, 630 (1937).

67. Vallance, R. H., *J. Chem. Soc. (London)* p. 1421 (1931).

68. Saddington, K., and Cahn, R. W., *J. Chem. Soc. (London)* p. 3526 (1950).

69. Sillén, L. G., *in* "Coordination Chemistry" (A. E. Martell, ed.), Vol. 1, p. 491. Van Nostrand-Reinhold, New York, 1971.

70. Harvey, J. F., Redfern, J. P., and Salmon, J. E., *J. Chem. Soc. (London)* p. 2861 (1963).

71. Cooper, M. K., and Salmon, J. E., *J. Chem. Soc. (London)* p. 2009 (1962).

72. Heitner-Wirguin, C., and Cohen, R., *J. Inorg. Nucl. Chem.* **26**, 161 (1964).

73. Stock, H. P., and Plewinsky, B., *Kolloid-Z. Z. Polym.* **249**, 1148 (1971).
74. Birkholz, E., Fuchs, J., Schiller, W., and Stock, H. P., *Z. Naturforsch. B* **26**, 365 (1971).
75. Boyer, M., and Souchay, P., *Rev. Chim. Miner.* **8**, 591 (1971).
76. Glemser, O., and Höltje, W., *Z. Naturforsch. B* **20**, 492 (1965).
77. Jahr, K. F., and Witzmann, H., *Z. Anorg. Allg. Chem.* **208**, 145 (1932).
78. Jander, G., and Jahr, K. F., *Kolloid-Beih.* **41**, 1 (1934).
79. Jander, G., and Ertel, D., *J. Inorg. Nuclear Chem.* **14**, 75 (1960).
80. Brintzinger, H., and Ratanarat, C., *Z. Anorg. Allg. Chem.* **224**, 97 (1935); **222**, 317 (1935).
81. Baker, L. C. W., and Pope, M. T., *J. Amer. Chem. Soc.* **82**, 4176 (1960).
82. von Kiss, A., and Acs, V., *Z. Anorg. Allg. Chem.* **247**, 190 (1941).
83. Tobias, R. S., *J. Inorg. Nucl. Chem.* **19**, 348 (1961).
84. Sillén, L. G., *Acta Chem. Scand.* **16**, 159 (1962); **18**, 1085 (1964).
85. Ingri, N., and Sillén, L. G., *Acta Chem. Scand.* **16**, 173 (1962).
86. Ingri, N., Sillén, L. G., *Ark. Kemi* **23**, 97 (1964).
87. Sasaki, Y., and Sillén, L. G., *Acta Chem. Scand.* **18**, 1014 (1964).
88. Tytko, K. H., and Glemser, O., *Z. Naturforsch. B* **25**, 429 (1970).
89. Tytko, K. H., unpublished results.
90. Lindqvist, *Ark. Kemi* **2**, 325 (1950).
91. Shimao, E., *Nature (London)* **214**, 170 (1967); *Bull. Chem. Soc. Jap.* **40**, 1609 (1967).
92. Evans, H. T., *J. Amer. Chem. Soc.* **90**, 3275 (1968).
93. Gatehouse, B. M., and Leverett, P., *Chem. Commun.* p. 901 (1968).
94. Sjöbom, K., and Hedman, B., *Acta Chem. Scand.* **27**, 3673 (1973).
95. Evans, H. T., Gatehouse, B. M., and Leverett, P., *J. Chem. Soc., Dalton Trans.* p. 505 (1975).
96. Lindqvist, I., *Acta Crystallogr.* **5**, 667 (1952).
97. Weiss, G., *Z. Anorg. Allg. Chem.* **368**, 279 (1969).
98. Allmann, R., *Acta Crystallogr., Sect. B* **27**, 1393 (1971).
99. d'Amour, H., and Allmann, R., *Z. Kristallogr.* **136**, 23 (1972); **138**, 5 (1973).
100. Tsay, Y. H., and Silverton, J. V., *Z. Kristallogr.* **137**, 256 (1973).
101. Fuchs, J., Hartl, H., and Schiller, W., *Angew. Chem.* **85**, 417 (1973); *Angew. Chem., Int. Ed. Engl.* **12**, 420 (1973).
102. Henning, G., and Hüllen, A., *Z. Kristallogr.* **130**, 162 (1969).
103. Allcock, H. R., Bissell, E. C., and Shawl, E. T., *Inorg. Chem.* **12**, 2963 (1973).
104. Böschen, I., Buss, B., and Krebs, B., *Acta Crystallogr., Sect. B* **30**, 48 (1974); Böschen, I., Buss, B., Krebs, B., and Glemser, O., *Angew. Chem.* **85**, 409 (1973); *Angew. Chem. Int. Ed. Engl.* **12**, 409 (1973).
105. Knöpnadel, I., Hartl, H., Hunnius, W. D., and Fuchs, J., *Angew. Chem.* **86**, 894 (1974); *Angew. Chem., Int. Ed. Engl.* **13**, 823 (1974).
106. Krebs, B., and Böschen, I., *Acta Crystallogr., Sect. B* **30**, 1795 (1974).
107. The computer program used was written by one of the authors. The program HALTAFALL [N. Ingri, W. Kakołowicz, L. G. Sillén, and B. Warnqvist, *Talanta* **14**, 1261 (1967)] can be used in the same way.
108. Tytko, K. H., unpublished results.
109. Jahr. K. F., Fuchs, J., Witte, P., and Flindt, E. P., *Chem. Ber.* **98**, 3588 (1965).
110. Jahr, K. F., and Fuchs, J., *Chem. Ber.* **96**, 2457 (1963).
111. Jahr, K. F., Fuchs, J., and Oberhauser, R., *Chem. Ber.* **101**, 477 (1968).
112. Fuchs, J., and Jahr, K. F., *Z. Naturforsch. B* **23**, 1380 (1968).

113. Schott, G., and Harzdorf, C., *Z. Anorg. Allg. Chem.* **288**, 15 (1956).
114. Kolli, I. D., Pirogova, G. N., and Spitsyn, V. I., *Z. Neorg. Chim.* **1**, 470 (1956).
115. Kiss, A. B., Gadó, P., and Hegedüs, A. J., *Acta Chim. Acad. Sci. Hung.* **72**, 371 (1972).
116. Babushkin, A. A., *Izv. Akad. Nauk. SSSR., Ser. Fiz.* **22**, 1131 (1958).
117. Babushkin, A. A., Yuknevich, G. N., Berezkina, Y. F., and Spitsyn, V. I., *Z. Neorg. Chim.* **4**, 823 (1959); *Russ. J. Inorg. Chem.* **4**, 373 (1959).
118. Pope, M. T., and Varga, G. M., *Chem. Commun.* p. 653 (1966).
119. Spitsyn, V. I., Lunk, H. J., Čuvaev, V. F. and Kolli, I. D., *Z. Anorg. Allg. Chem.* **370**, 191 (1969).
120. Signer, R., and Gross, H., *Helv. Chim. Acta* **17**, 1076 (1934).
121. Keggin, J. F., *Proc. Roy. Soc. (London) Ser. A* **144**, 75 (1934).
122. Mattes, R., Bierbüsse, H., and Fuchs, J., *Z. Anorg. Allg. Chem.* **385**, 230 (1971).
123. Kiss, B. A., Holly, S., and Hild, E., *Magy. Kem. Foly.* **77**, 418 (1971).
124. Fuchs, J., *Z. Naturforsch. B* **28**, 389 (1973).
125. Hunnius, W. D., *Z. Naturforsch. B* **29**, 599 (1974).
126. Kiss, B. A., *Acta Chim. Acad. Sci. Hung.* **75**, 351 (1973).
127. Spitsyn, V. I., *Acta Chim. Acad. Sci. Hung.* **12**, 119 (1957).
128. Spitsyn, V. I., and Torchenkova, E. A., *Dokl. Akad. Nauk SSSR* **95**, 289 (1954).
129. Ripan, R., and Marcu, G., *Acad. Rep. Pop. Rom., Fil. Cluj, Stud. Cercet. Chim.* **10**, 201 (1959).
130. Spitsyn, V. I., Aistova, R. I., and Vasil'ev, V. N., *Dokl. Akad. Nauk SSSR* **104**, 741 (1955).
131. Yukhnevich, G. V., *Z. Neorg. Chim.* **4**, 1459 (1959); *Russ. J. Inorg. Chem.* **4**, 656 (1959).
132. Kabanov, V. Ya., and Spitsyn, V. I., *Dokl. Akad. Nauk. SSSR* **148**, 109 (1963).
133. Spitsyn, V. I., and Berezkina, Y. F., *Dokl. Akad. Nauk. SSSR* **108**, 1088 (1956).
134. Arnek, R., and Szilard, I., *Acta Chem. Scand.* **22**, 1334 (1968).
135. Arnek, R., *Acta Chem. Scand.* **23**, 1986 (1969).
136. Pope, M. T., *Inorg. Chem.* **11**, 1973 (1972).
137. Launay, J. P., Souchay, P., and Boyer, M., *Collect. Czech. Chem. Commun.* **36**, 740 (1971).
138. Boyer, M., *J. Electroanal. Chem. Interfacial Electrochem.* **31**, 441 (1971).
139. Boyer, M., *C. R. Acad. Sci. Ser. C* **274**, 778 (1972).
140. Boyer, M., and Souchay, P., *J. Electroanal. Chem. Interfacial Electrochem.* **38**, 169 (1972).
141. Tourné, C., *Bull. Soc. Chim. Fr.* pp. 3196, 3199, 3214 (1967).
142. Mellström, R., and Ingri, N., *Acta Chem. Scand. Ser. A* **28**, 703 (1974).
143. Schwarzenbach, G., and Meier, J., *J. Inorg. Nucl. Chem.* **8**, 302 (1958).
144. Schwarzenbach, G., Geier, G., and Littler, J., *Helv. Chim. Acta* **45**, 2601 (1962).
145. Glemser, O., and Höltje, W., *Angew. Chem.* **78**, 756 (1966).
146. Asay, J., and Eyring, E. M., unpublished results.
147. Honig, D. S., and Kustin, K., *Inorg. Chem.* **11**, 65 (1972).
148. Wagner, G., Thesis, Göttingen, 1970.
149. Glemser, O., and Tytko, K. H., *Z. Naturforsch. B* **24**, 648 (1969).

150. Honig, D. S., Kustin, K., *J. Phys. Chem.* **76**, 1575 (1972).
151. Chauveau, F., Boyer, M., and Le Meur, B., *C. R. Acad. Sci., Ser. C* **268**, 479 (1969).
152. Fuchs, J., Jahr, K. F., and Nebelung, A., *Chem. Ber.* **100**, 2415 (1967).
153. Neu, F., and Schwing-Weill, M. J., *Bull. Soc. Chim. Fr.* p. 4821 (1968).
154. Cotton, F. A., and Wilkinson, G., "Anorganische Chemie," p. 875. Verlag Chemie, Weinheim (Bergstrasse), 1967.
155. Byé, J., *Ann. Chim.* [11] **20**, 463 (1945).
156. Chojnacki, J., and Hodorowicz, S., *Rocz. Chem.* **47**, 2213 (1973); **48**, 1399 (1974).
157. Rosenheim, A., and Felix, J., *Z. Anorg. Chem.* **79**, 292 (1913).
158. Svanberg, L., and Struve, H., *J. Prakt. Chem.* **44**, 282 (1848).
159. Hallada, C. J., *J. Less-Common Metals* **36**, 103 (1974).
160. Glemser, O., Wagner, G., Krebs, B., and Tytko, K. H., *Angew. Chem.* **82**, 639 (1970); *Angew. Chem., Int. Ed. Engl.* **9**, 639 (1970).
161. Foote, H. W., and Bradley, W. M., *J. Amer. Chem. Soc.* **58**, 930 (1936).
162. Hunnius, W. D., Thesis, Freie Universität, Berlin, 1970.
163. "Gmelins Handbuch der Anorganischen Chemie," 8th ed., Molybdän, System No. 53. Verlag Chemie, Berlin, 1935.
164. Haeringer, M., and Schwing, J. P., *Bull. Soc. Chim. Fr.* p. 708 (1967).
165. Goldstein, G., Wolff, C. M., and Schwing, J. P., *Bull. Soc. Chim. Fr.* p. 1201 (1971).
166. Souchay, P., Boyer, M., and Chauveau, F., *Kgl. Tek. Hoegsk. Handl. No.* **259** (1972) (Contributions to Coordination Chemistry in Solution, Stockholm, 1972, p. 159).
167. Lagrange, P., and Schwing, J. P., *Bull. Soc. Chim. Fr.* p. 1340 (1970).
168. Collin, J. P., and Lagrange, P., *Bull. Soc. Chim. Fr.* p. 777 (1974).
169. Baldwin, W. G., and Wiese, G., *Ark. Kemi* **31**, 419 (1968).
170. LeMeur, B., and Souchay, P., *Rev. Chim. Miner.* **9**, 501 (1972).
171. Angus, H. J. F., Briggs, J., and Weigel, H., *J. Inorg. Nucl. Chem.* **33**, 697 (1971).
172. Chojnacka, J., and Madejska, M., *Rocz. Chem.* **46**, 553 (1972).
173. Pauling, L., "General Chemistry." Edward Bros., Ann Arbor, Michigan, 1944.
174. Ricci, J. E., *J. Amer. Chem. Soc.* **70**, 109 (1948).
175. Rohwer, E. F. C. H., and Cruywagen, J. J., *J. South Afr. Chem. Inst.* **16**, 26 (1963); **17**, 145 (1964); **22**, 198 (1969).
176. Jain, D. V. S., *Indian J. Chem.* **8**, 945 (1970).
177. Tytko, K. H., unpublished results.
178. Schwing, J. P., Ph.D. Thesis, Strassbourg, 1961.
179. Wiese, G., *Z. Naturforsch. B* **27**, 616 (1972).
180. Tytko, K. H., and Petridis, G., unpublished results.
181. Sillén, L. G., *Pure Appl. Chem.* **17**, 55 (1968).
182. See Cotton and Wilkinson (154), p. 878.
183. Griffith, W. P., and Lesniak, P. J. B., *J. Chem. Soc., A* p. 1066 (1969).
184. Babko, A. K., and Gridchina, G. I., *Russ. J. Inorg. Chem.* **13**, 61 (1968).
185. Byé, J., Fischer, R., Krumenacker, L., Lagrange, J., and Vierling, F., *Kgl. Tek. Hoegsk. Handl. No.* **255** (1972) (Contributions to Coordination Chemistry in Solution, Stockholm, 1972, p. 99).
186. MacInnis, M. B., Kim, T. K., and Laferty, J. M., *J. Less-Common Metals* **36**, 111 (1974).

187. Ojo, J. F., Taylor, R. S., and Sykes, A. G., *J. Chem. Soc., Dalton Trans.* p. 500. (1975),

188. Collin, J. P., Lagrange, P., and Schwing, J. P., *J. Less-Common Metals* **36**, 117 (1974).

189. Collin, J. P. and Lagrange, P., *Bull. Soc. Chim. Fr.* p. 773 (1974).

190. Busey, R. H., and Keller, O. I., *J. Chem. Phys.* **41**, 215 (1964).

191. Hüllen, A., *Naturwissenschaften* **51**, 508 (1964); *Angew. Chem.* **76**, 588 (1964).

192. Borisov, S. V., Klevtsova, R. F., and Belov, N. V., *Kristallografija* **13**, 1980 (1968); *Sov. Phys. Crystallogr.* **13**, 852 (1969).

193. Lipscomb, W. N., *Inorg. Chem.* **4**, 132 (1965).

194. Ressel, H., Thesis, Technische Universität, Berlin, 1961.

195. Hähnert, M., *Z. Anorg. Allg. Chem.* **318**, 222 (1962).

196. Čuvaev, V. F., Lunk, H. J., and Spitsyn, V. I., *Dokl. Akad. Nauk SSSR* **180**, 133 (1968).

197. Tytko, K. H., unpublished results [cited in Tytko and Glemser (19)].

198. Wolff, C. M., and Schwing, J. P., *C. R. Acad. Sci., Ser. C* **272**, 1974 (1971).

199. Pan, K., and Hseu, T. M., *Bull. Chem. Soc. Jap.* **26**, 126 (1953); Pan, K., Lin, S. F., and Sheng, S. T., *ibid.* **26**, 131 (1953).

200. Freedman, M. L., *J. Amer. Chem. Soc.* **80**, 2072 (1958).

201. Arnek, R., and Sasaki, Y., *Acta Chem. Scand., Ser. A* **28**, 20 (1974).

202. Schulz, H., and Jander, G., *Z. Anorg. Allg. Chem.* **162**, 141 (1927); Jander, G., Mojert, D., and Aden, T., *Z. Anorg. Allg. Chem.* **180**, 129 (1929).

203. Johansson, G., *Acta Chem. Scand.* **14**, 771 (1960).

204. Cordis, V., unpublished results.

205. Boyer, M., and Souchay, P., *C. R. Acad. Sci., Ser. C* **268**, 2073 (1969).

206. Babad-Zakhryapin, A. A., and Gorbunov, N. S., *Izv. Akad. Nauk SSSR, Ser. Chim.* p. 1870 (1962).

207. Wolff, C. M., and Schwing, J. P., *C. R. Acad. Sci., Ser. C* **268**, 1339 (1969).

208. Glemser, O., Holznagel, W., and Höltje, W., *Z. Anorg. Allg. Chem.* **342**, 75 (1966).

209. Miolati, A., *J. Prakt. Chem.* **77**, 239, 417 (1908).

210. Rosenheim, A., *et al.*, *Z. Anorg. Chem.* **69**, 247 (1910); **70**, 73, 418 (1911); **75**, 141 (1912); **77**, 239 (1912); **89**, 224 (1914); **93**, 273 (1915); **96**, 139 (1916); **100**, 304 (1917); **101**, 215 (1917); **220**, 73 (1934).

211. Pauling, L., *J. Amer. Chem. Soc.* **51**, 2868 (1929).

212. Kepert, D. L., *Inorg. Chem.* **8**, 1556 (1969).

213. Tytko, K. H., unpublished results.

214. Babad-Zakhryapin, A. A., and Berezkina, Y. F., *Z. Strukt. Chim.* **4**, 346 (1963).

215. Chojnacki, J., *Bull. Acad. Pol. Sci., Ser. Sci. Chim.* **11**, 365 (1963).

216. Chojnacki, J., *Bull. Acad. Pol. Sci., Ser. Sci. Chim.* **11**, 369 (1963).

SUBJECT INDEX

A

Addition mechanisms of polymetalates, 296–305
  acid-base functions, 299, 300
  energetics, 300–304
  geometrical alternatives, 296–299
  kinetics, 304, 305
Aminosulfanuryl fluoride ions, 226, 227
Aminosulfonyl compounds, 209
Aminosulfur trifluorides, 192
Ammonia, labeled, exchange studies, 6–8
Archibald technique, 256

B

Benzophenone as sensitizer, 94
Bisaminosulfur difluorides, 192
Bis(fluorosulfonyl)amides, 197
Bis(fluorosulfonyl)aminoxenon fluoride, 198
Bis(imino)fluorosulfonylamide ions, 200, 201
Bis(sulfonyl)amides, 198, 200
S,S-Bis(sulfur oxydifluorideimidosulfonyl)-amide, 199
N-Bromacetamide, 56, 58
Bromine–nitrogen compounds, see Nitrogen–bromine compounds

C

Carbene–metal carbonyl complexes, 119
Carbonylation reaction, 111, 129
β-Carotene as quencher, 145
Catalytic reactions of organometallics, photoassisted, 114–118
Chlorine–nitrogen compounds, see Nitrogen–chlorine compounds
N-Chloropentafluorosulfanylamine, 190
Chlorophyll, 144–147
Chromium carbonyls
  flash photolysis, 82, 83
  infrared spectra, 85

photosubstitution, 119, 121, 122
  visible absorption spectra, 85
Circular dichroism spectra, 72, 75
Cobalamin, 151, 154
Cobaloximes, 154
  photochemistry of, 155–162
Cobinamide, 154
Concentration-proportional titrations of polymetalates, 250, 251, 254
Condensation mechanisms of polymetalates, 305–309
  acid-base functions, 306
  energetics, 306–308
  kinetics, 309
  octahedron formation, 305, 306
Conductivity titration of polymetalates, 251
Crystallization of polymetalates, 250
Cytochromes, 146, 147

D

Diene–metal carbonyl complexes, 76–79
N-Diiodoformamide, 31, 32
Diiodomethylamine, 22–26
  infrared spectra of, 23, 24
  iodine adducts of, 28, 29
  structure of, 22
Diiodomethylamine–0.5-pyridine, 23, 25

E

Electromigration analysis of polymetalates, 247, 248
Electronic absorption spectroscopy
  charge transfer transitions, 71
  d–d transitions, 70, 71
  intraligand transitions, 71–80
  of organometallics, 69–80
Elimination reactions of organometallics, 109–111, 126–129
Equilibrium curves for polymetalates, 253, 254

CONTENTS OF PREVIOUS VOLUMES

VOLUME 15

VOLUME 16

VOLUME 17

VOLUME 18

A 6
B 7
C 8
D 9
E 0
F 1
G 2
H 3
I 4
J 5